T0280994

The series Lecture Notes in Computer Science (LNCS), including its subseries Lecture Notes in Artificial Intelligence (LNAI) and Lecture Notes in Bioinformatics (LNBI), has established itself as a medium for the publication of new developments in computer science and information technology research, teaching, and education.

LNCS enjoys close cooperation with the computer science R & D community, the series counts many renowned academics among its volume editors and paper authors, and collaborates with prestigious societies. Its mission is to serve this international community by providing an invaluable service, mainly focused on the publication of conference and workshop proceedings and postproceedings. LNCS commenced publication in 1973.

Mohamed Mosbah · Tahar Kechadi ·
Ladjel Bellatreche · Faiez Gargouri
Editors

Model and Data Engineering

12th International Conference, MEDI 2023
Sousse, Tunisia, November 2–4, 2023
Proceedings

Editors
Mohamed Mosbah
Bordeaux INP
Talence, France

Tahar Kechadi
Dublin City University
Dublin, Ireland

Ladjel Bellatreche
ENSMA
Poitiers, France

Faiez Gargouri
University of Sfax
Sfax, Tunisia

ISSN 0302-9743 ISSN 1611-3349 (electronic)
Lecture Notes in Computer Science
ISBN 978-3-031-49332-4 ISBN 978-3-031-49333-1 (eBook)
https://doi.org/10.1007/978-3-031-49333-1

This Springer imprint is published by the registered company Springer Nature Switzerland AG
The registered company address is: Gewerbestrasse 11, 6330 Cham, Switzerland

Paper in this product is recyclable.

Preface

The Annual International Conference on Model and Data Engineering (MEDI) is a prominent platform for researchers and practitioners, enabling them to show-case the latest advances in modelling and data management. The conference encompasses a wide array of subjects, such as data models, machine learning and optimisations, advanced database and healthcare applications, and data analysis. Established by researchers from Euro-Mediterranean countries, MEDI has facilitated numerous global scientific collaborations, projects, and student/faculty exchanges. MEDI has been hosted in various countries, including Portugal (2011), France (2012 and 2019), Italy (2013), Cyprus (2014), Greece (2015), Spain (2016 and 2017), Morocco (2018), Estonia (2021), and Egypt (2022).

The 12th edition of MEDI took place in Sousse, Tunisia, from November 2–4, 2023. A total of 99 submissions were received. Each manuscript was rigorously evaluated and received three to five single-blinded reviews from Program Committee members chosen from 29 countries. Based on the evaluation results, we accepted 27 regular papers, representing an acceptance rate of 28%, and 12 short papers for short presentations. The regular papers are published in these proceedings, while short articles are published in a separate volume. The accepted papers were from authors from 17 countries. The conference program covers a wide range of topics, such as data modelling, data management (ontology and database systems), machine learning, model-driven engineering, image processing, natural language processing, optimisation, and advanced AI applications such as healthcare and security.

MEDI 2023 invited two distinguished keynote speakers. Khalil Drira from CNRS (French National Centre for Scientific Research), gave a talk entitled "Network services: design challenges for abstraction and autonomy". Said Boussakta from the Newcastle University, UK, gave a talk entitled "Fast Transforms: Powering Innovation in the Digital Age". The organisers would like to thank all authors who submitted research papers to MEDI 2023, as well as all members of the Program Committee and external reviewers, who carefully evaluated and gave feedback to all contributions. Moreover, we extend our special thanks to the Local Organizing Committee members who worked immensely hard to make the MEDI 2023 edition a great success.

November 2023

Tahar Kechadi
Mohamed Mosbah
Faiez Gargouri
Ladjel Bellatreche

Preface

Organization

General Chairs

Faiez Gargouri	University of Sfax, Tunisia
Ladjel Bellatreche	ISAE-ENSMA Poitiers, France

Program Committee Chairs

Mohamed Mosbah	LaBRI - Bordeaux INP, France
Tahar Kechadi	University College Dublin, Ireland

Proceedings Chairs

Mohamed Turki	University of Sfax, Tunisia
Sahbi Moalla	University of Sfax, Tunisia

Program Committee

El Hassan Abdelwahed	University Cadi Ayyad Marrakech, Morocco
Idir Ait Sadoune	Paris-Saclay University, France
Moulay Akhloufi	Université de Moncton, Canada
Sanaa Alwidian	Ontario Tech University, Canada
Ikram Amous	MIRACL, Tunisia
Heba Aslan	Nile University, Egypt
Christian Attiogbé	L2N - Université de Nantes, France
Ahmed Awad	University of Tartu, Estonia
Narjès Bellamine Ben Saoud	University of Manouba, Tunisia
Ladjel Bellatreche	LIAS/ENSMA, France
Orlando Belo	Universidade do Minho, Portugal
Antonio Corral	University of Almeria, Spain
Sahraoui Dhelim	University College Dublin, Ireland
Karima Dhouib	University of Sfax, Tunisia
Tai Dinh	Kyoto College of Graduate Studies for Informatics
Georgios Evangelidis	University of Macedonia, Greece
Flavio Ferrarotti	Software Competence Centre Hagenberg, Austria

Mamoun Filali-Amine	IRIT, France
Philippe Fournier-Viger	Shenzhen University, China
Jaroslav Frnda	University of Zilina, Slovakia
Enrico Gallinucci	University of Bologna, Italy
Faiez Gargouri	University of Sfax, Tunisia
Raju Halder	Indian Institute of Technology Patna, India
Ahmed Hassan	Nile University, Egypt
Irena Holubova	Charles University in Prague, Czech Republic
Luis Iribarne	University of Almería, Spain
Mohamed Jmaiel	University of Sfax, Tunisia
Rocheteau Jérôme	ICAM site de Nantes, France
Pinar Karagoz	Middle East Technical University, Turkey
Regine Laleau	Paris-Est Créteil University, France
Yves Ledru	Laboratoire d'Informatique de Grenoble - Université Grenoble Alpes, France
Ben Ayed Leila	ENSI, Tunisia
Ivan Luković	University of Belgrade, Serbia
Sofian Maabout	LaBRI. University of Bordeaux, France
Yannis Manolopoulos	Open University of Cyprus, Cyprus
Walaa Medhat	Nile University, Egypt
Dominique Mery	Université de Lorraine, LORIA, France
Mohamed Mhiri	University of Sfax, Tunisia
Chokri Mraidha	CEA LIST, France
Ahlem Nabli Chakroun	University of Sfax, Tunisia
Mourad Nouioua	Hunan University, China
Samir Ouchani	CESI Lineact, France
Milu Philip	University College Dublin, Ireland
Jaroslav Pokorný	Charles University in Prague, Czech Republic
Giuseppe Polese	University of Salerno, Italy
Elvinia Riccobene	University of Milan, Italy
Oscar Romero	Universitat Politècnica de Catalunya, Spain
Milos Savic	University of Novi Sad, Serbia
Arsalan Shahid	University College Dublin, Ireland
Neeraj Singh	INPT-ENSEEIHT/IRIT, University of Toulouse, France
Benkrid Soumia	ESI, Algeria
Goce Trajcevski	Iowa State University, USA
Mohamed Turki	Higher Institute of Computer Science and Multimedia of Sfax (ISIMS), Tunisia
Javier Tuya	Universidad de Oviedo, Spain
Panos Vassiliadis	University of Ioannina, Greece
Hala Zayed	Benha University, Egypt

Additional Reviewers

Alshawabkeh, Hamza
Baouya, Abdelhakim
Breve, Bernardo
Caruccio, Loredana
Cirillo, Stefano
Desiato, Domenico
Eid, Soha
Fischer, Bernhard
Gervais, Frédéric
Ghnaya, Imed
Kechadi, Tahar
Khaldi, Amine
Khamari, Sabri
Sochor, Hannes
Tasidou, Aimilia
Yousef, Ahmed Hassan Mohamed

Contents

Healthcare Applications

Applications and Security

Modelling

A Comparative Analysis of Time Series Prediction Techniques a Systematic Literature Review (SLR)

Sawssen Briki(✉), Nesrine Khabou, and Ismael Bouassida Rodriguez

ReDCAD, ENIS, University of Sfax, Sfax, Tunisia
sawssenbriki2019@gmail.com, {nesrine.khabou,bouassida}@redcad.org

Abstract. This paper highlights the significance of systematic literature reviews and explores the different techniques employed in these reviews, including statistical methods, machine learning, deep learning, and hybrid methods. The study aims to understand the performance and effectiveness of these techniques in the context of literature reviews. Statistical methods offer quantitative insights and analysis, while machine learning and deep learning techniques enable automation and uncover complex patterns in large volumes of data. However, hybrid methods, which integrate multiple techniques, have shown superior performance in systematic literature reviews, combining the strengths of different methodologies to achieve more comprehensive and accurate outcomes. Further development and refinement of hybrid methods can enhance the quality and effectiveness of literature review processes.

Keywords: Time series prediction · Systematic Literature Review · Machine Learning · Deep Learning

1 Introduction

Time series analysis plays a crucial role in understanding and predicting the behavior of data that evolves over time. It finds applications in various domains, including finance, economics, weather forecasting, stock market analysis, and many others. Accurate and reliable prediction of future values in a time series is of great importance for informed decision-making and proactive planning.

Over the years, a wide range of techniques have been developed for time series prediction, each with its own strengths and limitations. These techniques encompass traditional statistical methods, advanced machine learning algorithms, and more recently, deep learning models. The choice of the appropriate technique depends on the characteristics of the data, the complexity of patterns, interpretability requirements, and available computational resources.

A systematic literature review provides a comprehensive and objective analysis of the existing research in a specific area. In the context of time series

M. Mosbah et al. (Eds.): MEDI 2023, LNCS 14396, pp. 3–14, 2024.
https://doi.org/10.1007/978-3-031-49333-1_1

prediction, a systematic literature review enables us to identify the various techniques employed, understand their performance, and evaluate their suitability for different types of time series data.

The aim of this paper is to conduct a systematic literature review on time series prediction techniques and provide a comparative analysis of statistical methods, machine learning algorithms, deep learning models, and hybrid approaches. By analyzing and synthesizing the findings from a wide range of studies, we seek to identify the strengths and limitations of each technique and explore their performance in different scenarios.

This review will serve as a valuable resource for researchers, practitioners, and decision-makers seeking guidance on the selection and application of appropriate time series prediction techniques. By understanding the comparative merits of different approaches, stakeholders can make informed decisions and employ the most suitable techniques for their specific time series prediction tasks.

In the following sections, we will present the methodology employed for conducting the systematic literature review, followed by a detailed analysis of the identified studies. We will then discuss the findings, compare the techniques, and highlight the implications for future research and practical applications. Through this comprehensive review, we aim to contribute to the advancement of time series prediction methodologies and facilitate more accurate and reliable predictions in various domains.

2 Systematic Literature Review Planning

Systematic literature review planning involves developing clear research questions, creating a comprehensive search strategy, defining inclusion and exclusion criteria for screening articles, extracting relevant data, and analyzing and synthesizing the data. Careful planning is essential for ensuring that the review is rigorous, transparent, and comprehensive.

2.1 Research Questions

In order to find all relevant primary studies related to the study types of the models Existing prediction in the case of a series Temporal data, the following research questions (RQs) were generated:

1. RQ1:What are the prediction models in the case of data series?
2. RQ2: What are the existing implementations for prediction models ?

Next, we defined an initial search in the database. On the basis of keywords, four groups were created:

- Group1: ("predict","prediction","predicting")
- Group2: ("model","models")
- Group3: ("time serie","time series","data serie","data series")
- Group4: ("implementation","implementations")

2.2 Search Strategy

The research strategy combines the key concepts in our research question to achieve specific results. It is an organized structure of keywords, which are "Prediction", "Prediction models", "data series", "future value","Predictive Analysis", "Data Analysis", used to search a database. Subsequently, we added synonyms, variations, and related terms for each keyword. A Boolean operator (AND and OR) allow us to try different combinations of search terms. The final search string is ("predict" or "prediction" or "predicting") and "model" or "models") and ("time serie" or "time series" or "data serie" or "data series") and ("implementation" or "implementations")

2.3 Selection Criteria

After obtaining search results from various sources, we applied inclusion and exclusion criteria to identify relevant primary studies for our systematic literature review on models of prediction in the case of a series of temporal data. Inclusion criteria focused on studies that used machine learning or statistical techniques for prediction, reported quantitative measures of prediction accuracy, and used real-world datasets or simulations to evaluate the performance of prediction models. Exclusion criteria were used to remove irrelevant or unavailable studies, duplicates of included studies, and those that did not describe the use of machine learning or statistical techniques for prediction. By applying these criteria, we aimed to ensure a rigorous and comprehensive review process, including high-quality and informative primary studies relevant to our research question.

2.4 Data Collection

Table 1 summarizes the number of articles found in the search results. After filtering irrelevant, duplicate, and incomplete papers, a total of 61 papers in Table 3 were selected for the review process. Table 2 shows the filtering process. The distribution of selected papers based on resources is shown in Table 3.

2.5 Times Series Prediction by Model

Time series refers to a sequence of data points collected or recorded in chronological order, where each data point is associated with a specific time stamp or period. Time series data can be collected from various sources, such as stock prices, weather measurements, sensor readings, and social media posts, among others. Time series data often exhibit patterns, trends, and seasonality, and analyzing and forecasting these patterns can provide valuable insights and help make informed decisions in fields such as finance, economics, healthcare, and transportation, among others. Time series analysis is an important field in statistics, econometrics, and data science, with various tools and techniques available, including traditional statistical methods such as ARIMA, as well as more

Table 1. Search results by Resource.

Resource	Number of papers
Springer	122
IEE Xplore Digital Library	82
ACM Digital library	522
Science Direct	116
Hyper Articles en Ligne (HAL)	7
Total	849

Table 2. Filtered search results.

Irrelevant and duplicates	6
File not found	8
Excluded by reading title and abstract (Not related to RQ)	732
Total for Introduction reading	103
Excluded by reading Introduction (Not related to RQ)	42
Total	61

Table 3. Filtered search results by Resource.

Resource	Number of papers
Springer	4
IEE Xplore Digital Library	27
ACM Digital library	2
Science Direct	28
Hyper Articles en Ligne (HAL)	0
Total	61

advanced approaches such as machine learning and deep learning models like LSTM (Fig. 1).

In this study, we conducted a Systematic Literature Review (SLR) to identify relevant studies related to our research question. We found a total of 61 studies that utilized different statistical methods [1–4], machine learning methods [5–8], and deep learning methods [9,10]. Therefore, in this paper, we will focus on discussing the findings of the most relevant paper among the 61 studies identified through our SLR.

Statistical Techniques. Aman Swaraj et al. [11] proposed an ensemble model in 2021 that combines the autoregressive integrated moving average model (ARIMA) and the nonlinear autoregressive neural network (NAR) to improve the accuracy of predicting future COVID-19 cases (RMSE (16.23%), MAE (37.89%)

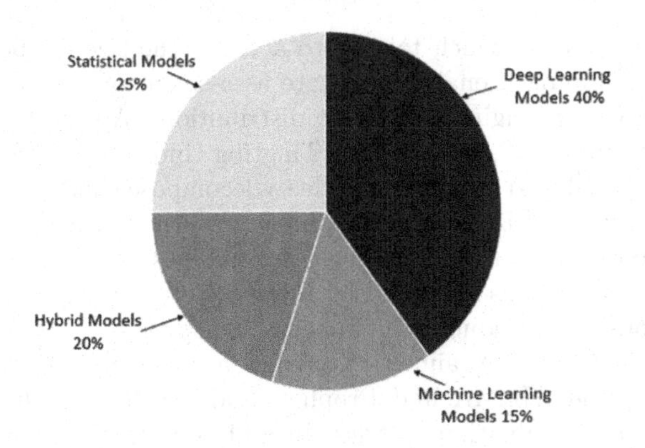

Fig. 1. Type of Approaches selected in this paper.

and MAPE (39.53%). The ARIMA models extract the linear correlations from the data, while the NAR neural network models the residuals containing nonlinear components. The combination of ARIMA-NAR models produced better prediction results than the single ARIMA model and other existing models applied to COVID-19 data from various countries. Hans Pratyaksa et al. [12] presented a study at the 1st International Conference on Biomedical Engineering in 2016, where they applied the ARIMA method to predict the usage of an antiseptic povidone-iodine at the Prof. Soeparwi Veterinary Hospital. Their analysis involved identifying patterns in the past behavior of the variable over time to forecast future values. Based on the historical dataset, the authors determined that ARIMA (1,0,1) is the most suitable model to represent the data. In their, Lester Marrero et al. [13] proposed a novel approach for predicting electric load one day ahead using an online implementation and self-control method. The proposed approach combines the ARIMA model to capture the temporal patterns and trends in the load data and particle swarm optimization (PSO) to optimize the model's parameters. They tested their approach using real-world load data and compared it with other commonly used forecasting methods. The results showed that their method outperformed other approaches, demonstrating its potential for improving the efficiency and reliability of electric power systems. Adhistya Erna proposed the use of the Seasonal Autoregressive Integrated Moving Average (SARIMA) model in 2013 [14] to forecast the incidence of a disease in humans. The SARIMA model incorporates seasonality into the analysis, building on the ARIMA model. The authors used time series data of Malaria occurrences in the United States to develop the forecasting model. The SARIMA (0,1,1)(1,1,1) model was selected, achieving a reasonable degree of accuracy with a Mean Absolute Percentage Error (MAPE) of 21.6%.

Jin Dong et al. [15] proposed a short-term probabilistic model for solar power output using historical inter-minute data. Their forecasting model for solar photovoltaic (PV) is useful for predicting solar energy production over short periods of time. Jie Ding et al. [16] proposes a new methodology that introduces the concept of predictability, which takes into account the size of the model class being used. This approach enables accurate assessment of the predictability of time series data with changing generating distributions. A new robust dynamic-wavelet-enabled approach is proposed by Tingting Guo et al. [17] for wind power smoothing. The developed approach is able to decompose wind power time series into self-adaptively optimized wavelet parameters without violating the physical constraints. A novel computational approach combines simulated annealing and agent-based simulation was proposed by Filippo Neri [18], in 2019, to simulate market conditions that produced the financial time series.

Sumit Mohan et al. [19] aimed to predict a spike in COVID-19 cases in India using a hybrid ARIMA and Prophet model, validated through various analytical tools and evaluation metrics. In addition, sentiment analysis using NLP libraries revealed negative sentiments in most analyzed articles and blogs related to the potential third wave of the pandemic in India. Overall, the study provided insights into the future trajectory of COVID-19 cases in India and public sentiments towards the situation. The model produced a root mean square error (RMSE) of 0.14 and a mean absolute percentage error (MAPE) of 0.06, indicating a high level of accuracy in the predictions.

Machine Learning Techniques. Tian [20] in 2020 proposed method that combines local mean decomposition and least squares support vector machine with a combined kernel function for short-term wind speed prediction. It achieves high prediction accuracy and strong robustness through simulations with short-term wind speed datasets. Hu Jingjing [21] proposed a proactive approach to service selection to avoid backlog and failure of service composition. It involves time series analysis and a negotiation process using least squares support vector learning algorithm and an acquaintance model, resulting in improved success rate and reduced execution time. In 2014, JinXing Che [22] proposed a multiple linear regression and support vector regression models for short-term electric load forecasting in the California electricity market. In 2015, Xinying Wang and Min Han [23] proposed an improved approach to extreme learning machine (ELM) for multivariate time series online sequential pre- diction. The proposed approach overcomes ELM's limitations by incorporating a weighted input selection method and a sliding window-based online sequential learning strategy. In 2021, Huang Chongyang et al. [24] presented a new analysis method that uses time series data and a backpropagation (BP) neural network to estimate the potential for electric energy substitution. This method can effectively identify the potential for electric energy substitution and provide valuable insights for energy planning and policy-making. Tesfamariam M. Abuhay et al. [25] proposed method involves analyzing the content of scientific papers by using topic modeling, specifically the Non-negative Matrix Factorization (NMF) algorithm.

In 2016 Wu Lijuan and Cao Guohua [26] proposed a novel approach called SFOASVR to forecast monthly inbound tourist flow, which combines SVR with FOA and seasonal index adjustment. The results show that the hybrid model is effective in handling nonlinear characteristics and seasonal tendencies, making it a viable option for tourism planning and administration. In 2017, Wu Xu et al. [27] proposed a new kernel function for remote sensing (RS) data, and uses the Least Squares Support Vector Regression (LSSVR) model to analyze spatio-temporal data. The resulting spatio-temporal kernel function is used to create a cellular automata (CA) model, which is used to simulate and predict changes in eco-environmental vulnerability. Experimental results show that the CA model based on LSSVR yields better simulation and prediction results.

Deep Learning Techniques. Mohd Rizman Sultan Mohd et al. [28] developed a non-linear NNARX model to predict solar radiation in Malaysia using meteorological and measured data. The model showed promising results with the lowest MSE of 0.0116, contributing to the advancement of reliable solar radiation prediction models. This research has practical implications for renewable energy stakeholders in Malaysia and worldwide. In 2022 Willian de Assis Pedrobon Ferreira et al. [29] proposed a fuzzy ARTMAP neural network to predict particulate matter sampled in a domestic bedroom environment, utilizing an online training architecture. The fuzzy ARTMAP network shows promise in predicting particulate matter time series data modeled in sliding windows with a 24-h ahead prediction and an MAE ranging from 0.26 to 7.65

IN 2022 Jaeseob Han et al. [30] Suggest a method for predicting a subset of Internet of Things (IoT) sensor readings using a Convolutional Neural Network (CNN) model. In addition, Raquel Espinosa and her team introduced a technique in 2021 [31] for assessing various pollutant prediction models based on accuracy and reliability criteria, using three years of hourly nitrogen oxides concentrations, traffic data, and meteorological information. The 1D CNN model surpassed other methods.

Femke Jansen et al. [32] presented a study that tackles the challenge of predicting equipment malfunctions in an industrial manufacturing process using multivariate time series data. They proposed a solution based on convolutional neural networks (CNN) and recurrent neural networks (RNN). Their findings demonstrate that RNN outperforms the CNN model in predicting machine failures.

Kareem Kamal et al. [33] implement a model to predict COVID-19 confirmed and death cases is based on Long Short-Term Memory (LSTM) and utilizes ten hidden units (neurons). Yun Jing et al. proposed a new approach for predicting passenger flow accurately using machine learning techniques. Their approach involves creating statistical features to capture the complex and non-linear relationships between passenger flow and various influencing factors, such as time, weather, and events. They applied their approach to real-world data and found that LSTM outperformed traditional prediction models, achieving high accuracy in predicting passenger flow. In 2021, Huiju Wang and al. [34] proposed the incremental ensemble LSTM model (IncLSTM), which uses ensemble learning and transfer learning for incremental updating of the model. The proposed method improves prediction accuracy by 15.6% and reduces training time by 18.8% compared to traditional methods, with even greater efficiency for larger training data sizes. Linglan Zhang et al. [28] proposed an LSTM neural networks to predict urban road diseases trends by analyzing historical data on road conditions and other relevant factors.

JAtin Bedi and Durga Toshniwal [35] proposes a deep learning-based hybrid approach to accurately estimate increasing electricity demand, addressing limitations of existing demand prediction models. The approach extracts meaningful sub-signals using VMD and Autoencoder models and uses LSTM networks to forecast electricity demand, incorporating historical, seasonal, and timestamp data dependencies. The proposed model outperforms other state-of-the-art demand forecasting models, achieving the lowest MAPE of 3.04% through experiments on an electricity consumption dataset of Himachal Pradesh, India. In 2019, Chao Luo et al. [36] propose an evolving recurrent interval type-2 intuitionistic fuzzy neural network (ERT2-IFNN) for online learning and time series prediction. The proposed network combines the advantages of interval type-2 fuzzy logic systems (IT2FLS), intuitionistic fuzzy sets (IFS), and recurrent neural networks (RNNs) to improve the accuracy and adaptability of time series prediction

Hybrid Approach. In 2017, Jeandro de M. Bezerra et al. [20] proposed a hybrid prediction model called FARIMA-RNN, which combines aspects of the FARIMA and Recurrent Neural Network (RNN) models. The FARIMA- RNN model aims to capture both the linear and nonlinear relationships in time series data to improve prediction accuracy. The results show that the FARIMA-RNN model outperformed other models in terms of prediction accuracy.

In 2014, Dominique Gay et al. [37] introduced a novel approach to time series classification by combining multiple representations of time series data to extract more discriminative features. This method achieved an impressive accuracy of 95.62.

In 2021 Salim Jibrin Danbatta and Asaf Varol [21] proposes a new approach to modeling and forecasting time series data related to tourism using an ANN-Fourier series model and Monte Carlo simulation. They introduce a hybrid model that combines artificial neural networks (ANN) and Fourier series analysis to capture both the nonlinear and cyclical patterns in the data. They then use Monte Carlo simulation to generate probabilistic forecasts that take into account the uncertainty and variability of the data, demonstrates the potential of combining ANN-Fourier series models with Monte Carlo simulation for modeling and forecasting time series data in the tourism industry, and provides useful insights for practitioners and policymakers seeking to make informed decisions based on these forecasts.

In 2021, Ruobin Gao et al. proposed a two-stage predictive algorithm that combines empirical wavelet transformation (EWT) and Echo state network (ESN) for improved predictive accuracy. EWT is used to extract features with predictability and eliminate noise in the data, while ESN is utilized for the overall predictive process. The proposed method outperforms other models on twelve public datasets with different mean-volatility features, as evidenced by improved out-of-sample forecasts compared to baseline persistence model and conventional benchmarks. Li-wei H. Lehman [38] proposes a novel approach to stratify mortality risks of intensive care units (ICU) patients receiving vasopressor treatment. The approach uses the switching autoregressive (SVAR) dynamics inferred from the multivariate vital sign time series to predict outcome. The results show that the bivariate HRiMAP dynamics contain additional prognostic information beyond the MAP values in mortality prediction, with an area under the curve (AUC) of 0.74. Furthermore, the HRiMAP dynamics achieved better performance among a subgroup of patients in a low MAP range (median MAP < 65 mmHg) while on pressors. In 2020, Harya Widiputra et al. [39] revisit, implements, and evaluates the multivariate transductive Neuro-Fuzzy Inference System model (mTNFI) for analyzing and modeling interrelated time-series data. Results confirm mTNFI's capability in recognizing patterns of relationship and modeling them in human-readable form, as well as its superiority in predicting future values compared to other time-series forecasting techniques.

3 Discussion

the discussion of the systematic literature review on time series prediction techniques reveals that statistical methods, machine learning algorithms, deep learning models, and hybrid approaches each have their strengths and limitations. Statistical methods offer interpretability and are suitable for simpler time series data, while machine learning algorithms provide flexibility and adaptability for complex datasets. Deep learning models excel in capturing temporal dependencies but require significant computational resources. Hybrid approaches, combining multiple techniques, have the potential to enhance prediction accuracy and robustness.

The selection of a time series prediction technique depends on factors such as data characteristics, interpretability requirements, computational resources, and forecasting goals. Researchers and practitioners must carefully consider these factors when choosing the most appropriate technique for their specific task.

Future research should focus on developing novel hybrid approaches that integrate statistical methods, machine learning, and deep learning techniques in more sophisticated ways. Comparative studies on larger and diverse datasets can provide deeper insights into the strengths and weaknesses of different techniques for time series prediction. Ultimately, the advancement of time series prediction techniques will contribute to more accurate and reliable forecasts in various domains.

4 Conclusions

The systematic review of time series prediction techniques reveals a diverse range of approaches, each with its unique strengths and limitations. Statistical methods offer interpretability and are suitable for simpler time series data, while machine learning and deep learning techniques provide flexibility for handling complex patterns and large datasets. Promisingly, hybrid approaches that combine multiple techniques show potential in improving prediction accuracy and robustness. Future research should prioritize the development of advanced hybrid methods that optimize the integration of statistical methods, machine learning, and deep learning techniques, and comparative studies on larger, more diverse datasets can deepen our understanding of their performance and limitations. Overall, the advancement of time series prediction techniques has far-reaching implications, enhancing the accuracy and reliability of forecasts across domains like finance, economics, healthcare, and climate science, thereby empowering researchers to make more informed decisions and extract valuable insights from time-dependent data.

Acknowledgments. This work was partially supported by the LABEX-TA project MeFoGL: "Méthodes Formelles pour le Génie Logiciel".

References

1. Xu, X., Hu, Z., Su, Q., Li, Y., Dai, J.: Multivariable grey prediction evolution algorithm: a new metaheuristic. Appl. Soft Comput. **89**, 106086 (2020)
2. Krupitzer, C., Pfannemüller, M., Kaddour, J., Becker, C.: Satisfy: towards a self-learning analyzer for time series forecasting in self-improving systems. In: 2018 IEEE 3rd International Workshops on Foundations and Applications of Self-Systems (FAS* W), pp. 182–189. IEEE (2018)
3. Rostam, N.A.P., Malim, N.H.A.H., Abdullah, R., Ahmad, A.L., Ooi, B.S., Chan, D.J.C.: A complete proposed framework for coastal water quality monitoring system with algae predictive model. IEEE Access **9**, 108249 (2021)
4. Rinchen, S., Yassine, A., Schwartzentruber, K., Ahmed, H., Armitage, A.: Integrating small scale green energy into smart grids: prediction for peak load reduction. In: 2018 International Conference on Computer and Applications (ICCA), pp. 104–109. IEEE (2018)
5. Roberts, L., Michalák, P., Heaps, S., Trenell, M., Wilkinson, D., Watson, P.: Automating the placement of time series models for IoT healthcare applications. In: 2018 IEEE 14th International Conference on e-Science (e-Science), pp. 290–291. IEEE (2018)
6. Bazine, H., Mabrouki, M.: Prediction of photovoltaic production for smart grid energy management using hidden Markov model: a study case. In: 2017 International Renewable and Sustainable Energy Conference (IRSEC), pp. 1–7. IEEE (2017)
7. Carvalho, J., Jr., Costa, C., Jr.: Identification method for fuzzy forecasting models of time series. Appl. Soft Comput. **50**, 166 (2017)
8. Tessoni, V., Amoretti, M.: Advanced statistical and machine learning methods for multi-step multivariate time series forecasting in predictive maintenance. Procedia Comput. Sci. **200**, 748 (2022)
9. Zhang, Q., et al.: Data-driven approaches for time series prediction of daily production in the Sulige tight gas field, China. Artif. Intell. Geosci. **2**, 165 (2021)
10. Farías, R.L., Flores, J.J., Puig, V.: Qualitative and quantitative multi-model forecasting with nonlinear noise filter applied to water demand. In: 2015 IEEE International Autumn Meeting on Power, Electronics and Computing (ROPEC), pp. 1–6. IEEE (2015)

11. Swaraj, A., Verma, K., Kaur, A., Singh, G., Kumar, A., de Sales, L.M.: Implementation of stacking based ARIMA model for prediction of COVID-19 cases in India. J. Biomed. Inform. **121**, 103887 (2021)
12. Pratyaksa, H., Permanasari, A. E., Fauziati, S., Fitriana, I.: Arima implementation to predict the amount of antiseptic medicine usage in veterinary hospital. In: 2016 1st International Conference on Biomedical Engineering (IBIOMED), pp. 1–4. IEEE (2016)
13. Marrero, L., García-Santander, L., Carrizo, D., Ulloa, F.: An application of load forecasting based on ARIMA models and particle swarm optimization. In: 2019 11th International Symposium on Advanced Topics in Electrical Engineering (ATEE), pp. 1–6. IEEE (2019)
14. Permanasari, A.E., Hidayah, I., Bustoni, I.A.: SARIMA (seasonal ARIMA) implementation on time series to forecast the number of malaria incidence. In: 2013 International Conference on Information Technology and Electrical Engineering (ICITEE), pp. 203–207. IEEE (2013)
15. Dong, J., Kuruganti, T., Djouadi, S.M.: Very short-term photovoltaic power forecasting using uncertain basis function. In: 2017 51st Annual Conference on Information Sciences and Systems (CISS), pp. 1–6. IEEE (2017)
16. Ding, J., Zhou, J., Tarokh, V.: Asymptotically optimal prediction for time-varying data generating processes. IEEE Trans. Inf. Theory **65**, 3034 (2018)
17. Guo, T., Liu, Y., Zhao, J., Zhu, Y., Liu, J.: A dynamic wavelet-based robust wind power smoothing approach using hybrid energy storage system. Int. J. Electr. Power Energy Syst. **116**, 105579 (2020)
18. Neri, F.: Combining machine learning and agent based modeling for gold price prediction. In: Cagnoni, S., Mordonini, M., Pecori, R., Roli, A., Villani, M. (eds.) WIVACE 2018. CCIS, vol. 900, pp. 91–100. Springer, Cham (2019). https://doi.org/10.1007/978-3-030-21733-4_7
19. Mohan, S., Solanki, A.K., Taluja, H.K., Singh, A., et al.: Predicting the impact of the third wave of COVID-19 in India using hybrid statistical machine learning models: a time series forecasting and sentiment analysis approach. Comput. Biol. Med. **144**, 105354 (2022)
20. Bezerra, J.D.M., Pinheiro, A.J., de Souza, C.P., Campelo, D.R.: Performance evaluation of elephant flow predictors in data center networking. Future Gener. Comput. Syst. **102**, 952–964 (2020)
21. Danbatta, S.J., Varol, A.: Modeling and forecasting of tourism time series data using ANN-Fourier series model and Monte Carlo simulation. In: 2021 9th International Symposium on Digital Forensics and Security (ISDFS), pp. 1–6. IEEE (2021)
22. Wang, X., Han, M.: Improved extreme learning machine for multivariate time series online sequential prediction. Eng. Appl. Artif. Intell. **40**, 28 (2015)
23. Soualhi, A., Medjaher, K., Celrc, G., Razik, H.: Prediction of bearing failures by the analysis of the time series. Mech. Syst. Signal Process. **139**, 106607 (2020)
24. Sinha, A., Jana, P.K.: MRF: MapReduce based forecasting algorithm for time series data. Procedia Comput. Sci. **132**, 92 (2018)
25. Abuhay, T.M., Nigatie, Y.G., Kovalchuk, S.V.: Towards predicting trend of scientific research topics using topic modeling. Procedia Comput. Sci. **136**, 304 (2018)
26. Lijuan, W., Guohua, C.: Seasonal SVR with FOA algorithm for single-step and multi-step ahead forecasting in monthly inbound tourist flow. Knowl.-Based Syst. **110**, 157 (2016)

27. Xu, W., Binbin, H., Xiao, Y., Cirenluobu, KanAike, Jinji, L.: A spatio-temporal series simulation and prediction method of geography based on SVR-CA model. In: Proceedings of the 2nd International Conference on Intelligent Information Processing, pp. 1–7 (2017)
28. Zhang, L., Meng, W., Chen, A., Mei, M., Liu, Y.: Application of LSTM neural network for urban road diseases trend forecasting. In: 2018 IEEE International Conference on Big Data (Big Data), pp. 4176–4181. IEEE (2018)
29. Ferreira, W.D.A.P., Grout, I., da Silva, A.C.R.: Application of a fuzzy ARTMAP neural network for indoor air quality prediction. In: 2022 International Electrical Engineering Congress (iEECON), pp. 1–4. IEEE (2022)
30. Han, J., Lee, G.H., Park, S., Lee, J., Choi, J.K.: A multivariate-time-series-prediction-based adaptive data transmission period control algorithm for IoT networks. IEEE Internet Things J. **9**, 419 (2021)
31. Espinosa, R., Palma, J., Jiménez, F., Kamińska, J., Sciavicco, G., Lucena-Sánchez, E.: A time series forecasting based multi-criteria methodology for air quality prediction. Appl. Soft Comput. **113**, 107850 (2021)
32. Jansen, F., Holenderski, M., Ozcelebi, T., Dam, P., Tijsma, B.: Predicting machine failures from industrial time series data. In: 2018 5th International Conference on Control, Decision and Information Technologies (CoDIT), pp. 1091–1096. IEEE (2018)
33. Ghany, K.K.A., Zawbaa, H.M., Sabri, H.M.: COVID-19 prediction using LSTM algorithm: GCC case study. Inform. Med. Unlocked **23**, 100566 (2021)
34. Wang, H., Li, M., Yue, X.: IncLSTM: incremental ensemble LSTM model towards time series data. Comput. Electr. Eng. **92**, 107156 (2021)
35. Bedi, J., Toshniwal, D.: Energy load time-series forecast using decomposition and autoencoder integrated memory network. Appl. Soft Comput. **93**, 106390 (2020)
36. Luo, C., Tan, C., Wang, X., Zheng, Y.: An evolving recurrent interval type-2 intuitionistic fuzzy neural network for online learning and time series prediction. Appl. Soft Comput. **78**, 150 (2019)
37. Gay, D., Guigourès, R., Boullé, M., Clérot, F.: Feature extraction over multiple representations for time series classification. In: Appice, A., Ceci, M., Loglisci, C., Manco, G., Masciari, E., Ras, Z.W. (eds.) NFMCP 2013. LNCS (LNAI), vol. 8399, pp. 18–34. Springer, Cham (2014). https://doi.org/10.1007/978-3-319-08407-7_2
38. Li-wei, H.L., Nemati, S., Mark, R.G.: Hemodynamic monitoring using switching autoregressive dynamics of multivariate vital sign time series. In: 2015 Computing in Cardiology Conference (CinC), pp. 1065–1068. IEEE (2015)
39. Widiputra, H.: Evaluation of multivariate transductive neuro-fuzzy inference system for multivariate time-series analysis and modelling. In: Proceedings of the 5th International Conference on Sustainable Information Engineering and Technology, pp. 45–50 (2020)

A Formal Metamodel for Software Architectures with Composite Components

James Baak, Quentin Rouland(✉), and Jason Jaskolka

Systems and Computer Engineering, Carleton University, Ottawa, ON, Canada
{james.baak,quentin.rouland,jason.jaskolka}@carleton.ca

Abstract. Formal component-based modeling has been shown to be invaluable for verifying the compatibility of specified components, discovering flaws early in design stages, and enabling the reuse of components, across multiple projects and teams. However, complex system specifications are large and difficult to reason with which has limited the adoption of formal approaches. In this paper, we use a formal language to build a metamodel to represent software architectures consisting of composite components. First, we propose a metamodel to describe the high-level concepts of software architectures in a component-port-connector fashion. We focus on providing hierarchical modeling capabilities by considering the construction of composite components from existing ones. Second, using Alloy as a tooled formal language, we formalize the metamodel concepts to build a reusable framework for modeling complex systems consisting of composite component structures that can be automatically constructed and checked for architectural conformance. We use a smart metering system to demonstrate the use our formal metamodel.

Keywords: composite component · component-based software engineering · metamodeling · formalization

1 Introduction

Increasingly complex software systems [12] are prevalent in all types of embedded, cloud, and industrial systems and the need to ensure they behave as expected is more apparent. To manage the increasing complexity, systems are often built using components with the intention of reusing them. Using formal methods to specify components enables unambiguous interpretations of the services provided and required by the component making the composition of components well-formed and easy to follow. However, formal component-based design still has limited adoption due to the increasing complexity of modern software-dependent systems that often lead to a *state explosion problem* [3]. Therefore, methods are needed to support component-based system design and specification of components into logical groups, or *composite components*, that

Supported by the Ericsson-Carleton Partnership 5G Fellowship Program.

M. Mosbah et al. (Eds.): MEDI 2023, LNCS 14396, pp. 15–29, 2024.
https://doi.org/10.1007/978-3-031-49333-1_2

represent higher-level entities that can be used in the verification properties of a system's architecture and behavior.

In this work, we propose a formal component-port-connector metamodel enabling the specification of component-based software architectures as hierarchical structures of components organized into logical groups. The metamodel extends the one provided by Rouland et al. [14,16] by introducing the concept of composite components to the metamodel. The contribution of this paper is a reusable framework for modelling complex systems with composite component structures that can be checked for architectural conformance including: (1) a formal component-port-connector metamodel for hierarchical component-based software architectures, and (2) a novel idea to enable the automatic construction of composite components from existing ones. To demonstrate the formal metamodel, we formalize it using Alloy and we use a smart metering system to illustrate the specification and verification of composite structures.

The rest of this paper is organized as follows. Section 2 presents our component-port-connector metamodel. Section 3 presents our approach for formalizing and verifying the metamodel concepts using Alloy. Section 4 demonstrates how to use the formal metamodel to specify and verify composite components and connector behaviors for a smart metering system. Section 5 positions our contributions within the related literature. Lastly, Sect. 6 concludes and provides an outlook for several future research directions.

2 Defining the Software Architecture Metamodel

In this section, we propose a metamodel that adopts a UML-like vocabulary to define software architecture models that are conceptually close to industrial practice. The metamodel is visualized as a class diagram with UML notations in Fig. 1. It provides concepts for describing software architectures in terms different views [11]: (1) the *logical view* which is concerned with capturing the functional architecture of the system in terms of components, and (2) the *scenario view* which is concerned with the representing the communication behavior between components. We describe the concepts of our metamodel below.

Component. A component is an abstract artifact representing all types of independently deployable and reusable units of composition that provides and requires services to and from its environment [5]. To arrange components in a hierarchical structure, all components have one parent which is the composite component in which they are contained or is the root system node. A component can be one of the following two types:

- **Atomic Component.** An atomic component is the lowest abstraction level of a component in the system and represents a unit that cannot, or does not need to be, divided into smaller components. Atomic components realize the functional system properties by requiring services and provide services.
- **Composite Component.** A composite component is a logical group of one or more components (atomic or composite) that can interact with other system components. Composite components only use the ports of the

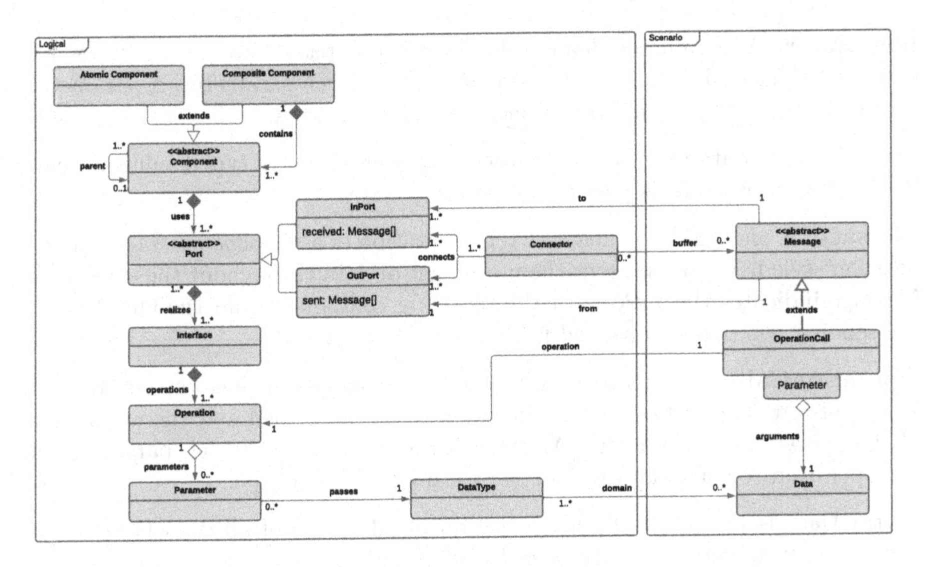

Fig. 1. Component-port-connector metamodel supporting composite components.

components they contain that interact with other components outside the composite component. These ports are called the *external ports* as they reach out to external components outside those of the composite component.

Port. Components use ports to interact with their environment by exposing or exploiting a known interface. A port may realize more than one interface to support grouping interfaces of a particular nature dealt with by the port.

- **InPort.** An InPort represents a required service of its component where the receive action of a communication style that enables data to be passed to the component to which the port belongs.
- **OutPort.** An OutPort represents a provided service of its component where the send action of a communication style that enables data to be passed to another port realizing the same interface.

Connector. A connector is a communication link enabling messages to be passed from OutPorts to InPorts. In this paper, we only consider a *message passing* communication style. Connectors buffer any abstract Message type. They can connect more than one InPort and OutPort, but each interface used must have at least one InPort and OutPort that realizes the interface.

Interface. An interface defines the set of operations required on an InPort or provided on an OutPort to establish a kind of contract between components using ports that realize the same interface.

Operation. An operation promises that a certain action will occur based on the input from the OutPort. An operation will either have no parameters or a set of typed parameters. They represent the methods of interacting with a component in a contractual manner.

Parameter. A parameter defines the data types passed to the operation to which it belongs. Multiple data types can be passed as a parameter. A parameter can be specified to accept data of one or more data types.

DataType. A data type is the abstract representation of a typed value system. It defines all data that is part of the data type.

Message. A message is an abstract representation of a *message passing* communication style used to represent the movement of data throughout the system. A Message indicates the InPorts of the receiving component and the OutPort on the sending component involved in the communication.

OperationCall. An OperationCall defines the message payload structure consisting of a representation of the operation to be performed and the arguments of the operation's parameters. Arguments are a mapping of each parameter of the operation to some data within the parameter's data type.

Data. Data is the abstract representation of all values of all data types (e.g., the set of all integers for an Integer data type).

3 Formalizing the Metamodel with Alloy

Next, we formalize the principal concepts of our software architecture metamodel. The formal language should meet the following requirements: (1) enable the creation of a formal component-port-connector metamodel; (2) enable the creation of a component-port-connector-based system architecture model, describing the target application model according to the metamodel; (3) enable the verification of static properties and dynamic properties (e.g., behavioral aspects) and (4) have tool support.

While any suitable formal language with tool support (e.g., nuSMV, SPIN, UPPAAL) could be used, we choose to use Alloy [7] because it additionally: (1) enables the incremental construction of models, allowing rapid iterations between modeling and analysis; (2) offers a simple and straightforward usage of its analyzer; (3) and supports visualizing models. Additionally, a recent article [8] has further highlighted the strengths of Alloy for software design.

3.1 A Brief Introduction to Alloy

Alloy is a lightweight formal modeling language based on first-order relational logic [7]. An Alloy model is composed of a set of signatures each defining a set of atoms. Atoms may contain fields that specify the relationships between them. Additionally, signatures act as types, and subtyping can be described as an extension of a signature. A model's constraints can be expressed as *facts* that must always be true, *predicates* that are defined as parameterized formulas that can be used elsewhere, or *assertions* that are supposed to result from the facts of the model. The specification and verification processes may advantage from the usage of *functions* that take the form of parameterized expressions.

In the context of our work, we use Alloy to formalize our metamodel and verify that the theory produces component-based systems that satisfy the metamodel. Once the metamodel framework is defined, it can be used to specify component-based system architectures, behaviors, and functional and non-functional properties, and explore model instances that satisfy the metamodel and the system model's objectives. Alloy offers automated model-checking using several different solvers and includes a visualizer to investigate model instances and property counterexamples which makes it an ideal candidate to implement a metamodel or system quickly and providing informative feedback to designers.

In the following, we formalize in Alloy both the structural part (i.e., logical view) and the behavioral part (i.e., scenario view) presented in Sect. 2.

3.2 Structural Model in Alloy

Component. The `Component` signature in Alloy represents a node in our metamodel's hierarchical tree-like structure and therefore contains a `parent` relation which defines the parent of a component. The `lone` keyword specifies that every component will have one or no parent. In our metamodel, all defined instances of components will have a parent and be part of the system tree, except for the root `System` node (Line 4). The `Component` signature also contains a `uses` relation which defines a set of ports used by the component defining its interface. The `CompositeComponent` signature extends the abstract `Component` and defines another relation named `contains` that specifies the components contained within the composite component. Line 3 in Listing 1.1 specifies an Alloy fact on the composite component stating the `uses` relation in the component signature is equal to the composite component's external ports. The final signature Line 5 is the `AtomicComponent` which defines the leaf nodes of the tree structure. An atomic component must have only one parent in the component tree and must have some ports in the `uses` field, otherwise the component would be unable to interact with its environment. Lines 8–10 define the *external ports* as the ports of the components contained within the composite component that are connected to components outside the composite component.

```
1  abstract sig Component { parent: lone Component, uses: set Port }
2  sig CompositeComponent extends Component { contains: some Component }{
3    uses = this.external_ports[] }
4  one sig System in CompositeComponent {}{no parent }
5  sig AtomicComponent extends Component {}{
6    one parent
7    some uses }
8  fun CompositeComponent.external_ports[] : set Port {{ p : Port |
9    p in this.ports[]
10   and p.connectors[].components[] not in this.^contains }}
11 fun CompositeComponent.ports[] : set Port {{ p : Port |
12   p in this.contains.uses }}
13 fact { all cc: CompositeComponent, c: cc.contains | c.parent = cc }
14 fact { Component in System.*contains }
15 fact { all c: Component | c.uses.connectors[].components[] != c }
```

Listing 1.1. Formalization of Components

To further constrain the metamodel, *facts* are used define invariants of the system metamodel or specific signatures. Line 13 states all composite components are the parent of the components they contain. Line 14 ensures that all components are contained within the system tree by using the reflexive-transitive operation *. Lastly, Line 15 ensures that no component is connected to itself. This is an assumption ensuring no component relies on itself to provide its own services.

Port. Listing 1.2 shows the formalization of ports. The `realizes` field of the `Port` signature defines a set of interfaces the port realizes, providing or requiring the operations in the interface as a service depending on the port type. As in our metamodel in Fig. 1, the `Port` signature is abstract and is extended by two different types of ports: `InPort` and `OutPort`. The InPort has a variable `received` field where messages addressed to the `InPort` are received from a connecting connector buffer. The `OutPort` has a variable `sent` field where messages are sent to buffers of connecting connectors to be received by the message recipients.

```
1 abstract sig Port { realizes: set Interface }
2 sig InPort extends Port { var received: set Message }
3 sig OutPort extends Port { var sent: set Message }
```

Listing 1.2. Formalization of Ports

Connector. Listing 1.3 defines the `Connector` signature. The signature has a `connects` field to connect `OutPorts` to `InPorts`. A connector must be connected to at least one `OutPort` and one `InPort` (Lines 2–3), but there is no limit on the number of ports that can be connected to the connector. The variable `buffer` field holds a set of messages that are buffered by the connector. When a message is sent to a connector, it can be dropped before it reaches the buffer emulating a message being lost in transmission. Once a message is in a connector's buffer, it can be received by an attached `InPort` for processing.

```
1 sig Connector { connects: some Port, var buffer: set Message}
2 fact ConnectorHasAtLeastOneInAndOutPort { all c: Connector |
3   some (InPort & c.connects) and some (OutPort & c.connects) }
```

Listing 1.3. Formalization of Connectors

Interface, Operation, Parameter, and Datatype. Listing 1.4 provides the formalization of interfaces, operations, and parameters. The `operations` field of the `Interface` signature defines the set of operations provided or required. The `Operation` signature defines an operation that is available on an `Interface`. An operation has a set of parameters specified by the `params` field. The `Parameter` signature defines the type of the parameter using the `passes` field which specifies the allowable values, variable and/or constant, that can be passed to the parameter. The abstract `Variable` and `Constant` signatures represent mutable and immutable *DataTypes* respectively from the metamodel in Fig. 1. We differentiate these *DataTypes* in Alloy because variable values can be used to create data structures that change during the execution, whereas constants need to remain the same throughout all time steps.

```
1 abstract sig Interface { operations: set Operation }
2 abstract sig Operation { params: set Parameter }
3 abstract sig Parameter { var passes: set (Variable + Constant) }
4 abstract var sig Variable {}
5 abstract sig Constant {}
```

Listing 1.4. Formalization of Interfaces, Operations, and Parameters

3.3 Behavioral Model in Alloy

Message. Listing 1.5 specifies `Message` as an abstract signature representing any type of message that can be sent or received with the system. A `Message` has two variable fields which are set by the sending `OutPort` when the message is created: `from`, which states the origin `OutPort` of the message, and `to`, which states the destination `InPort` of the message.

```
1 var abstract sig Message { var from: one OutPort, var to: one InPort }
2 fact toFromStatic { always all m: Msg |
3   (m.to' = m.to and m.from' = m.from) or m.release[] }
```

Listing 1.5. Formalization of Messages

We use facts to constrain sending and receiving messages through connectors. Listing 1.6 shows some of the facts in Alloy for message movement focusing on message sending. Lines 1–2 state that a message does not exist before it is sent on a time step. Lines 3–4 ensure that a message is always sent by the origin port of the message. Lines 5–8 state all messages are eventually sent and there is only one unique sending of that message. Lines 9–10 ensure all messages are not in any connector buffer until the sending of the message by an `OutPort`. Lines 11–12 ensure a message is not sent over a connector that is not shared between the message's origin and destination ports as the message would never reach its destination. Lastly, Line 13–15 specifies that all messages sent to all connectors shared between the origin `OutPort` and destination `InPort` and will either be in those connector's buffers or have been dropped by the connector. Further facts are then defined for receiving the message at the `InPort` along with facts to release a message that has been dropped or received.

```
1  fact MessageCreatedOnSend { always all m:Message |
2      (m not in Message until m.sentOn) or once m.sentOn }
3  fact MessageOriginCorrect { always all m:Message |
4    m.sentOn => m.sentBy[] = m.from }
5  fact UniqueSendInstant { all m:Message |
6    eventually m.sentOn always { all m:Message {
7      (m.sentOn => after always no sent.m)
8      (lone m.sentBy[])}}}
9  fact MessageNotBufferBeforeSend { always all m: Message |
10     (m not in Connector.buffer until m.sentOn) or once m.sentOn }
11 fact MessageNotInNonSharedBuffer { always all m: Message |
12   m not in (Connector - sharedConnector[m.from,m.to]).buffer }
13 fact MessageInBufferAfterSend { always { all m:Message |
14   all sb:sharedConnector[m.from,m.to] | m.sentOn[]
15   => m not in sb.buffer and after (m in sb.buffer or sb.drop[m])}}
```

Listing 1.6. Specification of Message sending invariants

Then, as depicted in Listing 1.7, we write predicates to specify sending and receiving operations of messages over component ports. Furthermore, we can provide an abstraction of the sending and receiving abstract predicates for `OutPorts` and `InPorts` by using the port's components as points of sending and receiving messages as in Lines 12–16. Then, we ensure that messages have the expect behavior conform to our metamodel presented in Sect. 2. Lines 17–18 ensure that a message does not exist before it was sent. Lines 19–20 ensure that all messages in buffers were once sent by an `OutPort`. Lastly, Lines 21–22 ensure all messages sent by an `OutPort` are either buffered by a connector or dropped.

```
1  pred OutPort.send[ip:InPort] {
2    some m:Message {
3      m.from = this
4      m.to = ip
5      m in this.sent }}
6  pred OutPort.multiSend[ips: set InPort] { all ip:ips | this.send[ip] }
7  pred InPort.receive[] {
8    some this.connectors[].buffer
9    some c:this.connectors[] | some m: c.buffer {
10     m.to = this
11     m in this.received }}
12 pred AtomicComponent.send[c:Component, o:Operation] {
13   this.outOperationPort[o].send[c.inOperationPort[o]]
14   some m:OperationCall { m.operation = o}}
15 pred AtomicComponent.receive[o: Operation] {
16   this.inOperationPort[o].receive[] }
17 check MessageNotInBufferBeforeSend { always all m:Message |
18  m.sentOn[] => historically no m.~buffer } for 4
19 check MessageInBufferSent { always all m: Message |
20   some m.~buffer => once m.sentOn[] } for 4
21 check AllSentMessageInBufferOrDropped { always all m:Message | m.sentOn[]
22   => after (some buffer.m or Connector.drop[m]) } for 4
```

Listing 1.7. Predicates specifying message sending and receiving

OperationCall and Data. Listing 1.8 shows the specification of the `Operation Call` which provides the ports with a method to use the interfaces and operations they realize by allowing the passing of messages containing an operation and arguments and invoking an action on the `InPort`'s component. The `operation` field of the `OperationCall` signature defines the operation that is performed to send data to the receiving `InPort` and an `arguments` field which maps the operation Parameters to data values. Line 6 states that the arguments of an `OperationCall` will not contain parameters that are not part of the operation's parameters. Lines 7–8 ensure there is one argument for every parameter of the `OperationCall`'s operation and the value of the parameter's argument is within the parameters data type or passable values (Variable and/or Constant). Line 9 and 10 state that the `OperationCall`'s operation should be included in the sender's and receiver's realized interfaces.

```
1  var abstract sig OperationCall extends Message{
2    var operation: one Operation,
3    var arguments: Parameter -> (Variable + Constant) }{
4    always {(operation' = operation and arguments' = arguments)
5      or this.release[]}
6    always no arguments[Parameter - operation.params]
7    always all p: operation.params |
8      one arguments[p] and arguments[p] in p.passes
9    always operation in from.realizes.operations
10   always operation in to.realizes.operations}
```

Listing 1.8. Formalization of OperationCall

Using the Alloy formalization of our metamodel[1], models can be built to study a system's functional and non-functional properties and the metamodel's components, connectors, ports, and messages can be extended to define additional structures and behaviors. When developing software architecture models, the use of composite components allows us to group together components into high-level entities which can help increase the readability of system requirements and manage the complexity of the natural hierarchical structure of a system's subcomponents. Architecture specifications can use composite component structures to specify and verify properties that encompass the components contained within a composite component to ensure the internal structures satisfy the requirements of the high-level entities represented by the composite components.

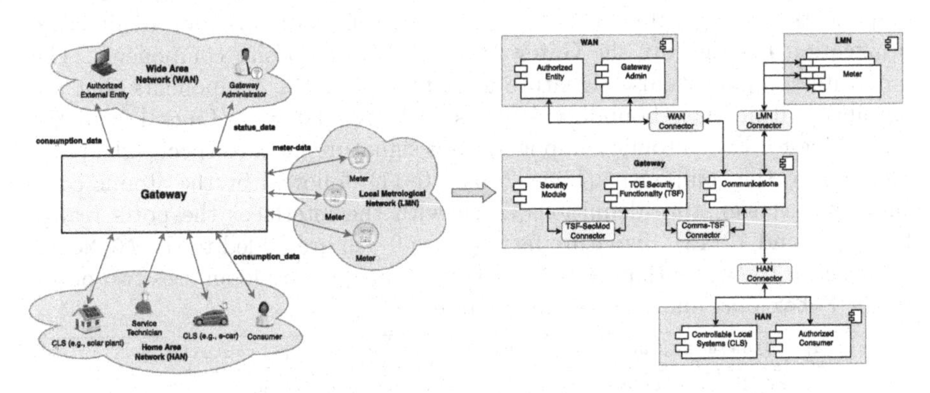

Fig. 2. Overview of the smart metering system (left) and its architecture (right), adapted from [2], organized into atomic (white) and composite (blue) components.

4 Illustrating the Use of the Formal Metamodel

To demonstrate how to adopt our formal reusable metamodel to model a software architecture, we specify a smart metering system model using Alloy. The smart metering system, as depicted in Fig. 2, is a simplified version of a system from a real Gateway Protection Profile for a smart metering system [2]. It connects a Local Metrological Network (LMN) which consists of several electricity meters that communicate measurements with the Gateway. It also communicates data with remote entities in a Wide Area Network (WAN) and a Home Area Network (HAN). The primary function of this system is to ensure that the measurement information is processed in the gateway and exchanged with (1) authorized external entities in the WAN, and (2) with authorized consumers in the HAN. For the specification, we adopt the naming convention from [2] to support traceability to the system requirements in the Gateway Protection Profile. In addition, we

[1] Available online at https://gitlab.com/CyberSEA-Public/CC-Metamodel.

use prefixes to indicate the type of entity described by the metamodel; Atomic Component (AC), Composite Component (CC), InPort (IN), OutPort (OUT), Connector (CON), Interface (IF), Operation (OP), Parameter (P).

For the purposes of this paper, we use the scenario of raw meter data being sent from the meters in the LMN for processing by the Gateway, and where the processed meter data is subsequently sent from the Gateway to the Authorized External Entities in the WAN. We also assume that all network (LMN, WAN, HAN) connectors are reliable to ensure all messages are received.

4.1 Structural Model of the Smart Metering System

The first step in specifying the smart metering system architecture using our formal metamodel is to translate system entities into atomic and composite components. Listing 1.9 shows the partial smart metering system architecture specification highlighting the Gateway and LMN composite components. The composite components use signature facts to specify the components that are contained within their boundary, such as with the Gateway (Lines 1) and the LMN (Line 8). Each atomic component uses signature facts to specify the ports belonging to the component (Lines 2 and 10–11), followed by the atomic component's port signature definitions along with the interfaces the ports realize (Lines 5–6 and 12–13). Signature facts (Lines 9–11) are added to the `AC_Meter` specification to ensure that each `AC_Meter` of our system model uses two ports: one `IN_Meter` port and one `OUT_Meter` port.

```
1  one sig CC_Gateway_TOE extends CompositeComponent {}{ contains =
       AC_Gateway_Comms + AC_Gateway_TSF + AC_Security_Module }
2  one sig AC_Gateway_Comms extends AtomicComponent {}{ uses = IN_GW_LMN +
       OUT_GW_LMN + IN_Comms_TSF + OUT_Comms_TSF + IN_GW_WAN + OUT_GW_WAN }
3  one sig IN_GW_WAN extends InPort {}{ realizes = IF_ExtEnt2GW +
       IF_Admin2GW }
4  one sig OUT_GW_WAN extends OutPort {}{ realizes = IF_GW2ExtEnt +
       IF_GW2Admin }
5  one sig IN_GW_LMN extends InPort {}{ realizes = IF_MTR2GW }
6  one sig OUT_GW_LMN extends OutPort {}{ realizes = IF_GW2MTR }
7  ...
8  one sig CC_LMN extends CompositeComponent {}{ contains = AC_Meter }
9  some sig AC_Meter extends AtomicComponent {}{ #uses = 2
10   one (uses & IN_Meter)
11   one (uses & OUT_Meter)}
12 some sig IN_Meter extends InPort {}{ realizes = IF_GW2MTR }
13 some sig OUT_Meter extends OutPort {}{ realizes = IF_MTR2GW }
14 one sig CON_LMN_NET extends ReliableConnector {}{ connects = IN_Meter +
       OUT_Meter + IN_GW_LMN + OUT_GW_LMN }
```

Listing 1.9. Specification of the Gateway and LMN as Composite Components

Next, we need to specify the operations and parameters for the interfaces realized by the atomic component's ports. Listing 1.10 provides an example of the interface from the Meters to the Gateway with the operation `OP_send_raw_data` specifying the meter is sending unprocessed meter data to the Gateway for processing. An architect would specify the operation needed for an interface and then all parameters that are a part of the operation and the data types for the parameters. In this case, the parameter `P_MeterData` can be any member of the `MeterData` set generated by Alloy in our system model during model-checking.

```
1 one sig IF_MTR2GW extends Interface {}{ operations = OP_send_raw_data }
2 one sig OP_send_raw_data extends Operation {}{ params = P_RawMeterData }
3 one sig P_RawMeterData extends Parameter {}{ passes = MeterData }
```

Listing 1.10. Meter-to-Gateway Interface Definition

4.2 Behavioral Model of the Smart Metering System

After specifying the structural elements of the smart metering system architecture, we specify the individual atomic component and connector behaviors.

Listing 1.11 demonstrates an action-based specification of the `AC_Meter` atomic component. The behavioural constraint on Lines 1–5 specifies the valid transition invariants of the `AC_Meter` by using the `always` temporal operation in Alloy to ensure the following Boolean expression is always true across traces. The transition invariant ensures all `AC_Meters` of the system is carrying out one of the three defined actions; `AC_Meter.generateMeterData[]`, `AC_Meter .sendMeterData[md:MeterData]`, or `AC_Meter.doNothing[]`. Each action is prepended with the Component signature name and specifies how the variable fields of the Atomic Component change with the action. The `AC_Meter. generateMeterData[]` action generates one MeterData and adds that MeterData to its mdata field which holds the MeterData to be sent to the Gateway. The `AC_Meter.sendMeterData[md:MeterData]` action sends some MeterData that is within the `AC_Meter`'s mdata field. Finally, the `AC_Meter.doNothing[]` enables the `AC_Meters` to do nothing for a time step of the system trace.

```
1  fact AC_Meter_trans { always all m:AC_Meter {
2    (m.generateMeterData or (some md:MeterData |
3        m.sendMeterData[md]) or m.doNothing)
4    all md:m.mdata | eventually m.sendMeterData[md]
5    some md:MeterData | some m.output[].sent => m.sendMeterData[md] }}
6  pred AC_Meter.generateMeterData[] { after one md:MeterData {
7    before md not in MeterData before this.mdata' = this.mdata + md
8    md.generated_by = this}}
9  pred AC_Meter.sendMeterData[md:MeterData] {
10   md.generated_by = this
11   md in this.mdata
12   this.send[AC_Gateway_Comms,OP_send_raw_data]
13   (arguments.md.P_RawMeterData & this.output[].sent).sentOn[]
14   one this.output[].sent
15   this.mdata' = this.mdata - md}
16 pred AC_Meter.doNothing[] { this.mdata' = this.mdata }
```

Listing 1.11. Meter Behaviour Specification

Similar to Listing 1.11, the transitions and actions of the Gateway's Communication Components are defined in Listing 1.12.

```
1  fact AC_Gateway_Comms_trans { always AC_Gateway_Comms.Comms_In_Guarantee}
2  pred AC_Gateway_Comms.Comms_In_Guarantee {
3    all i:this.input[].received { (i.operation = OP_send_raw_data and
4        this.forwardMeterData[i.arguments[P_RawMeterData]]) or
5      (i.operation = OP_send_p_mdata and this.forwardProcessedMeterData
6        [i.arguments[P_ProcessedMeterData]])}}
7  pred AC_Gateway_Comms.forwardMeterData[md:MeterData] {
8      after this.send[AC_Gateway_TSF,OP_process_mdata]
9      md in this.output[].sent'.arguments'[P_MeterDataComms]}
10 pred AC_Gateway_Comms.forwardProcessedMeterData[pmd:ProcessedMeterData]{
11     one oc:this.output[].sent' { pmd.dest in oc.to'.user[]
12       pmd in oc.arguments'[P_ExtEntPData]}}
```

Listing 1.12. Communication Component's Behavioural Specification

4.3 Verification of the Smart Metering System Model

Now that we have specified the smart metering system's architecture and behavior using our formal metamodel, we use the Alloy Analyzer to verify the desired properties of the model.

The two assertions in Listing 1.13 are derived from [2] and ensure all meter data is received, processed, and delivered to the WAN. The first assertion on Lines 1–4 states that all meter data shall leave the LMN in the operation call parameter `P_MeterData` and shall only be received by the Gateway. The second assertion on Lines 5–8 specifies that all processed meter data shall be received by the WAN, where the authorized entity is located, and the processed meter data shall have originated from the LMN.

```
1 assert AllMeterDataFromLMNisDeliveredToTheGateway {
2   always all md:MeterData | once md in CC_LMN.output[].sent
3     .arguments[P_MeterData] and md.opcalls[].receivedOn[]
4       => md.opcalls[].receivedBy[] in CC_Gateway_TOE.input[] }
5 assert WANOnlyReceivesProcessedMeterDataFromLMN {
6   always all pmd:ProcessedMeterData, wan_rcv: CC_WAN.input[].received |
7     some lmn_sent:CC_LMN.output[].sent.arguments[P_RawMeterData] |
8       arguments.pmd.P_ExtEntPData in wan_rcv => once pmd.p_meter_data in
            lmn_sent }
```

Listing 1.13. Composite Assertions

Both assertions generate no counterexample indicating the assertions are consistent with the specified model. The satisfaction of these properties demonstrates how we can verify properties specified over only the higher-level composite components which helps to manage the complexity of the specification and verification of system properties.

5 Related Work

Our work extends the work of Rouland et al. [16] which developed a component-port-connector metamodel defining the elements of a component-based system and for verifying functional requirements and security properties [14,15] on the specified architecture models. Specifically, our metamodel extends the existing metamodel to support the concept of composite components by include a hierarchical component tree structure. Further, our metamodel extends the message passing communication style considered in [16] with the *OperationCall* concept. This enables more fine-grained specification and verification of system properties in the context of specific operations and parameters, which is closer to practical designs of component and connector behaviors.

Wong et al. [18] provide an Alloy implementation of their approach for verifying multi-styled architectures by dividing complex systems into submodules defined by their architectural styles. However, each submodule has its own Alloy specification leading to difficulty in the ability to automate the approach. Wong et al. also use a component-port-connector metamodel with an action-based behavioral model to track events within system processes. In contrast, our contributions enable further definitions of component and connector behaviors and use ports to define interfaces of the component architecture to ensure conformance. Also, instead of manually dividing our system model into separate formal submodule specifications, our concept of a composite component automatically inherits the external ports of the components of which it is comprised.

The concept of composite components has also been proposed in the rCOS [9,13] method. In this work, the composition is accomplished by merging the contracts (interfaces semantic information) of the components. As a result, it differs from our solution where composite components automatically inherit ports. It offers a similar concept of interface, but it does not have a specific connector concept. Further, our work enables possibilities to further define reusable interaction types (i.e., connector behaviors).

A similar notion called compound components has been considered in the BIP Framework [1]. It allows to define a new component instance from existing components (atomic or not), but it requires specifying additional connectors in the process. Therefore, it is distinct from the automatic inheritance of ports proposed by our approach. On the other hand, it takes a similar approach for defining interaction types by defining connector behavior, but it lacks the concept of interface; instead, operations are specified directly in the interactions.

The notion of composite components has also been considered in the INVEST framework [10]. Similar to what we proposed in this paper, the INVEST framework groups components into 'complex components,' otherwise known as composite components, to form a hierarchical component structure. However, the INVEST framework lacks a suitable means to define a system's structure and operational interfaces, something that is handled by our formal metamodel as demonstrated in the illustrative example in Sect. 4.

Other approaches for developing hierarchical component systems have considered assume-guarantee (e.g., [17]) or contract-based design logic (e.g., [4]), but within these settings there is no agreed upon method to decomposing system compositionally. Some approaches (e.g., [6]) use specific connectors to group components, while others (e.g., [4,17]) do not consider group components, but verify changes of assumptions on system changes. Instead, our approach aims to provide architects with capabilities to model composite components and define the line between which components can be viewed as a black box with its own assumptions and guarantees at different levels of the component tree hierarchy.

6 Concluding Remarks

In this work, we proposed a metamodel to describe the high-level concepts of software architectures in a component-port-connector fashion. Specifically, we focused on providing hierarchical modeling capabilities by enabling the automatic construction of composite components from existing ones. The proposed metamodel was formalized using Alloy as a tooled formal language. With the formalization, we developed a reusable framework for modeling complex systems consisting of composite component structures that can be checked for architectural conformance as demonstrate on an illustrative smart metering system example. Such modeling capabilities help manage system complexity by enabling the specification and verification of properties on high-level entities while ensuring that internal structures satisfy the requirements of the high-level entities.

In future work, we aim to explore the use of compositional verification methods to further reduce the burden of verifying complex component-based systems by skipping the re-verification of certain properties when changes are localized to specific component structures. We also seek to develop a Domain Specific Language similar to [16], to simplify the specification of software architectures conforming to our proposed metamodel.

References

1. Basu, A., et al.: Rigorous component-based system design using the BIP framework. IEEE Softw. **28**(3), 41–48 (2011)
2. BSI: Protection Profile for the Gateway of a Smart Metering System (Smart Meter Gateway PP). Common Criteria Protection Profile BSI-CC-PP-0073, Bundesamt für Sicherheit in der Informationstechnik (2014)
3. Clarke, E.M., Long, D.E., McMillan, K.L.: Compositional model checking. In: Fourth IEEE Symposium on Logic in Computer Science (1989)
4. Collet, P., Malenfant, J., Ozanne, A., Rivierre, N.: Composite contract enforcement in hierarchical component systems. In: Lumpe, M., Vanderperren, W. (eds.) SC 2007. LNCS, vol. 4829, pp. 18–33. Springer, Heidelberg (2007). https://doi.org/10.1007/978-3-540-77351-1_3
5. Crnkovic, I., Larsson, M. (eds.): Building Reliable Component-based Software Systems. Artech House, New York City (2002)
6. He, N., et al.: Component-based design and verification in X-MAN. In: Embedded Real Time Software and Systems (ERTS2012). Toulouse, France, February 2012
7. Jackson, D.: Software Abstractions: Logic, Language, and Analysis. The MIT Press, Cambridge (2006)
8. Jackson, D.: Alloy: a language and tool for exploring software designs. Commun. ACM **62**(9), 66–76 (2019)
9. Jifeng, H., Li, X., Liu, Z.: rCOS: a refinement calculus of object systems. Theor. Comput. Sci. **365**(1–2), 109–142 (2006)
10. Johnson, K., Calinescu, R., Kikuchi, S.: An incremental verification framework for component-based software systems. In: 16th International ACM Sigsoft Symposium on Component-Based Software Engineering, pp. 33–42. CBSE '13 (2013)
11. Kruchten, P.: Architectural blueprints–the "4+1" view model of software architecture. IEEE Softw. **12**(6), 42–50 (1995)
12. Lehman, M.M.: Programs, life cycles, and laws of software evolution. IEEE **68**(9), 1060–1076 (1980)
13. Liu, Z., Jifeng, H., Li, X.: Contract oriented development of component software. In: Levy, J.-J., Mayr, E.W., Mitchell, J.C. (eds.) TCS 2004. IIFIP, vol. 155, pp. 349–366. Springer, Boston, MA (2004). https://doi.org/10.1007/1-4020-8141-3_28
14. Rouland, Q., Hamid, B., Jaskolka, J.: Specification, detection, and treatment of STRIDE threats for software components: modeling, formal methods, and tool support. J. Syst. Architect. **117**, 102073 (2021)
15. Rouland, Q., Hamid, B., Bodeveix, J.P., Filali, M.: A formal methods approach to security requirements specification and verification. In: 2019 24th International Conference on Engineering of Complex Computer Systems (ICECCS), pp. 236–241. IEEE (2019)

16. Rouland, Q., Hamid, B., Jaskolka, J.: Formal specification and verification of reusable communication models for distributed systems architecture. Futur. Gener. Comput. Syst. **108**, 178–197 (2020)
17. Saoud, A., Girard, A., Fribourg, L.: Assume-guarantee contracts for continuous-time systems. Automatica **134**, 109910 (2021)
18. Wong, S., Sun, J., Warren, I., Sun, J.: A scalable approach to multi-style architectural modeling and verification. In: 13th IEEE International Conference on Engineering of Complex Computer Systems (ICECCS 2008), pp. 25–34 (2008)

A Floating-Point Numbers Theory for Event-B

Idir Ait-Sadoune(✉)

Paris-Saclay University, CentraleSupelec, LMF Laboratory Plateau de Saclay, Gif-Sur-Yvette, France
idir.aitsadoune@centralesuplec.fr

Abstract. Static type checking helps catch errors in manipulating variables values early on, and most specification languages, like Event-B, are strongly typed. However, the type system of Event-B language is relatively simple and provides only a way to specify discrete behaviour using *Integer* type. There is no possibility to model continuous behaviour, which would have helped analyse hybrid systems. More precisely, the Event-B language doesn't consider in its type-checking system the possibility of defining such behaviours and checking the correctness of the values of the continuous variables within the Event-B models. In this article, we propose to extend the type-checking system of Event-B to include *Float* variables by specifying a floating point numbers theory using the theory plugin.

Keywords: Hybrid systems · Event-B · Type checking · Floating-point numbers

1 Introduction

Since its invention, the use of the Event-B formal method [2] has continued to increase, and it has been applied to various applications and domains [6]. The Event-B method is practical and adapted to analyse discrete systems, and its type system offers the possibility of modelling discrete behaviours. Today, with the need to model and analyse hybrid systems to include different types of complex and cyber-physical systems, extending the Event-B type-checking systems becomes necessary to specify and analyse continuous behaviours or represent numerical algorithms in Event-B. This need involves considering the definition of real numbers, more concretely, floating-point numbers.

The interest and motivation for using the floating-point arithmetic in the case of the classical B method [1] was discussed in [12]. Today, with classical B language, it is possible to specify a treatment with real numbers and to ensure that its floating point implementation is "close" to its specification. In the case

This work was supported by a grant from the French national research agency ANR ANR-19-CE25-0010 (EBRP Project https://www.irit.fr/EBRP/).

M. Mosbah et al. (Eds.): MEDI 2023, LNCS 14396, pp. 30–43, 2024.
https://doi.org/10.1007/978-3-031-49333-1_3

of the Event-B method, we can use discretisation like in [5] and other works cited in the same paper to formalise continuous behaviours. Another known tool around the classical B and Event-B methods is the ProB model-checker [13]. Currently, *Leuschel et al.* work on integrating the floating-point arithmetic in ProB [14].

The Rodin platform [3] is the most used développement environment in the Event-B ecosystem. Since it is an Eclipse product, it can be extended by adding plugins. Among all the plugins developed for the Rodin platform, the theory plugin [7,10] is mainly used to extend the Event-B modelling possibilities by defining theories. The theory plugin provides the facility to define mathematical and prover extensions. Mathematical extensions are new operator definitions and new datatype definitions, and axiomatic definitions. In this article, we propose to develop a floating-point numbers theory using the theory plugin to extend the Event-B type-checking system with the possibility of handling floating-point numbers. We note that there is a theory for reals in the "Standard Library" of theories[1].

This paper is organized as follows: Sect. 2 presents the main concepts of the Event-B method. Section 3 gives an example to illustrate why there is a need to use floating-point arithmetic. Section 4 details the proposed approach, and Sect. 5 shows how the proposed theories improve the motivating example. The paper concludes with a summary and outlook in Sect. 6.

2 The Event-B Method

The Event-B method [2] is an evolution of the classical B method [1]. This method is based on the notions of pre-conditions and post-conditions [11], weakest pre-condition [9], and the calculus of substitution [1]. It is a formal method based on first-order logic and set theory.

2.1 The Event-B Model

An Event-B model is made of several components of two kinds: machines and contexts. The machines contain a model's dynamic parts (states and transitions), whereas the contexts contain the static parts (axiomatization and theories). A machine can be refined by another machine, and a context can be extended by another. Moreover, a machine can see one or several contexts (see Listings 1.1 and 1.2).

A context is defined by a set of clauses (see Listing 1.1) as follows:

- `SETS` describes a set of abstract and enumerated types.
- `CONSTANTS` represents the constants used by a model.
- `AXIOMS` describes, in first-order logic expressions, the properties of the attributes defined in the `CONSTANTS` clause. Types and constraints are described in this clause as well.

[1] https://sourceforge.net/projects/rodin-b-sharp/files/Theory_StdLib/.

- THEOREMS are logical expressions that can be deduced from the axioms.

An Event-B machine is defined by a set of variables, described in the VARIABLES clause, that evolves thanks to events depicted in the EVENTS clause. It encodes a state transition system where the variables represent the state, and the events represent the transitions from one state to another.

Listing 1.1. The Event-B context

```
CONTEXT ctx1
EXTENDS ctx2

SETS s
CONSTANTS c
AXIOMS
    A(s,c)
THEOREMS
    T(s,c)
END
```

Listing 1.2. The Event-B machine

```
MACHINE mch1
REFINES mch2
SEES ctxi

VARIABLES v
INVARIANTS
    I(s,c,v)
THEOREMS
    T(s,c,v)
EVENTS
    < events_list >
```

Similarly to contexts, a machine is defined by a set of clauses (see Listing 1.2). Briefly, the clauses mean.

- VARIABLES represents the state variables of the specification model.
- INVARIANTS describes, by first-order logic expressions, the properties of the variables defined in the VARIABLES clause. Typing information and functional and safety properties are usually expressed in this clause. These properties need to be preserved by events.
- THEOREMS defines a set of logical expressions that can be deduced from the invariants.
- EVENTS defines all the events that occur in a given model. Each event is characterized by its guard and the actions performed when the guard is true. Each machine must contain an "*Initialisation*" event.

The refinement operation offered by Event-B encodes model decomposition. A transition system is decomposed into another transition system with more and more design decisions while moving from an abstract level to a less abstract one. A refined machine is defined by adding new events, new state variables and a glueing invariant. Each event of the abstract model is refined in the concrete model by adding new information expressing how the new set of variables and the new events evolve.

2.2 The Proof Obligations (PO)

Proof obligations (PO) are associated with any Event-B model. They define the formal semantics associated with each Event-B component. PO are automatically generated, and *the PO generator plugin* in the Rodin platform [3] is in charge of generating them. These PO need to be proved to ensure the correctness of developments and refinements. The obtained PO can be proved automatically or interactively by *the prover plugin* in the Rodin platform.

The rules for generating PO follow the substitutions calculus [1] close to the weakest precondition calculus [9]. To define some PO rules, we use the notations defined in Listings 1.1 and 1.2 where s denotes the seen sets, c the seen constants, and v the variables. Seen axioms are represented by $A(s, c)$ and theorems by $T(s, c)$, whereas invariants are denoted by $I(s, c, v)$ and local theorems by $T(s, c, v)$. For an event, the guard is denoted by $G(s, c, v, x)$ and the action is represented by the before-after predicate $BA(s, c, v, x, v')$ (a predicate expressing the relationship between the variable contents before and after an event triggering). Here we give a list of the most used/generated PO rules :

- *The theorem PO rule*: ensures that proposed theorems of a context or machine are provable.

$$A(s, c) \Rightarrow T(s, c)$$

$$A(s, c) \wedge I(s, c, v) \Rightarrow T(s, c, v)$$

- *Invariant preservation PO rule*: ensures that each event preserves each invariant in a machine.

$$A(s, c) \wedge I(s, c, v) \wedge G(s, c, v, x) \wedge BA(s, c, v, x, v') \Rightarrow I(s, c, v')$$

- *Feasibility PO rule*: ensures that a non-deterministic action is feasible.

$$A(s, c) \wedge I(s, c, v) \wedge G(s, c, v, x) \Rightarrow \exists v'.BA(s, c, v, x, v')$$

There are other rules for generating PO to prove the correctness of variables construction and using operators (Well definedness - WD) and refinement listed in [2].

2.3 The Theory Plugin

To extend the Event-B modelling possibilities with new mathematical objects, the theory plugin [7,10] extends the Rodin platform by providing a new syntax to define mathematical and prover extensions with the theory component. A theory can contain new datatype definitions, new polymorphic operator definitions, axiomatic definitions, theorems and associated rewrite and inference rules. The installation for the theory plug-in is available under the main Rodin Update site[2] under the category “Modelling Extensions”. If you have never used the theory plugin, consult the user manual available in this link[3].

In this work, we use the theory Plugin to define the power operator (with its axiomatic definitions, theorems and inference rules) and the floating-point numbers data type (with all its operators, theorems and associated rewrite and inference rules). The justification of why we need these two theories will be given in the following sections.

---

[2] http://rodin-b-sharp.sourceforge.net/updates.

[3] https://wiki.event-b.org/images/Theory_Plugin.pdf.

3 The Motivating Example

To illustrate our approach for extending the Event-B core with the floating-point numbers data type, we propose to model a system that continuously calculates a moving object's speed (cf. listing 1.3). The main objective of this example is to show some modelling and validation problems that we can face when we analyse physical phenomena, mainly when we use integer variables to handle small values and expressions that come from the laws of physics. For simplicity reasons, we ignore in this study all the problems related to the units of measurement and which will be treated in our future work.

Listing 1.3. An Event-B model calculating a moving object's speed.

```
MACHINE mch_integer_version
...
INVARIANTS
    @inv1: traveled_distance ∈ ℕ
    @inv2: measured_time ∈ ℕ₁
    @inv3: speed ∈ ℕ
    @inv4: starting_position ∈ ℕ
    @inv5: starting_time ∈ ℕ
    @inv6: speed = travelled_distance ÷ measured_time
    @inv7: traveled_distance > 0 ⇒ speed > 0

EVENTS
...
get_speed ≙
    any v t
    where
        @grd1: v ∈ ℕ₁ ∧ v > starting_position
        @grd2: t ∈ ℕ₁ ∧ t > starting_time
    then
        @act1: travelled_distance := v - starting_position
        @act2: measured_time := t - starting_time
        @act3: speed := (v - starting_position) ÷ (t - starting_time)
    end
END
```

The proposed Event-B model formalises two functional properties: **PROP 1 -** the speed of the moving object is equal to the *travelled_distance* divided by the *measured_time* ($v = d/t$), and **PROP 2 -** when the *travelled_distance* is strictly positive, the *speed* of the moving object must also be strictly positive (the object moves when its speed is different from zero). These two properties are formalised by the invariants @$inv6$ and @$inv7$ of listing 1.3.

The main event of the proposed Event-B model is called *get_speed*. It captures the new position of the moving object and calculates the new values of the *measured_time*, *travelled_distance*, and *speed* variables. These new values depend on the initial position stored in the *starting_time* and *starting_position* variables captured by another event that doesn't interest us in this study.

From the model validation point of view, we encountered a problem with the invariant preservation proof obligation of the @*inv*7 invariant generated for the *get_speed* event (cf. Fig. 1, all the OPs are green except the one maintaining the @*inv*7 invariant by the *get_speed* event). For recall, this invariant formalises the **PROP 2 -** property of our system (if the value of the *travelled_distance* variable is strictly positive, the *speed* variable must also be strictly positive). However, in the *get_speed* event, the value of the expression "*v* − *starting_position*" can be less than that of "*t* − *starting_time*". In this case, the new value of the *speed* variable becomes equal to zero while the one of the *travelled_distance* variable is not because all variables of our model are integer variables, and the "÷" Event-B operator makes an integer division. For these reasons, the `get_speed/inv7/INV` PO cannot be proved. Conceptually, our model correctly specifies our requirements; on the other hand, the basic types and operators of the Event-B language are not adapted to our needs and do not allow us to validate continuous behaviours requirements and manipulate small and big values simultaneously.

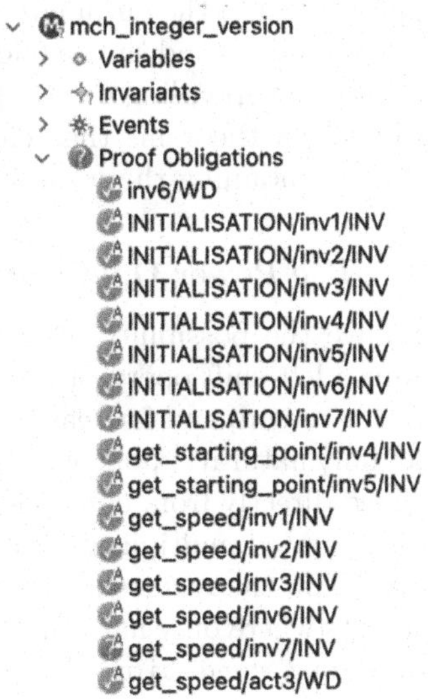

mch_integer_version
Variables
Invariants
Events
Proof Obligations
inv6/WD
INITIALISATION/inv1/INV
INITIALISATION/inv2/INV
INITIALISATION/inv3/INV
INITIALISATION/inv4/INV
INITIALISATION/inv5/INV
INITIALISATION/inv6/INV
INITIALISATION/inv7/INV
get_starting_point/inv4/INV
get_starting_point/inv5/INV
get_speed/inv1/INV
get_speed/inv2/INV
get_speed/inv3/INV
get_speed/inv6/INV
get_speed/inv7/INV
get_speed/act3/WD

Fig. 1. The summary of the generated and proven (or not) proof obligations.

For these reasons and those discussed in [12], we propose to develop a floating-point numbers theory using the theory plugin to extend the Event-B type-checking system with the possibility of handling floating-point numbers.

4 The Proposed Approach

As known, the floating point is the most used method for representing and approximating real numbers in computer-based arithmetic. Therefore, we propose to represent real numbers by using floating-point arithmetic. This approach represents floating-point numbers using an integer called the **significand**, scaled by an integer **exponent** of a fixed **base**. We have chosen that the base always equals ten in our models (see the following example).

$$x = 3.14159265359 = \underbrace{314159265359}_{\text{significand}} \times \underbrace{10}_{\text{base}}{}^{\overbrace{-11}^{\text{exponent}}}$$

To allow the Event-B language to embed this floating-point representation, we need to define two theories: the first one formalises the power operator that isn't included in the Event-B language (the "^" caret Event-B operator is not implemented in the automated proofs supported by the Rodin platform, besides power 0 and 1), and the second one formalises floating-point numbers by specifying the corresponding data type, the supported arithmetic operators, and some axioms and theorems that characterise the proposed modelling. The Event-B project containing these theories can be downloaded from this link[4].

4.1 The Power Operator

To have the possibility of comparing two floating point numbers, we need to define a left-shift operator that uses multiplication by powers of ten. In our case, for this reason, and for reasons we explain later, we define a power operator that uses only natural exponents. The power operation with natural exponents may be defined directly from multiplication operations. The definition of exponentiation as an iterated multiplication can be formalized using induction. The base case is $x^0 = 1$ and the recurrence is $x^n = x \times x^{n-1}$ (cf. Listing 1.4). For the case 0^0, in contexts where only natural powers are considered, 0 to the power 0 is undefined (see the `wd condition` defined for the `pow` operator in the Listing 1.4).

Listing 1.4. The theory defining the power operator

```
THEORY thy_power_operator

AXIOMATIC DEFINITIONS
   operators
   pow(x ∈ ℤ, n ∈ ℕ) : ℤ INFIX
       wd condition : ¬ (x = 0 ∧ n = 0)
   axioms
       @axm1: ∀ n. n ∈ ℕ₁ ⇒ 0 pow n = 0
       @axm2: ∀ x. x ∈ ℤ ∧ x≠0 ⇒ x pow 0 = 1
       @axm3: ∀ x,n. x ∈ ℤ ∧ x≠0 ∧ n ∈ ℕ₁ ⇒ x pow n = x × (x pow (n-1))
       ...
THEOREMS
   @thm1: ∀ x,n,m. ... ⇒ (x pow n) × (x pow m) = x pow (n+m)
   @thm2: ∀ x,n,m. ... ⇒ (x pow n) pow m = x pow (n×m)
   @thm3: ∀ x,y,n. ... ⇒ (x×y) pow n = (x pow n)×(y pow n)
   ...
END
```

The proposed theory also contains some proven theorems formalising some exponent rules (the product rule, the power rule, the multiplying exponents rule, ...). The proofs of all these theorems were made by induction following the rules defined in [8]. Notice that we have chosen to define the `pow` operator in a single theory to offer the possibility of reusing this operator in other Event-B components (theories, machines, or contexts) using the theory path mechanism available in the theory plugin [7,10].

4.2 The Floating-Point Numbers Theory

The proposed theory formalizes a floating-point number by defining a new data type called `FLOAT_Type`. This new data type provides the `NEW_FLOAT` constructor

[4] https://www.idiraitsadoune.com/recherche/modeles/eventb.theories.zip.

that allows creating a floating-point number by following the definition $x = s \times 10^e$ (with s representing the significand part and e the exponent part). This way, it is possible to create constants like 0 and 1 ($F0 = 0 \times 10^0$ or $F1 = 1 \times 10^0$) (cf. Listing 1.5). The proposed theory does not model limited precision floating point numbers. This implies that the operators defined in the theory involve no precision loss. This choice is made in order to allow the user to refine the proposed theory towards any implementation, the IEEE Standard 754, for example. This also allows us to remain compliant with the definition of the Event-B integer type, which is independent of any implementation. This theory provides an operator (FLOAT) to convert any Event-B integer to FLOAT_Type. We consider that the abstract numbers are those defined in the Event-B theory, and the concrete ones are those described by the IEEE Standard.

Listing 1.5. The theory defining the floating-point numbers (part 1)

```
THEORY thy_floating_point_numbers

DATATYPES
    FLOAT_Type ≙ NEW_FLOAT(s ∈ ℤ, e ∈ ℤ)

OPERATORS
    F0 ≙ NEW_FLOAT(0,0)
    F1 ≙ NEW_FLOAT(1,0)
    FLOAT(x ∈ ℤ) ≙ NEW_FLOAT(x,0)

    l_shift(x ∈ FLOAT_Type, offset ∈ ℕ) ≙
        NEW_FLOAT(s(x) × (10 pow offset), e(x)-offset)

    eq(x ∈ FLOAT_Type, y ∈ FLOAT_Type) INFIX ≙
        s(l_shift(x, e(x)-min({e(x),e(y)}))) =
        s(l_shift(y, e(y)-min({e(x),e(y)})))

    gt(x ∈ FLOAT_Type, y ∈ FLOAT_Type) INFIX ≙
        s(l_shift(x, e(x)-min({e(x),e(y)}))) >
        s(l_shift(y, e(y)-min({e(x),e(y)})))

    geq(x ∈ FLOAT_Type, y ∈ FLOAT_Type) INFIX ≙
        x eq y ∨ x gt y

    lt(x ∈ FLOAT_Type, y ∈ FLOAT_Type) INFIX ≙
        ¬(x geq y)

    leq(x ∈ FLOAT_Type, y ∈ FLOAT_Type) INFIX ≙
        ¬(x gt y)
    ...
END
```

The floating-point theory redefines all essential numeric operators (comparison and calculation operators), and the operator we have to define to overload all the numeric operators is the left shift operator (`l_shift` operator). This operator uses a positive offset to perform a left shift by multiplying the significand part by powers of ten[5]. To compare two numbers, we left-shift the number containing the biggest exponent to have the same exponent as the other number. Then, when two numbers have the same exponents, it's possible to compare them by comparing their significand parts. In this way, we have defined the operators `eq`, `gt`, `geq`, `lt`, and `leq` ($=$, $>$, $\geq$, $<$, $\leq$) comparing two floating-point numbers (cf. Listing 1.5).

Listing 1.6. The theory defining the floating-point numbers (part 2)

```
THEORY thy_floating_point_numbers
...
OPERATORS
    ...
    plus(x ∈ FLOAT_Type, y ∈ FLOAT_Type) INFIX ≙
        NEW_FLOAT(s(l_shift(x, e(x)-min({e(x),e(y)}))) +
        s(l_shift(y, e(y)-min({e(x),e(y)}))) , min({e(x),e(y)}))

    neg(x ∈ FLOAT_Type) ≙
        NEW_FLOAT(−1 × s(x), e(x))

    minus(x ∈ FLOAT_Type, y ∈ FLOAT_Type) INFIX ≙
        x plus neg(y)

    mult(x ∈ FLOAT_Type, y ∈ FLOAT_Type) INFIX ≙
        NEW_FLOAT(s(x) × s(y) , e(x) + e(y))

    f_pow(x ∈ FLOAT_Type, n ∈ ℕ) INFIX ≙
        NEW_FLOAT(s(x) pow n, n × e(x))
    ...
END
```

Using the same reasoning for the comparison, we have generalized the idea to the addition and subtraction operators. A left-shift of one of the two operands is necessary to perform the addition and subtraction operations (cf. Listing 1.6). However, the multiplication operation is performed by multiplying the significand parts of the two operands, and the resulting exponent is obtained by adding the exponent parts of the two operands (cf. Listing 1.6). The `f_pow` operator generalises the `pow` operator for the floating-point numbers.

While the proposed theory involves no precision loss for multiplication and addition, division sometimes induces a precision loss. For example, we cannot precisely represent the result of 1/3 or 2/3. That is why, for the case of the division and inverse operators, we have firstly defined the well-defined conditions (by the `inv_WD` and `div_WD` operators in listing 1.7). To calculate the inverse of x, we must find a z, which we multiply by the significand part of x to obtain a power of ten (The value of z corresponds to the significand part of the result of the inverse of x) . For example, to calculate the inverse of 2, 5 corresponds to

[5] This is why we have defined a power operator with only natural exponents.

z in our case, which does not exist for the inverse of 3. The same reasoning is done for the division operator.

Listing 1.7. The theory defining the floating-point numbers (part 3)

```
THEORY thy_floating_point_numbers
...
OPERATORS
    ...
    inv_WD(a ∈ FLOAT1_Type) ≙
        ∃ n,z. n ∈ℕ ∧ z ∈ℤ ∧ 10 pow n = s(a) × z

    div_WD(a ∈ FLOAT_Type , b ∈ FLOAT1_Type) ≙
        ∃ n,z. n ∈ℕ ∧ z ∈ℤ ∧ s(a) × (10 pow n) = s(b) × z

AXIOMATIC DEFINITIONS
    operators
    inv(x ∈ FLOAT_Type) : FLOAT1_Type
        wd condition : inv_WD(x)
    axioms
        @inv_1: ∀ x,y.(... ⇒ ((x mult y) = F1 ⇔ inv(x) = y))
        @inv_2: ∀ x,y.(... ⇒ ((x mult y) eq F1 ⇔ inv(x) eq y))

    operators
    div(x ∈ FLOAT_Type , y ∈ FLOAT_Type) : FLOAT_Type INFIX
        wd condition : div_WD(x,y)
    axioms
        @div_1: ∀ x,y,z.(... ⇒ ((y mult z) = x ⇔ (x div y) = z))
        @div_2: ∀ x,y,z.(... ⇒ ((y mult z) eq x ⇔ (x div y) eq z))
        @div_3: ∀ x,y.(... ⇒ x mult inv(y) = x div y)
    ...
END
```

The last basic arithmetic operations, inverse and division, are formalized by axiomatic definitions, and both are invocable if their well-defined conditions are true (defined in the `wd condition` clause). The inverse of x is y if and only if y is the number we multiply by x to obtain $F1$, and the result of dividing x by y is z, if and only if z is the number we multiply by y to obtain x (cf. Listing 1.7). We must prove the `WD PO` generated from the `wd condition` for both operators. The axiom @*div_3* gives the relationship between the inverse operator and the division operator.

Finally, the floating-point data type is often used in laws of physics and scientific calculations. Functions calculating the integer part, the fractional part, the floor function and the ceiling function are very useful. This theory provides all these operators, and due to the page number limitations, these operators are not presented in this article. The reader may consult them by downloading this theory from this link[6].

The last part of the proposed theory contains a set of theorems that we have proved, and that correspond to laws defining properties of arithmetic operators

[6] https://www.idiraitsadoune.com/recherche/modeles/eventb.theories.zip.

(equality, addition, and multiplication are commutative, the order is total, reflexive, anti-symmetric and transitive, addition and multiplication have an inverse, ...) and others theorems combining the comparison operators and the arithmetic operators (cf. Listing 1.8).

Listing 1.8. The theory defining the floating-point numbers (part 4)

```
THEORY thy_floating_point_numbers
...
THEOREMS
    @thm1: ∀ x,y.(... ⇒ x eq y ⇔ y eq x)
    @thm2: ∀ x.(... ⇒ x geq x ∧ x leq x)
    @thm3: ∀ x,y.(... x leq y ∧ y leq x ⇒ x eq y)
    @thm4: ∀ x,y.(... ⇒ x leq y ∨ y leq x)
    @thm5: ∀ x,y,z.(... x leq y ∧ y leq z ⇒ x leq z)
    @thm6: ∀ x,y,z.(... x leq y ⇒ (x plus z) leq (y plus z))
    @thm7: ∀ x,y,z.(... x leq y ⇒ (x mult z) leq (y mult z))

    @thm8: ∀ x.(... ⇒ x plus F0 eq x)
    @thm9: ∀ x,y.(... ⇒ x plus y = y plus x)
    @thm10: ∀ x,y.(... ⇒ x plus neg(y) = y minus x)
    @thm11: ∀ x.(... ⇒ x minus F0 eq x)
    @thm12: ∀ x.(... ⇒ x minus x eq F0)

    @thm13: ∀ x.(... ⇒ x mult F0 eq F0)
    @thm14: ∀ x.(... ⇒ x mult F1 = x)
    @thm15: ∀ x,y.(... ⇒ x mult y = y mult x)

    @thm16: ∀ x.(... ⇒ inv(x) = F1 div x)
    @thm17: ∀ x.(... ⇒ x div F1 = x)
    @thm18: ∀ x.(... ⇒ x div x = F1)
    @thm19: ∀ x.(... ⇒ x mult inv(x) = F1)
    ...
END
```

Due to our choice to formalise unlimited precision floating-point numbers (the operators defined in the proposed theory involve no precision loss), we can deduce some properties that are not true in the floating-point numbers world (the associativity of addition and multiplication, for example). When this theory is refined towards any implementation (the IEEE Standard 754, for example), the developer must pay attention to this point.

5 Revisiting the Motivating Example

The example presented in Sect. 3 is updated to use the floating-point numbers theory. All NATURAL variables are typed by PFLOAT_Type set containing positive floating-point numbers (cf. Listing 1.9), and the rest of the model was adapted using the equivalent operators from the proposed theory. The obtained Event-B machine contains almost the same invariants and the same events (cf. Listing 1.10). The only difference is the addition of the invariant @*inv*6 concerning the well-defined condition of the division operator used to formalise the speed of the moving object (**PROP 1** of the motivating example). Thus, **PROP 1** and **PROP 2** of the initial model are formalised by the invariants @*inv*7 and @*inv*8 (cf. Listing 1.10).

Listing 1.9. The definition of the positive floating-point numbers

```
THEORY thy_floating_point_numbers
    ...
    PFLOAT_Type = { x · x ∈ FLOAT_Type ∧ s(x) ≥ 0 | x }
END
```

Listing 1.10. The new version of the model calculating the speed of a moving object

```
MACHINE mch_floating_point_version
...
INVARIANTS
    @inv1: traveled_distance ∈ PFLOAT_Type
    @inv2: measured_time ∈ PFLOAT_Type ∧ s(measured_time) ≠ 0
    @inv3: speed ∈ PFLOAT_Type
    @inv4: starting_position ∈ PFLOAT_Type
    @inv5: starting_time ∈ PFLOAT_Type
    @inv6: div_WD(traveled_distance, measured_time)
    @inv7: speed eq traveled_distance div measured_time
    @inv8: traveled_distance gt F0 ⇒ speed gt F0

EVENTS
...
get_speed ≙
    any v t
    where
        @grd1: v ∈ PFLOAT_Type ∧ v gt starting_position
        @grd2: t ∈ PFLOAT_Type ∧ t gt starting_time
        @grd3: div_WD(v minus starting_position, t minus starting_time)
    then
        @act1: traveled_distance := v minus starting_position
        @act2: measured_time := t minus starting_time
        @act3: speed := (v minus starting_position) div (t minus starting_time)
    end
END
```

From the model validation point of view, contrary to the initial model, all generated proof obligations have been proven. The problem with the invariant linked to the integer division operator no longer arises. As shown in the Fig. 2, the `get_speed/inv8/INV` PO becomes green, and it has been proven using the interactive prover of the Rodin platform. As we have said, the @*inv*8 formalises the following property : if the value of the *travelled_distance* variable is strictly positive, the *speed* variable must also be strictly positive. Even if in the *get_speed* event, the value of the expression "*v minus starting_position*" can be less than that of "*t minus starting_time*", the new value of the *speed* variable is never equal to zero because the value of "*v minus starting_position*" is also never equal to zero (thanks to the guard @*grd*1 of the *get_speed* event). All this is possible thanks to the new `div` operator specification, which acts on the floating-point numbers.

This is one of the reasons that allow us to conclude that our floating-point numbers theory is more suitable than the basic integers of Event-B in modelling hybrid systems and continuous behaviours.

Fig. 2. The summary of the generated and proven POs of the new Event-B machine.

6 Conclusion

In this article, we have proposed an approach using the theory plugin to extend the Event-B type-checking system with the possibility of handling floating-point numbers. We have developed a floating-point numbers theory that formalises a floating-point number using an integer called the **significand**, scaled by an integer **exponent** of a fixed **base** (equals ten in our theory). Our proposition includes an extension of the Event-B power operator to handle powers of ten more than 0 and 1. We have proposed an abstract representation of the floating-point numbers to offer the possibility to refine the proposed theory to any more concrete implementation (the IEEE standard, for example).

For the next step of our work, we consider the floating-point numbers theory as the first step before developing a more general theory that will formalise the standard units of measurement defined by the International System of Units (SI). Such theories will be helpful in modelling cyber-physical, and these works will be integrated into our framework [4] for generating the Event-B model from ontologies that can define concepts in the context of hybrid systems.

References

1. Abrial, J.: The B-book - assigning programs to meanings. Cambridge University Press, Cambridge (1996). https://doi.org/10.1017/CBO9780511624162
2. Abrial, J.R.: Modeling in Event-B: System and Software Engineering. Cambridge University Press, Cambridge (2010). https://doi.org/10.1017/CBO9781139195881
3. Abrial, J., Butler, M.J., Hallerstede, S., Hoang, T.S., Mehta, F., Voisin, L.: Rodin: an open toolset for modelling and reasoning in Event-B. Int. J. Softw. Tools Technol. Transf. **12**(6), 447–466 (2010). https://doi.org/10.1007/s10009-010-0145-y
4. Ait-Sadoune, I., Mohand-Oussaid, L.: Building formal semantic domain model: an Event-B based approach. In: Schewe, K.-D., Singh, N.K. (eds.) MEDI 2019. LNCS, vol. 11815, pp. 140–155. Springer, Cham (2019). https://doi.org/10.1007/978-3-030-32065-2_10
5. Babin, G., Aït-Ameur, Y., Singh, N.K., Pantel, M.: Handling continuous functions in hybrid systems reconfigurations: a formal Event-B development. In: Butler, M., Schewe, K.-D., Mashkoor, A., Biro, M. (eds.) ABZ 2016. LNCS, vol. 9675, pp. 290–296. Springer, Cham (2016). https://doi.org/10.1007/978-3-319-33600-8_23
6. Butler, M.: The first twenty-five years of industrial use of the B-method. In: ter Beek, M.H., Ničković, D. (eds.) FMICS 2020. LNCS, vol. 12327, pp. 189–209. Springer, Cham (2020). https://doi.org/10.1007/978-3-030-58298-2_8
7. Butler, M., Maamria, I.: Practical theory extension in Event-B. In: Liu, Z., Woodcock, J., Zhu, H. (eds.) Theories of Programming and Formal Methods. LNCS, vol. 8051, pp. 67–81. Springer, Heidelberg (2013). https://doi.org/10.1007/978-3-642-39698-4_5
8. Cervelle, J., Gervais, F.: Introducing inductive construction in B with the theory plugin. In: Glässer, U., Creissac Campos, J., Méry, D., Palanque, P. (eds.) ABZ 2023. LNCS, vol. 14010, pp. 43–58. Springer, Cham (2023). https://doi.org/10.1007/978-3-031-33163-3_4
9. Dijkstra, E.W.: A Discipline of Programming. Prentice-Hall, Hoboken (1976)
10. Hoang, T.S., Voisin, L., Salehi, A., Butler, M.J., Wilkinson, T., Beauger, N.: Theory Plug-in for Rodin 3.x. CoRR abs/1701.08625 (2017). http://arxiv.org/abs/1701.08625
11. Hoare, C.A.R.: An axiomatic basis for computer programming. Commun. ACM **12**(10), 576–580 (1969). https://doi.org/10.1145/363235.363259
12. Lecomte, T., Burdy, L., Dufour, J.L.: The B method takes up floating-point numbers. In: Embedded Real Time Software and Systems (ERTS2012) (2012)
13. Leuschel, M., Butler, M.: ProB: a model checker for B. In: Araki, K., Gnesi, S., Mandrioli, D. (eds.) FME 2003. LNCS, vol. 2805, pp. 855–874. Springer, Heidelberg (2003). https://doi.org/10.1007/978-3-540-45236-2_46
14. Rutenkolk, K.: Extending modelchecking with ProB to floating-point numbers and hybrid systems. In: Glässer, U., Creissac Campos, J., Méry, D., Palanque, P. (eds.) ABZ 2023. LNCS, vol. 14010, pp. 366–370. Springer, Cham (2023). https://doi.org/10.1007/978-3-031-33163-3_27

Model-Based Testing Approach for EIP-1559 Ethereum Smart Contracts

Mohamed Amin Hammami and Mariam Lahami(✉)

ReDCAD Laboratory, National School of Engineering of Sfax, University of Sfax, Sokra Road km 4, 1173 Sfax, Tunisia
mariam.lahami@redcad.org

Abstract. Smart contracts are computer programs that are deployed and executed on the blockchain without the need of third parties. They are characterized by their immutability because once deployed, they cannot be modified. Thus, it is highly demanded to verify and validate them at development phase before their deployment. This work introduces a Model-Based Testing (MBT) approach for checking functional and execution related properties of Ethereum smart contracts. Our MBT solution supports the transaction pricing mechanism set by the Ethereum Improvement Proposal EIP-1559. It consists of four steps: (1) modelling the smart contract and its blockchain environment as UPPAAL Timed Automata while defining the contract gas usage regarding the EIP-1559 proposal, (2) generating abstract test cases, (3) executing dynamically the obtained tests, and at the end (4) analyzing and reporting the obtained test results. To illustrate the feasibility of our MBT approach, tests for the smart banking case study are generated and executed.

1 Introduction

Blockchain technology has gained a lot of attention during the last decade from academic researchers and several industries [1], including supply chain management, intelligent transportation, e-health, etc. As a decentralized system architecture initially introduced by Satochi Nakamoto [2], it is characterized with a linked chain of blocks in which transactions are securely stored. The most important features which have boosted the interest in this technology are security, decentralization and immutability. For example, the immutability feature is supplied by sharing identical copies of the ledger among several peer-to-peer nodes, while security is ensured through the use of cryptographic algorithms.

Recently, the emergence of smart contracts has extended these features. In fact, Ethereum platform is growing rapidly and according to the Ethereum statics[1], the total number of created smart contracts in 2022 have reached 1.45 million. A smart contract is defined as an immutable software program which is deployed and executed on the blockchain infrastructure. Nevertheless, multiple functional and security issues may occur during the design and the development

[1] https://www.alchemy.com/overviews/ethereum-statistics.

M. Mosbah et al. (Eds.): MEDI 2023, LNCS 14396, pp. 44–57, 2024.
https://doi.org/10.1007/978-3-031-49333-1_4

of these smart contracts. For instance, 3.6 million of Ether, around 50 million dollars, were lost in the well-known "DAO attack", due to the famous reentrancy vulnerability [3]. To avoid such attacks and the potential loss of funds due to smart contract failures, it highly required to verify and check their correctness.

For this reason, a recent branch of work has adopted Verification and Validation (V&V) techniques to ensure the trustworthiness and the correctness of Blockchain oriented Software (BoS) [4,5]. The most used V&V techniques in this context are model checking [6–8], theorem proving [9,10] and software testing [11–14]. However, proposing Model-based Testing (MBT) approaches for BoS that automate the generation of abstract test suites from abstract models and also perform test execution and test reporting has been rarely addressed [15,16] and without taking into consideration the modelling of the gas mechanism following the EIP-1559.

To overcome this limitation, we introduce an extension of our previous model-based testing approach for BoS, called *MBT4BoS*, that tests Ethereum smart contracts for detecting functional bugs [16]. The novelty in this paper is that we take into account the EIP-1559 standard while modeling transactions and gas related properties. To do so, we make use of UPPAAL model checker and its timed automata formalism to model smart contracts and their blockchain environment. Furthermore, we exploit especially its model-based testing module (UPPAAL Yggdrasil) [17] to generate test cases since the UPPAAL Co$\surd$er tool used in our previous work has not been updated anymore. The major contribution here is that obtained tests check functional aspects and also the gas related properties of ethereum transactions following the EIP-1559 standard. Thus, transaction modelling is enhanced to support such improvement protocol.

The rest of this paper is organized as follows. Section 2 provides background materials on blockchain technology, the gas mechanism and the EIP-1559 standard. Subsequently, the proposed approach is outlined in Sect. 3. Afterward, its application to a small banking system is highlighted in Sect. 4. At the end, Sect. 5 concludes the paper while giving a summary about our main contributions, and identifying possible areas of future research.

2 Theoretical Background and Definitions

To properly comprehend our contribution in the next sections, it is crucial to provide briefly some theoretical key concepts related to Blockchain (BC), Smart Contracts (SCs), the gas mechanism of Ethereum and EIP-1559 standard.

2.1 Blockchain and Smart Contracts

The Blockchain. It is a distributed and decentralized register of transactions. It is stored and updated simultaneously on a peer-to-peer network, each node keeping in permanently the most recent version of the register. It offers the possibility of recording, simultaneously for each user, an operation, transaction or event without the need of third parties. These irreversible transactions are ordered and grouped into blocks. For each transaction, the blockchain records the address of the sender, the address of the recipient and the data transferred to the whole network. The blockchain stores one or more transactions in a block and encrypts the contents of the block by the use of cryptographic functions into a single value called a hash. This hash can be viewed at any time by anyone on the blockchain. The executed transactions cannot be modified or deleted from the distributed ledger.

Smart Contracts. The concept of smart contract (SC) first appeared in 1997 by the American computer scientist Nick Szabo [18]. It has gained more and more attention thanks to the emergence of public blockchains, such as Ethereum. SC is a computer program executed by a network of peer-to-peer nodes, guaranteed not by a central authority, but by cryptography and blockchain technology. It provides a coordination and enforcement framework for agreements between network participants, without the need for traditional legal contracts. In blockchain, smart contracts are deployed and executed by specific types of transactions and can be used to transfer digital currency, record information and also interact to other systems. In Ethereum, smart contracts are commonly written in the Solidity language and then they are compiled to the Ethereum Virtual Machine (EVM) bytecode. A SC is publicly accessible, transparent, and immutable. Therefore, the immutability feature makes its code tamper-proof. It is extremely expensive to fix an issue once it has been deployed on the blockchain since a new smart contract needs to be created. Thus, it is essential to validate smart contract reliability and safety before deploying it on the blockchain infrastructure.

2.2 The Gas Mechanism

In Ethereum, a single cryptographically signed instruction created by an externally owned account is referred to as a transaction. This transaction object includes mainly two fields: a gasLimit and a gasPrice. The gasPrice displays the unit's current market price in Wei. In fact, a gas is a unit that describes basic computing operations. The execution of one atomic instruction, or bytecode, equals one unit of gas. For instance, obtaining the balance of a specific account takes 400 gas but multiplying is a simple operation that only needs a small number of processing units (5 gas). The gasLimit is the maximum amount of gas that may be burned in order to complete the transaction. The total amount of gas required for the execution of a given smart contract relies on the number of instructions run by the EVM and also their types. Prior to the London upgrade, the total transaction fee is calculated as follows:

$$txFee = Gas\ unit(limits) * gasPrice\ per\ unit. \tag{1}$$

This gas mechanism proposed by Ethereum accomplishes two main goals: it controls resource usage and pays miners for their labor. The creator of a transaction has to pay this fee to the miner that validates and commits the transaction and includes it into a block [19]. After the London hard fork update, EIP-1559 has been proposed in order to make transactions fees less volatile and more predictable.

2.3 EIP-1559

The major problem with the historical gas mechanism is that prices can fluctuate very wildly based on sudden spikes in demand for Ethereum's limited free block space. Users are always uncertain about the right price level when they submit a transaction and often have to overpay to be sure that it will be included in the next block. To address these problems, a novel gas fee mechanism was introduced and implemented as an Ethereum improvement called EIP-1559.

With this new mechanism, variable-sized blocks are now required instead of fixed-sized blocks. Consequently, it proposes a new transaction fee calculation as given in the following equation:

$$txFee = Gas\ units(limit) * (Basefee + tip) \tag{2}$$

where;

- The *Base fee*: it is the block's network fee per gas determined by the network itself and it will be burnt. The base fee per gas increases when blocks are above the gas target (i.e., block gas limit divided by a given elasticity multiplier), it decreases when blocks are below the gas target. In other words, the base fee is sensitive to the size of the previous block [20].
- The *max priority fee (tip)*: is specified by the creator of the transaction to be paid to the miner of the block that includes the transaction. Although the tip is optional, it is included to speed up transactions.
- The *max fee per gas*: is the maximum fee per gas unit that users specify and they are willing to pay in order to get their transactions included into a block. A given transaction will be included in a block only if the max fee per gas is greater than or equal to the base fee [20,21].

In our work, we make use of this novel standard to model transactions in Ethereum blockchain and its gas fee mechanism.

3 MBT Approach for Ethereum Smart Contracts

Model-based Testing (MBT) is an automated approach which consists on generating abstract test cases on the basis of abstract model of the System Under Test (SUT). The primary justification for choosing model-based testing is that its main goal is to automate manual processes by decreasing the cost of producing models for coverage and minimizing the time and effort required to create and build test cases. Therefore, we apply this black-box testing technique in the context of Ethereum smart contracts to speed up and automate the testing activities.

The proposed approach is highlighted in Fig. 1 that outlines an overview of its different constituents. The first module is used to model the system under test, from the functional requirements or from a specification file of the system under test. In our case, we adopt UPPAAL's timed automata to formally model smart contracts. The second module consists in generating test cases from the smart contract model we have designed. Then, the third module is used to translate the generated abstract test cases into concrete and executable tests. The last one focuses on the generation of the test report containing the test results. Deeper discussion of these modules is provided in the next subsections.

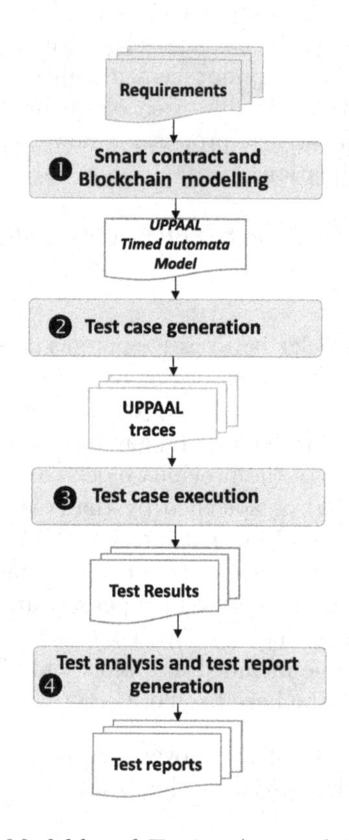

Fig. 1. Model-based Testing Approach for BoS.

3.1 Smart Contract and Blockchain Modelling

First of all, we conceive a formal test model that specifies the expected SUT behaviours with reference to its requirements. To that aim, we make use of the most popular and widespread formalism for specifying real-time and critical systems, named Timed Automata (TA). Indeed, we adopt the UPPAAL's timed automata formalism to model not only the smart contract but also its blockchain environment by producing a network of timed automata.

Regarding the smart contract model, a timed automata is defined by the tuple

$$(S, s_0, \mathcal{A}\text{ct}, \mathcal{C}, \mathcal{V}, \mathcal{T}), \text{ where:}$$

- S: a finite set of states.
- $s_0 \in S$: the initial state and $i_0 \in \mathcal{I}$ indicates the initial input action corresponding to the smart contract's constructor.
- $\mathcal{A}ct$: a finite set of Input and Output actions. The Input actions are related to smart contract function calls.
- $\mathcal{C}$: a finite set of clocks defined to model temporal constraints.
- $\mathcal{V}$: the collection of state variables. Each variable $x \in \mathcal{V}$ is seen as a global variable that may be accessed at any state $s \in S$.
- $\mathcal{T}$: a finite set of transitions, where $e = \langle l, g, r, a, l' \rangle \in \mathcal{T}$ corresponds to the transition from l to l', g is the guard associated to e, r is the set of clock to be reset and a is a label of e. We note $l \xrightarrow{g,r,a} l'$.

Regarding the blockchain modelling, our approach is specific to Ethereum Blockchain and we consider only accounts, transactions and gas mechanism following the EIP-1559 improvement. Modelling blocks, consensus algorithms and mining process are out the scope of this work. As presented in the Ethereum Yellow paper [22], a smart contract account or an externally owned account are both possible types of Ethereum accounts. Both of them have a unique identifier named *address* as well as other fields like a balance which indicates how many Wei belong to this address, a codeHash, an EVM code of this account and a storageRoot which represents the root node of a Merkle Patricia tree that encodes the account's storage contents.

We assume that an ethereum transaction has four[2] states *created, confirmed, reverted and rejected.* Moreover, the transaction fee (txFee) is calculated following the EIP-1559 as shown in the Eq. (2) introduced in Sect. 2.3:

- A given transaction is **created** when the constructor of the smart contract is called and the creator has enough ether in his account to execute such deployment transaction: $Balance >= txFee$.
- It is **confirmed** when the sender of the transaction has enough ether in his account to perform it and this requirement is met: $maxFee >= txFee$.
- It can be **rejected** if transaction fee exceeds the maximum fee: $maxFee < txFee$.
- It can be **reverted** if the user's account balance is insufficient to cover the transaction fee: $Balance < txFee$.

[2] Note that the pending state in which transaction in the pool waiting for minor validation is out the scope of this paper.

3.2 Test Case Generation

In our work, the test generation process is fully automated since we are based on a model-based testing approach that generates the required number of test cases from the abstract test model. Each produced abstract test case generally consists of a sequence of high-level SUT actions, each of which has associated input parameters and expected results.

In our case, the test suites were generated from the model using the UPPAAL Test Generator (Yggdrasil) [17]. The Yggdrasil tab includes an offline test-case generating tool with the aim of enhancing edge coverage in order to produce test cases. It generates traces from the test model, and translates them into test cases based on test code entered into the model on edges and locations.

3.3 Test Case Execution

To execute the generated tests, we have implemented a test tool named *BC Test Runner* which makes it possible to automate test execution by stimulating the smart contracts deployed locally on the Ganache blockchain, as well as the generation of test reports. As shown in Fig. 2, this test tool is composed of two parts including a front-end and a server-side backend. The front-end, allows testers to put two inputs as follows: a set of test cases generated from the test model given by UPPAAL Test Generator (Yggdrasil) and after compiling the smart contract, we obtain smart contract artefact as a Json file. This file contains all the specifications of the smart contract. The back-end has many modules: such as *Test Executor*, *Test result analyzer* and *Report generator*. Through the Web3.js library, we can communicate within the deployed smart contract.

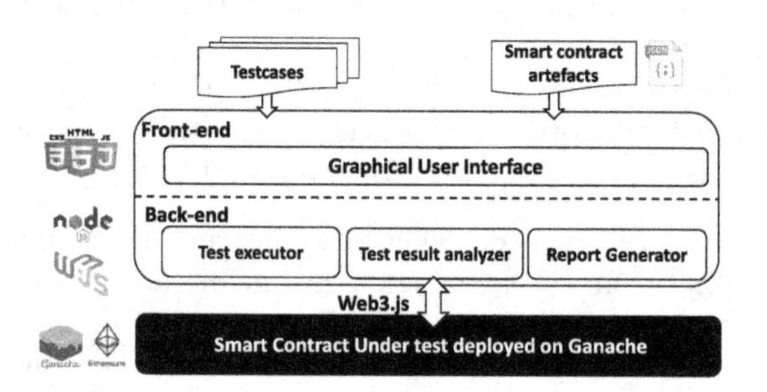

Fig. 2. Architecture of BC Test Runner tool.

The *Test Executor* module serves the purpose of executing test cases and interacting with the smart contract. It retrieves essential information, such as the contract's address and ABI (Application Binary Interface), from the Json file. The ABI provides a detailed description of the smart contract's functions, including their names, parameters, return types, and other relevant specifications. Using the test cases stored in a separate Text file, where each test case consists of input values and expected results separated by (/), the *Test Executor* module sends the input values to the deployed smart contract. Then, it captures the generated results and compares them against the expected results.

Based on this comparison, this module generates a verdict for each test case, indicating whether it has passed or failed. This crucial assessment ensures that the smart contract performs as intended and produces the expected outcomes, allowing for effective testing and validation of its functionality.

3.4 Test Analysis and Test Report Generation

This process involves the examination of the obtained test results, which are recorded into log files during the test execution, and the generation of test reports. To do such task, BC Test Runner tool incorporates the module *Test Result Analyzer* that calculates the percentage of Pass verdicts and Fail verdicts. Subsequently, the *Report Generator* module generates test reports in the form of trace text files.

4 Prototype Implementation

Before showing the feasibility of our approach and its fault detection capability, we introduce the prototype implementation details.

4.1 Development Tools

In this subsection, we present the development tools, that we used for the implementation of our test tool.

Ganache[3] is a local blockchain that allows developers to develop, deploy and test their distributed applications in a safe and deterministic environment. This tool is mainly used to test Ethereum contracts locally. It creates a simulation of a blockchain that allows anyone to use multiple accounts.

Truffle[4] is a very familiar tool for developers to create a smart contract project. It provides us with a project structure, files and folders that facilitate deployment and testing of Ethereum smart contract.

web3.js[5] is a library that allows users to interact with the blockchain. Additionally, web3.js is a collection of libraries for performing actions like sending Ether from one account to another, and reading and writing data from smart contracts.

[3] https://trufflesuite.com/ganache/.
[4] https://trufflesuite.com/.
[5] https://web3js.readthedocs.io/en/v1.10.0/.

4.2 Test Tool Implementation

This section provides an introduction to *BC Test Runner*, our testing tool developed using JavaScript and HTML. BC Test Runner is designed to seamlessly connect with the local blockchain, specifically Ganache, utilizing the *Web3.js* library. By leveraging this tool, testers can easily invoke smart contracts deployed on the local blockchain by providing their specifications, such as address and ABI. It features a user interface that encompasses three sub-interfaces, as illustrated in Fig. 3.

In the first sub-interface (1) of BC Test Runner, testers are given the ability to select the smart contract specification file (.json) and the test cases file (.txt). They can then initiate the test process by clicking on the *Start Test* button or generate test reports using the *Generate Report* button. The second sub-interface (2) provides a comprehensive display of important metrics, including the number of executed test cases, their respective verdicts, and the duration of each test. The third sub-interface (3) presents the test results visually, utilizing a pie chart format. This graphical representation effectively highlights the outcomes of the tests, providing a concise overview for analysis and evaluation.

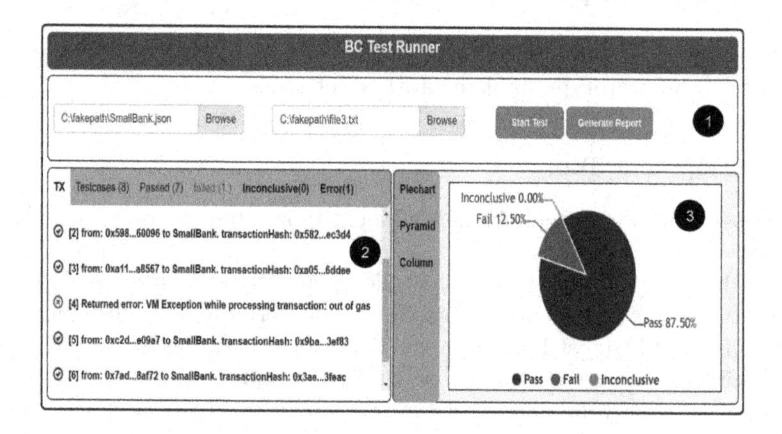

Fig. 3. The user interface of BC Test Runner.

5 Illustration

This section presents the case study that we utilized to demonstrate the application of our MBT approach in the context of EIP-1559 smart contracts.

5.1 Case Study Description

Today, blockchain technology is widely used in various sectors of the global economy, and one of its most popular applications is in the banking sector. This is primarily because blockchain has the capability to reduce costs, expedite money transfers, improve workflow efficiency, and protect confidential bank and customer data. Our idea is to create a smart contract that empowers users to create individual bank accounts and initiate fund transfers directly from their accounts. A smart contract, called *SmallBank*, is highlighted in the Listing 1.1.

```
pragma solidity ^0.5.3;
 contract SmallBank{
 address[] users;
function addUsers(address newUser) public {
 users.push(newUser);
 }
 function addInterest(uint interest) public {
 //Heavy code to compute interest per user
 for(uint i = 0; i < users.length; i++){
 users[i].call.value(interest)();
 } }}
```

Listing 1.1. Code snippet of The Small Bank smart contract.

5.2 Modelling the Small Bank System

The subsequent section provides the timed automaton specification of the Small Bank smart contract, which will be utilized as a reference in our approach.

The Small Bank Smart Contract Automaton. The Small Bank smart contract automaton described in Fig. 4 comprises three states. The initial state, labeled as A1 and represented by a double circle, serves as the starting point. The model evolves based on the received requests, resulting in transitions that lead either to state A3 or state A2. For example, the enabled transition $Tx\_addUsers[i]?$ allows the model to transition to state A2. Ultimately, the model returns to its initial state A1 via the transition $user\_added[i]!$.

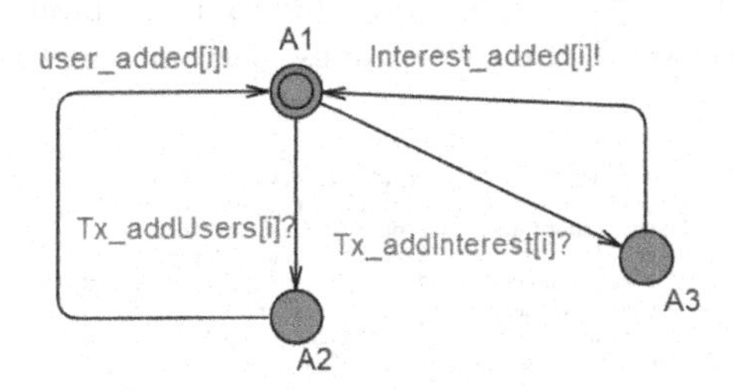

Fig. 4. Small Bank smart contract automaton.

Transaction Automaton. As depicted in Fig. 5, the Transaction Automaton consists of three states: T0, T1, and T2. The initial state T0, serves as the starting point for the model. Depending on the received request, the model can evolve either to state T1 or state T2 from the initial state. For instance, the transition $addUsers[i]?$ enables the model to move to state T1.

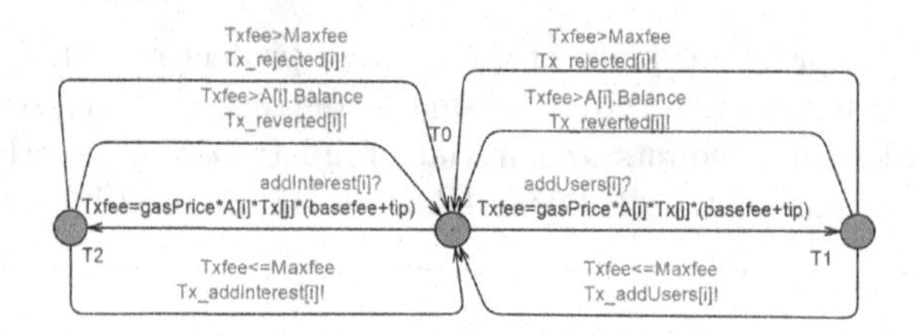

Fig. 5. Transaction automaton.

The model also has transitions that allow it to return to the initial state, T0, under certain conditions. If the transaction fee (txFee) exceeds the maximum fee (maxFee), the model follows the transition $Tx\_rejected[i]!$, indicating a rejected transaction, and returns to state T0. Similarly, if the user's account balance is insufficient to cover the transaction fee, the model follows the transition $Tx\_reverted[i]!$, representing a reverted transaction, and returns to state T0. Alternatively, if the transaction fee is less than the maximum fee, the model enables the transition $Tx\_addUsers[i]!$, signifying the invocation of the *addUsers* function of the smart contract. In this case, the transaction cost is deducted from the user's balance, and the model progresses accordingly.

Overall, the model demonstrates the flow of transactions and the conditions that determine the state transitions, allowing for proper handling of rejected, reverted, and confirmed transactions.

5.3 Test Case Generation

After modeling and compiling our test model, we were able to generate the test cases as a text file, as shown in Fig. 6. Each test case is composed by the function name of the smart contract that the sender invoked, the input parameters of the invoked function, the expected output values, and the sender's address.

```
addUsers/input:address=1/output:accepted/from:1
addUsers/input:address=1/output:accepted/from:2
addUsers/input:address=2/output:accepted/from:3
addUsers/input:address=3/output:accepted/from:4
addInterest/input:int=222/output:out of gas/from:10
addUsers/input:address=4/output:accepted/from:5
addUsers/input:address=5/output:accepted/from:6
addUsers/input:address=6/output:accepted/from:7
```

Fig. 6. Test cases.

6 Related Work

Most of the existing testing approaches and tools focus on the security of smart contracts and make use of black-box, White-box and grey-box testing techniques to detect functional and security issues [4]. Since our concern in this work is to propose a black-box and model based testing approach for BoS, we address all works similar to ours dealing with black-box fuzzing, MBT approaches, etc.

In fact, Black-box fuzzing is a fundamental technique that generates random test data based on a distribution for various inputs [23]. This technique shows its efficiency in detecting essentially security problems in smart contracts. For instance, the ContractFuzzer [15] detects well-known security vulnerabilities in Ethereum smart contracts. For this purpose, it takes as input the ABI[6] specification of the smart contract under test and proceeds to the generation of test inputs. After that, it proposes test oracles for increasing the vulnerability detection capabilities. Similar to ContractFuzzer, Pan et al. [24] adopt a black-box fuzzer engine to generate inputs in order to detect reentrancy vulnerability. Called ReDefender, the proposed framework would send transactions while gathering runtime data through fuzzing input. Then, ReDefender can detect the reentrancy issue and track the vulnerable functions by looking at the execution log. It demonstrates its ability to detect efficiently reentrancy bugs in real world smart contracts. However, we notice that functional correctness of smart contracts are not taken in to consideration as well as gas related issues.

Another interesting study was introduced in [11], called SolAnalyser, it offers a vulnerability detection tool with a three-phase process. In the first phase, SolAnalyser analyzes statically Solidity source code of smart contracts under test with the purpose of assessing locations prone to vulnerabilities and then instrumenting it with assertions. In the second step, an *inputGenerator* module has been implemented to automatically generate inputs for all transactions and functions in the instrumented contract. At the last phase, vulnerabilities are detected when the property checks are violated while executing smart contracts on the Ethereum Virtual Machine (EVM). Similarly, Grieco et al. in [25] introduce an open-source and black-box fuzzer for smart contracts that automatically generates tests to detect assertion violations and some custom properties. Called Echidna, this tool creates test inputs depending on user-supplied predicates or test functions. However, the major problem within it is that it may need a great knowledge to define the predicates and test methods.

The closest approach to our work is ModCon [26]. Indeed, ModCon is an MBT solution that enables the generation of test cases for enterprise smart contracts and it supports both permissioned and consortium blockchains. To do so, it makes use of an explicit abstract model of the target smart contract and allows users to define test oracles, and customize the testing process by choosing from different coverage strategies and test prioritization options. Compared to our solution, ModCon did not model blockchain environment and gas related issues, it focused only on modelling and testing functional aspects of smart contracts.

Regarding our previous work [16], it introduces model-based testing approach to automate the generation and the execution of test cases for blockchain oriented software. Similar to this paper, it ensures the modelling of both smart contracts and the blockchain environment through the use of UPPAAL time automata but without taking into consideration the novel gas mechanism. Moreover, the major problem within the older version is that it makes use of an obsolete test case

[6] Application Binary Interface.

generator called UPPAAL CO$\sqrt{}$ER, an old extension of the UPPAAL model checker which is no longer updated.

7 Conclusion

This paper proposed a model-based testing approach for EIP-1559 Ethereum Smart contracts. Our approach ensured the modelling both of smart contracts and the blockchain environment while considering essentially Ethereum gas mechanism according to the new Ethereum Improvement Proposal, EIP-1559. To do so, UPPAAL Timed Automata were used to elaborate test models. Afterwards, new abstract test cases were generated by using the UPPAAL Test Generator (Yggdrasil). We also reused our tool BC Test Runner to execute tests, analyze test results and generate test reports. As a proof of concept, our work was illustrated through the Small Bank smart contract.

As future work, we aim to extend our MBT approach to support security testing and to detect several vulnerability issues in the case of Ethereum smart contracts. The key idea here is to study firstly security properties like confidentiality, integrity, authentication, authorization, availability, and non-repudiation. Secondly, we investigate security modelling and the automatic security test cases and test suites generation.

References

1. Krichen, M., Ammi, M., Mihoub, A., Almutiq, M.: Blockchain for modern applications: a survey. Sensors **22**(14), 5274 (2022)
2. Nakamoto, S., et al.: Bitcoin: a peer-to-peer electronic cash system (2008)
3. Finley, K.: A $50 million hack just showed that the DAO was all too human (2016)
4. Lahami, M., Maâlej, A.J., Krichen, M., Hammami, M.A.: A comprehensive review of testing blockchain oriented software. In: Proceedings of the 17th International Conference on Evaluation of Novel Approaches to Software Engineering, ENASE 2022, Online Streaming, 25–26 April 2022, pp. 355–362. SCITEPRESS (2022)
5. Krichen, M., Lahami, M., Al-Haija, Q.A.: Formal methods for the verification of smart contracts: a review. In: 15th International Conference on Security of Information and Networks, SIN 2022, pp. 1–8. IEEE (2022)
6. Nelaturu, K., Mavridou, A., Veneris, A., Laszka, A.: Verified development and deployment of multiple interacting smart contracts with VeriSolid. In: Proceedings of the 2nd IEEE International Conference on Blockchain and Cryptocurrency (ICBC) (2020)
7. Ben Fekih, R., Lahami, M., Jmaiel, M., Ben Ali, A., Genestier, P.: Towards model checking approach for smart contract validation in the EIP-1559 Ethereum. In: Proceeding of the 46th IEEE Annual Computers, Software, and Applications Conference, (COMPSAC), pp. 83–88 (2022)
8. Ben Fekih, R., Lahami, M., Jmaiel, M., Bradai, S.: Formal modeling and verification of ERC smart contracts: application to NFT. In: The proceeding of IEEE Symposium on Computers and Communications (ISCC). IEEE (2023)
9. Amani, S., Bégel, M., Bortin, M., Staples, M.: Towards verifying Ethereum smart contract bytecode in Isabelle/HOL. In: Proceedings of the 7th ACM SIGPLAN International Conference on Certified Programs and Proofs, pp. 66–77 (2018)

10. Annenkov, D., Milo, M., Nielsen, J.B., Spitters, B.: Extracting smart contracts tested and verified in Coq. In: Proceedings of the 10th ACM SIGPLAN International Conference on Certified Programs and Proofs, pp. 105–121 (2021)
11. Akca, S., Rajan, A., Peng, C.: SolAnalyser: a framework for analysing and testing smart contracts. In: Proceeding of the 26th Asia-Pacific Software Engineering Conference (APSEC), pp. 482–489 (2019)
12. Sánchez-Gómez, N., Torres-Valderrama, J., García-García, J.A., Gutiérrez, J.J., Escalona, M.J.: Model-based software design and testing in blockchain smart contracts: a systematic literature review. IEEE Access **8**, 164556–164569 (2020)
13. Andesta, E., Faghih, F., Fooladgar, M.: Testing smart contracts gets smarter. In: Proceeding of the 10th International Conference on Computer and Knowledge Engineering (ICCKE 2020), pp. 405–412 (2020)
14. Wang, H., Li, Y., Lin, S.W., Artho, C., Ma, L., Liu, Y.: Oracle-supported dynamic exploit generation for smart contracts (2019)
15. Jiang, B., Liu, Y., Chan, W.K.: ContractFuzzer: fuzzing smart contracts for vulnerability detection. In: Proceedings of the 33rd ACM/IEEE International Conference on Automated Software Engineering, pp. 259–269 (2018)
16. Hammami, M.A., Lahami, M., Maâlej, A.J.: Towards a dynamic testing approach for checking the correctness of Ethereum smart contracts. In: 17th International Conference of Risks and Security of Internet and Systems, (CRiSIS) (2022)
17. Kim, J.H., Larsen, K.G., Nielsen, B., Mikučionis, M., Olsen, P.: Formal analysis and testing of real-time automotive systems using UPPAAL tools. In: Núñez, M., Güdemann, M. (eds.) FMICS 2015. LNCS, vol. 9128, pp. 47–61. Springer, Cham (2015). https://doi.org/10.1007/978-3-319-19458-5_4
18. Szabo, N.: Formalizing and securing relationships on public networks. **2**(9) (1997)
19. Liu, Y., Lu, Y., Nayak, K., Zhang, F., Zhang, L., Zhao, Y.: Empirical analysis of EIP-1559: transaction fees, waiting time, and consensus security. In: CCS '22: 2022 ACM SIGSAC Conference on Computer and Communications Security Los Angeles CA USA 7–11 November 2022, pp. 2099–2113. IEEE (2022)
20. Buterin, V., Conner, E., Dudley, R., Slipper, M., Norden, I., Bakhta, A.: EIP-1559: Fee market change for eth 1.0 chain. https://eips.ethereum.org/eips/eip-1559. Accessed May 2023
21. Azouvi, S., Goren, G., Heimbach, L., Hicks, A.: Base fee manipulation in ethereum's EIP-1559 transaction fee mechanism (2023)
22. Wood, G., et al.: Ethereum: a secure decentralised generalised transaction ledger. Ethereum Proj. Yellow Paper **151**(2014), 1–32 (2014)
23. Felderer, M., Büchler, M., Johns, M., Brucker, A.D., Breu, R., Pretschner, A.: Chapter one - security testing: a survey. Adv. Comput. **101**, 1–51 (2016)
24. Pan, Z., Hu, T., Qian, C., Li, B.: ReDefender: a tool for detecting reentrancy vulnerabilities in smart contracts effectively. In: Proceedings of the IEEE 21st International Conference on Software Quality, Reliability and Security (QRS), pp. 915–925 (2021)
25. Grieco, G., Song, W., Cygan, A., Feist, J., Groce, A.: Echidna: effective, usable, and fast fuzzing for smart contracts. In: Proceedings of the 29th ACM SIGSOFT International Symposium on Software Testing and Analysis, pp. 557–560 (2020)
26. Liu, Y., Li, Y., Lin, S.W., Yan, Q.: ModCon: a model-based testing platform for smart contracts. In: Proceedings of the 28th ACM Joint Meeting on European Software Engineering Conference and Symposium on the Foundations of Software Engineering, pp. 1601–1605 (2020)

Execution Planning for Aggregated Search in the Web of Data: A Free-Metadata Approach

Ahmed Rabhi[1](✉), Rachida Fissoune[2], Mohamed Tabaa[3], and Hassan Badir[2]

[1] LSIA laboratory, EMSI, Tangier, Morocco
a.rabhi@emsi.ma

[2] Data Engineering and System Team (IDS) - National School of Applied Sciences, Abdelmalek Essaadi University, Tangier, Morocco
{rfissoune,badir.hassan}@uae.ac.ma

[3] LPRI Lab, LSIA Laboratory, EMSI Casablanca, EMSI Tanger, Casablanca, Morocco
m.tabaa@emsi.ma

Abstract. An aggregate search system in the web of data makes it possible to search for pieces of information that may not exist entirely in a single source. It provides a unified query interface allowing access to several data sources. One of the major concerns while setting up an aggregate search system is adopting an adequate execution planning strategy, indeed, thanks to execution planning, the system optimizes network traffic while collecting fragments of data, avoids unnecessary executions, and reduces the waiting time for the user. Execution planning strategies in the state of the art often include using metadata or checking the existence of resources before execution. In this paper, we present a solution for execution planning in aggregated search systems without using metadata or consulting the sources. This solution aims to optimize network traffic by avoiding duplicate results and exaggerated size of intermediate data, our method is based only on the analysis of the input query.

Keywords: Aggregated search · SPARQL · Distributed queries · Query analysis · Execution planning

1 Introduction

Retrieving information from distributed data graphs is a subject of great importance in data integration field. Actually, the graphical representation of data opens up many possibilities to facilitate the integration of information from data belonging to different and independent graphs, and this by taking into consideration common nodes between graphs or nodes having the same semantic value.

The web of data is based on the standards of the semantic web to create an environment containing several data graphs that can be linked to each other [1].

M. Mosbah et al. (Eds.): MEDI 2023, LNCS 14396, pp. 58–68, 2024.
https://doi.org/10.1007/978-3-031-49333-1_5

Thus, the web of data nowadays contains a large number of graph-oriented datasets covering different domains, these data are presented in RDF format which is a graph-oriented format. Data sets of the web of data are accessible via sources to facilitate their interrogation, these sources are called SPARQL endpoints.

The general objective of this work is to set up an aggregated research system allowing to collect the necessary data from several sources and to answer a SPARQL query via an interface unifying several datasets. one of the major problems that arises in this kind of system is the execution time consumed during the collection of intermediate data, in other words, interaction with external data sources and the collection of intermediate data is the most expensive processing step in terms of runtime according to our previous work [2]. For this, adopting of an efficient execution plan is essential to optimize runtime.

According to the state of the art, some of the strategies adopted use indexes and metadata to select the relevant data sources, other strategies consist of consulting the sources before the executions to estimate the cardinalities and verify the existence of the necessary resources. In this paper, we present an execution planning strategy based only on query analysis. Our solution does not refer to an index and does not use any meta data, it analysis query parts in order to minimise interactions with data sources and optimise network traffic by avoiding unnecessary intermediate data, however, our solution does not neglect any data that may be important.

The rest of this paper is organized as follows: Sect. 2 presents background and basic concepts. Section 3 presents related work. We present in Sect. 4 our solution for execution planning without using meta-data. In Sect. 5 we present experimental evaluation of our solution. Finally, Sect. 6 concludes the paper.

2 Background and Basic Concepts

The work presented in this paper is the continuation of our previous work on aggregated research in the web of data, we presented our aggregated search system called WODII [16] which aims to maximize intermediate results for a SPARQL query based on a local index to select relevant sources and optimize network traffic. Our strategy of preparing queries is well illustrated in our paper [17] which presents our method of analysing SPARQL queries for aggregated search.

In this paper, a solution to optimise network traffic in aggregated search systems without using meta-data, so, the solution we propose in this paper consists in carefully analysing the query and setting up an execution plan to optimize runtime based on the query only.

As presented in Fig. 1, a SPARQL query is formed of a set of triple patterns, these triple patterns are linked to each other to represent the schema of an information. Hence, we can consider a SPARQL query as a linked graph representing the information sought by the user. In SPARQL, a star-group pattern denotes a set of triples having a common resource, for example, in Fig. 1, triple patterns having "?x" as a common resource form a star-group pattern, thanks to star-group patterns the system can identify links between triple patterns.

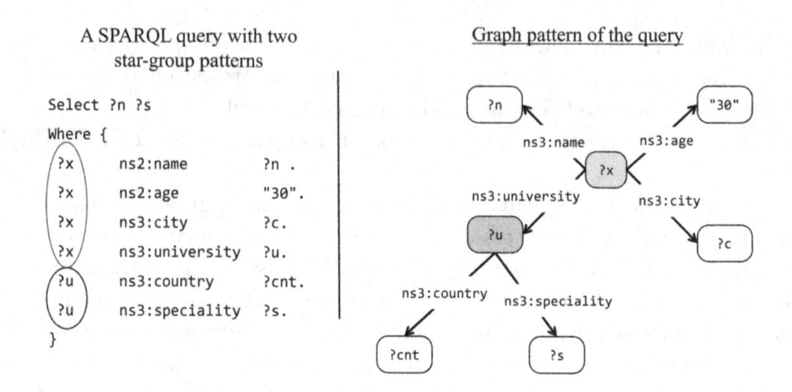

Fig. 1. Example of a SPARQL query and its graph pattern.

3 Related Works

According to T. H. Le et al. [3] a single data graph may not satisfy all the data to answer a query, thus, the authors consider aggregate research as an approach to seek the response to a query by collecting its fragments from several graphs. S. Sushmita [4] define aggregate search an approach to search for objects that do not exist entirely in a single data source, in other words, it allows to collect fragments belonging to the same information from distributed sources and then integrate them into a complete information.

B. Quilitz and U. Leser [5] introduce DARQ as a solution providing to the user a unique interface to query multiple sources, their solution is based on predicates of the query as well as primary statistics about the data sources to set up an execution plan. The solution proposed by A. Schwarte et al. [6] uses SPARQL ASK queries to check the existence of each triple pattern of the query. C. Başca and A. Bernstein [7] propose Avalanche as a solution to query the web of data, their system set up an execution plan using statistics and meta data about sources to select relevant sources based on their processing speed. The solution proposed by O. Görlitz and S. Staab [8] uses Void stores to explore and select relevant data sources, Void stores contain meta data about the content of data sources. M. Saleem et al. [9] introduce DAW as an index-based solution to select relevant data sources for federated SPARQL queries, their solution is based on a local index, cardinality estimation and predicates of the query to plan executions. Z. Akar et al. [10] propose a solution based on triple pattern of the query and external indexes to select relevant sources. The LHD system [11] uses query predicates, meta-data and ASK queries to select relevant data sources and set up an optimal execution plan.

Costfed [12] is a solution based on cardinality estimation and star-group patterns of the query for execution planning, M. Vidal et al. [13] propose an execution planning solution that is based on star-group patterns of the query as well as cardinality estimation to optimize joining intermediate results. The use of star-group patterns is also adopted by G. Montoya et al. [14] in their solution to optimize query planning in SPARQL federated queries. M. Meimaris et al. [15] propose a solution based on characteristic sets as well as star-group patterns for federated SPARQL queries optimization.

4 Our Proposed Solution

In this section we present a general view on the workflow of our system, and we explain our execution planning solution that does not need meta-data to optimize network traffic.

4.1 Workflow of Our System

Our system workflow can be summarized in three main steps: query preparing, execution planning and results processing, Fig. 2 illustrates these steps.

In the query preparing step, the system decomposes the user query by forming a sub query for each triple pattern, thus, the system looks for intermediate results corresponding to each triple pattern, this step is explained in mor details in our previous work [16]. The execution planning step is the main step in this paper, it allows filtering candidate sub-queries by excluding those leading to duplicate results and those leading to an exaggerated and unnecessary intermediate data size, this step is explained with more details in Sect. 4.2. The last step in our system is results processing, during this step, the system execute sub-queries, retrieves intermediate data and finally prepare the final answer to the user, our method to prepare final answers is explained with details in our previous paper [17].

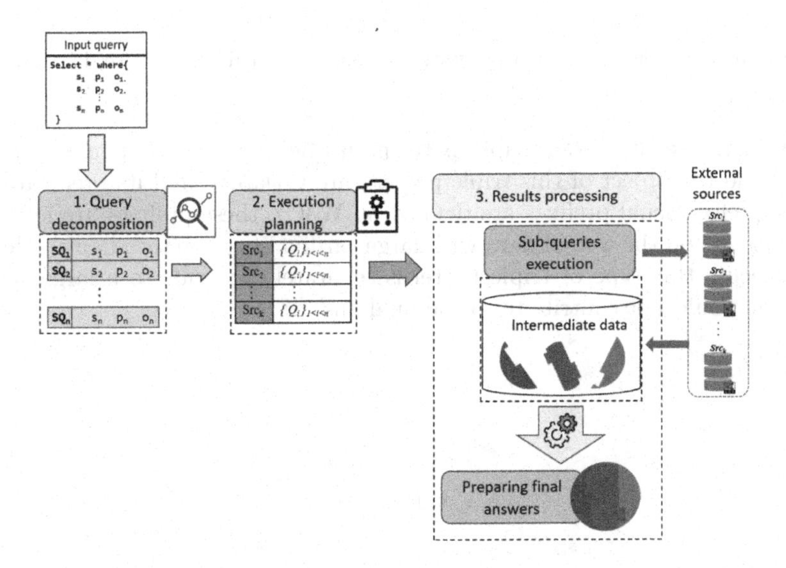

Fig. 2. A global view of our system's workflow.

4.2 Execution Planning

The proposed execution planning method is based on filtering candidate sub-queries, and this filtering goes through two main steps: The first step is identifying usual triple patterns, it means, triple patterns that are mostly found in all data sets of the web of data. The second step is identifying similar triple patterns; two similar triple patterns in our solution are triple patterns leading to duplicate results. We explain in the following our candidate sub-queries filtering mechanism with illustrative examples.

A. Filtering Usual Triple Patterns
As asserted above, we consider a triple pattern that is mostly found in all RDF datasets as usual triple pattern. Indeed, results for a usual triple pattern are generally found in all RDF datasets of the web of data, and if we try to retrieve results of a usual triple pattern, this engenders in an exaggerated intermediate data size. For example, the triple pattern **TP4** in Fig. 3, if we execute this triple pattern as it is (regardless of other triple patterns), it returns the entire dataset which is not relevant in aggregated search.

```
SELECT ?protein ?OMA_link
WHERE {
    ?protein a orth:Protein.                                  (TP1)
    ?protein orth:organism ?organism.                         (TP2)
    ?inTaxon rdfs:label 'in taxon'@en.                        (TP3)
    ?organism ?inTaxon ?taxon.                                (TP4)
    ?taxon up:scientificName 'Rattus norvegicus'.             (TP5)
    ?protein rdfs:seeAlso ?OMA_link.                          (TP6)
}
```

Fig. 3. Example of usual triple pattern that may return the entire dataset.

Another case of similar triple patterns is the one presented in Fig. 4; the subject and the object of this triple pattern are variables, and its predicate has a usual prefix. Usual prefixes are defined by W3C[1], these prefixes are found in most datasets on the web of data with large cardinalities (up to billions). Hence, if we execute this type of triple patterns separately, it leads to a large size of data that would not contribute to the final answer.

[1] https://www.w3.org/wiki/TheUsualPrefixes.

```
SELECT *
WHERE {
    ?group      rdf:type                        orthology:OrthologsCluster.
    ?group      orthology:hasHomologous         ?x.
    ?x          mbgd:gene                       ?y
    ?y          mbgd:uniprot                    uniprot:K9Z723.
    ?group      orthology:hasHomologous         ?a.
    ?a          mbgd:gene                       ?b.
    ?b          mbgd:uniprot                    ?uniprot.
    ?group      void:inDataset                  mbgdr:default.
    ?uniprot    rdfs:seeAlso                    ?xref.          (Usual)
    ?xref       up:database                     updb:PDB.
}
```

Fig. 4. Example of usual triple pattern that is mostly found in all datasets.

Our filtering mechanism begin by temporarily excluding usual triple patterns to avoid unnecessary data transfer. These triple patterns must be rewritten to look for their intermediate data. The search for intermediate data for the usual triple patterns is explained in the demonstration section (5).

B. Similar Triple Patterns

In our solution two similar triple patterns are two triple patterns leading to the same set of intermediate results (see more details in our paper [17]). As example, triple patterns **TP2**, and **TP5** of the query presented in Fig. 5 are two similar triple pattern because according to the SPARQL syntax, executing these two triple patterns independently returns the same set of results.

```
SELECT *
WHERE {
    ?group      rdf:type                        orthology:OrthologsCluster.
    ?group      orthology:hasHomologous         ?x.             (Similar)
    ?x          mbgd:gene                       ?y              (Similar)
    ?y          mbgd:uniprot                    uniprot:K9Z723.
    ?group      orthology:hasHomologous         ?a.             (Similar)
    ?a          mbgd:gene                       ?b.             (Similar)
    ?b          mbgd:uniprot                    ?uniprot.
    ?group      void:inDataset                  mbgdr:default.
    ?uniprot    rdfs:seeAlso                    ?xref.
    ?xref       up:database                     updb:PDB.
}
```

Fig. 5. Examples of similar triple patterns.

It must be noted that usual triple patterns are not taken into consideration in this step, in other words, two usual triple patterns cannot be considered similar in our solution since the system excludes them before identifying similar triple patterns.

5 Demonstration and Experimental Results

This section presents the experimental evaluation to demonstrate the effectiveness of our solution.

5.1 Set-Up Environment

A. Software and Hardware Environment
This experimental evaluation was conducted on a machine equipped with the hardware specifications presented in Table 1, and the development of our system was done using the software environment presented in Table 2.

Table 1. Hardware environment.

CPU	Name	Intel® Core™ i5-10310U
	Cores	4
	Threads	8
	Base frequency	1,7 GHz
	Turbo frequency	4,4 GHz
RAM	Capacity	16 Go
	Speed	2667 MHz
	Type	DDR4

Table 2. Software environment.

Operating system	Windows 10 Professional 64 bits
Programming language	Java 8 SE
IDE	Eclipse
Framework/API	RDF4J

B. Datasets and the Query
In this evaluation, we use the query in Fig. 6, this query contains 10 triple patterns and we executed it on the five different datasets: Uniprot, DisGeNET and MBGD are life science datasets, DBpedia contains cross-domain data, finally, DBLP contains scientific publications data. We know beforehand that the answers to the query are found in Uniprot is MBGD but we have added other datasets for the test.

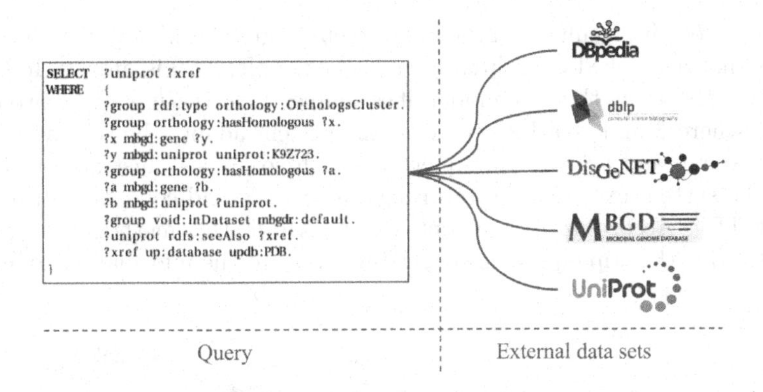

Fig. 6. The query and datasets used for the evaluation.

5.2 Executing Candidate Sub-queries

After filtering candidate triple patterns, the remaining triple patterns are: TP1, TP2, TP3, TP4, TP7, TP8 and TP10. The system excludes TP5 because it is similar to TP2, and TP6 was excluded because it is similar to TP3. The histogram in Fig. 7 presents results size (number of RDF triples) returned by each source for each triple pattern, the black line in Fig. 7 presents runtime of each triple pattern execution. According to these results, MBGD returned results for TP1, TP2, TP3, TP4, TP5, TP6, TP7 and TP8, while TP10 was found in Uniprot.

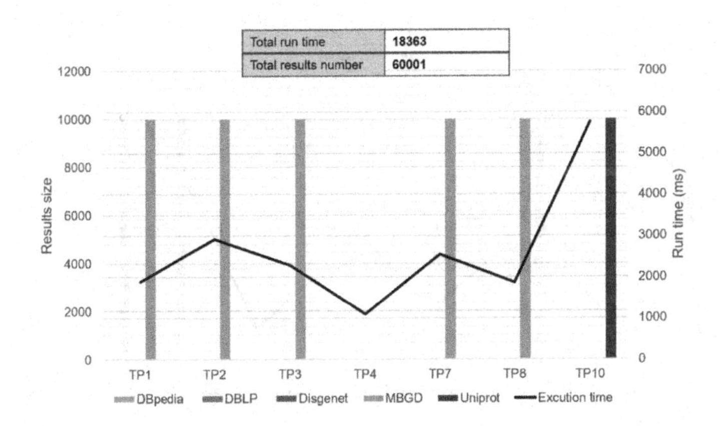

Fig. 7. Evaluation of candidate sub-queries execution.

By observing the chart of Fig. 7, we deduce that Uniprot and MBGD are the only sources that contributed to the query, hence, results for TP9 will be found in either MBGD or Uniprot (or both of datasets).

To look for the results of TP9, the system begins by identifying the triple patterns that are linked to it directly, in this example, TP9 is directly linked to TP7 and TP10, then, the system identifies sources where each of these two triple patterns where found (see Fig. 8), then, the system adapts the schema of TP9 for each source and writes a sub-query for each source as illustrated in Fig. 8, finally, the system executes the resulting sub-queries and collects intermediate data for TP9. According to the syntax of SPARQL this method narrows the results size for the triple pattern by getting closer to the information sought.

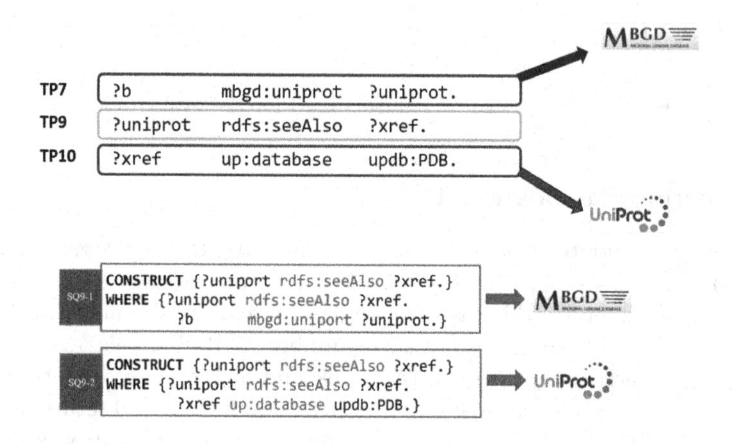

Fig. 8. The two versions of the corresponding subquery to TP9.

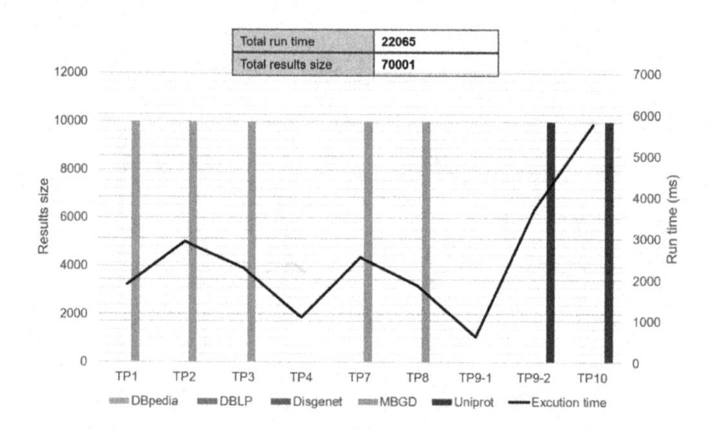

Fig. 9. Evaluation of candidate and usual sub-queries execution.

The histogram of Fig. 9 presents a summary results size (number of RDF triples) returned by each source for each candidate triple pattern additionally to the two schemas of the usual triple pattern. According to this evaluation, the relevant TP9 results for this query are found only in Uniprot, because the schema of TP9 that is intended for MBGD returned null results.

According to the experimental results, our sub-queries filtering solution allowed to avoid 10 unnecessary executions by avoiding duplicate results. Moreover, the triple pattern TP9 is found in the five the sources used, which would have generated an exaggerated size of the intermediate results, and thanks to our solution, we were able to narrow the number of intermediate results for TP9.

6 Conclusion

We presented in this paper a solution for execution planning in our aggregated search system in the web of data, this solution aims to minimize non necessary execution to avoid retrieving intermediate data with exaggerated sizes and to avoid duplicate intermediate data.

The proposed solution does not use metadata and does not consult data sources before executions. Our method is only based on analyzing the input. Our strategy identifies parts of the query leading to an exaggerated intermediate data size, the system reformulates the scheme of these parts to narrow the size of the intermediate results, in addition, the proposed strategy identifies query parts leading to duplicate results to avoid them.

According to the experimental results, our solution allows avoiding duplicate results. Besides, the processing method that we propose for usual triple patterns makes it possible to optimize runtime and memory consumption by avoiding unnecessary executions and minimizing unnecessary data, this method allows finding results as close as possible to the final answer.

References

1. Yu, L.: A Developer's Guide to the Semantic Web. Springer, Heidelberg (2014). https://doi.org/10.1007/978-3-662-43796-4
2. Rabhi, A., Fissoune, R., Tabaa, M., Badir, H.: A parallel processing architecture to optimize runtime in aggregated SPARQL queries. In: Proceedings of the 14th International Conference on Management of Digital EcoSystems, pp. 9–15 (2022)
3. Le, T.-H., Elghazel, H., Hacid, M.-S.: A relational-based approach for aggregated search in graph databases. In: Lee, S., Peng, Z., Zhou, X., Moon, Y.-S., Unland, R., Yoo, J. (eds.) DASFAA 2012. LNCS, vol. 7238, pp. 33–47. Springer, Heidelberg (2012). https://doi.org/10.1007/978-3-642-29038-1_5
4. Sushmita, S., Joho, H., Lalmas, M., Villa, R.: Factors affecting click-through behavior in aggregated search interfaces. In: Proceedings of the 19th ACM International Conference on Information and Knowledge Management, pp. 519–528 (2010)
5. Quilitz, B., Leser, U.: Querying distributed RDF data sources with SPARQL. In: Bechhofer, S., Hauswirth, M., Hoffmann, J., Koubarakis, M. (eds.) ESWC 2008. LNCS, vol. 5021, pp. 524–538. Springer, Heidelberg (2008). https://doi.org/10.1007/978-3-540-68234-9_39

6. Schwarte, A., Haase, P., Hose, K., Schenkel, R., Schmidt, M.: FedX: optimization techniques for federated query processing on linked data. In: The Semantic Web-ISWC 2011: 10th International Semantic Web Conference, Bonn, Germany, October, Proceedings, Part I 10, pp. 601–616 (2011)
7. Basca, C., Bernstein, A.: Querying a messy web of data with avalanche. J. Web Semant. **26**, 1–28 (2014)
8. Görlitz, O., Staab, S.: SPLENDID: SPARQL endpoint federation exploiting void descriptions. Proc. Second Int. Conf. Consuming Linked Data **782**, 3–24 (2011)
9. Saleem, M., Ngonga Ngomo, A.C., Xavier Parreira, J., Deus, H.F., Hauswirth, M.: DAW: duplicate-aware federated query processing over the web of data. In: The Semantic Web-ISWC 2013: 12th International Semantic Web Conference, Sydney, NSW, Australia, pp. 574–590 (2013)
10. Akar, Z., Halaç, T.G., Ekinci, E.E., Dikenelli, O.: Querying the web of interlinked datasets using VOID descriptions. LDOW **937** (2012)
11. Wang, X., Tiropanis, T., Davis, H.C.: LHD: Optimising linked data query processing using parallelisation. In: LDOW (2013)
12. Saleem, M., Potocki, A., Soru, T., Hartig, O., Ngomo, A.C.N.: CostFed: cost-based query optimization for SPARQL endpoint federation. Procedia Comput. Sci. **137**, 163–174 (2018)
13. Vidal, M.-E., Ruckhaus, E., Lampo, T., Martínez, A., Sierra, J., Polleres, A.: Efficiently joining group patterns in SPARQL queries. In: Aroyo, L., et al. (eds.) ESWC 2010. LNCS, vol. 6088, pp. 228–242. Springer, Heidelberg (2010). https://doi.org/10.1007/978-3-642-13486-9_16
14. Montoya, G., Skaf-Molli, H., Hose, K.: The *Odyssey* approach for optimizing federated SPARQL queries. In: d'Amato, C., et al. (eds.) ISWC 2017. LNCS, vol. 10587, pp. 471–489. Springer, Cham (2017). https://doi.org/10.1007/978-3-319-68288-4_28
15. Meimaris, M., Papastefanatos, G., Mamoulis, N., Anagnostopoulos, I.: Extended characteristic sets: graph indexing for SPARQL query optimization. In: 2017 IEEE 33rd International Conference on Data Engineering (ICDE), pp. 497–508 (2017)
16. Rabhi, A., Fissoune, R.: WODII: a solution to process SPARQL queries over distributed data sources. Cluster Comput. **23**(3), 2315–2322 (2019). https://doi.org/10.1007/s10586-019-03004-1
17. Rabhi, A., Fissoune, R., Tabaa, M., Badir, H: Intermediate results processing for aggregated SPARQL queries. In: 2021 IEEE/ACS 18th International Conference on Computer Systems and Applications (AICCSA), pp. 1–8. IEEE (2021)

Discovering Relationships Between Heterogeneous Declarative Mappings for RDF Knowledge Graph

Amel Belmaksene(✉) and Selma Khouri

Laboratoire de la Communication dans les Systemes Informatiques, Ecole nationale Supérieure d'Informatique (ESI), Algiers, Algeria
{a_belmaksene,s_khouri}@esi.dz

Abstract. Nowadays, Knowledge Graphs (KGs) are extensively used in companies, they are created using different techniques, mapping languages among them, and involve various designers teams. Mappings languages have been proposed to explicitly define mappings assertions between heterogeneous (semi)-structured datasets and the KG. The wide variety of mapping languages and their associated systems and the lack of a generic mapping language that covers all the existing languages, lead each expert/design team to use the mapping language they master and that fits their requirements, to enrich the company's KG. This situation creates new sources of heterogeneous mapping assertions. A straightforward analysis of these mappings is very complex because of the strong differences of their syntaxes. In this article, we propose an approach to detect relationships between heterogeneous mappings from the analysis of their generated graphs, following a reverse ETL approach. This analysis is relevant for analyzing and maintaining the mappings, and may provide relevant insights in the datasets usage and design patterns defined within the global KG project, where the mappings play a central role. A set of experiments is provided to show the feasibility of the approach.

Keywords: Heterogeneous declarative mappings · RDF KG · Comparative relationships

1 Introduction

Knowledge Graphs became an essential technology for many issues related to data processing, integration and analysis, encountered in different companies. A large number of techniques for creating KGs have been proposed, for delivering data from a set of heterogeneous data sources into a materialized RDF KG using a set of mappings following an ETL-based approach. A declarative definition of mappings is important to enhance reusability, transparency and maintenance of the KG creation process [1]. Following these goals, different mapping languages

M. Mosbah et al. (Eds.): MEDI 2023, LNCS 14396, pp. 69–83, 2024.
https://doi.org/10.1007/978-3-031-49333-1_6

have been proposed in literature such as R2RML [2], RML [3], Shexml [4], Sparql-Generate [5], etc. Each mapping language has its own syntax and grammar, and one or more mapping systems (supporting the mapping language) which allows the execution of the defined mappings to automatically create the RDF KG.

Different studies have considered technical issues related to the definition and management of declarative mappings, but less studies have considered the organizational context related to their usage in companies. A typical scenario in such context assumes that different users (eg. KG designers or teams of designers) contribute to the KG development. Such teams may have different requirements [6], they may use overlapping input datasets of the company (where each team master its used dataset, but not necessarily all the input data of the company used by other teams), and consequently, each team may produce different heterogeneous "*sources of declarative mappings*" that meets their requirements and skills.

In this context, each design expert/team benefits from a great autonomy, but this situation will lead to the generation of a plethora of mappings assertions in different languages, with different overlapping, equivalent and/or complementary parts. This situation makes the analysis and maintenance of these mappings a difficult task, where mappings may become like a "spaghetti dish". In our study, we consider that the autonomy of sources of mappings has to be maintained. In order to master the sources of mappings generated independently, we consider that the global KG generated from these mappings offers a global view that can be used for detecting new insights useful for each source of mappings. That said, we can say that our context is close to a "reverse ETL" approach, a current trend in ETL process issue [7], where we treat the KG as a new source of information that uses the generated data and insights, back into the mapping sources. For this study, we consider the set of all the attempts of heterogeneous mappings correctly executed by the designers through a mapping system, i.e. the mappings that have generated an RDF graph whether they are retained for alimenting the enterprise KG or not. We aim to identify different relationships (equivalence, inclusion/containment and disjointness) between these sources of mappings, using a backward analysis, i.e. by analyzing these relationships in their generated graphs. We consider the discovering of such relationships as a first set of key indicators for analyzing the mappings and consequently the global KG process. By analyzing the inclusion and containment between mappings of different teams, the designers may discover the potential of the datasets of the company, and how they have been used by other teams. Equivalent and disjoint mappings can also serve for identifying common patterns and anti-patterns of heterogeneous mappings defined. In this way, our approach considers the mappings as a focal point for mastering and monitoring the KG eco-system, the datasets usages and the design tasks of involved actors.

Discovering such relationships involves many challenges related to the heterogeneity of the mapping languages and their systems, which generates heterogeneous formats of RDF graphs even though equivalent mappings are defined. This heterogeneity is related to the use of different encoding, data type definitions, different serialization formats, the generation of special characters by some languages and not by others, etc. The discovering of relationships between mappings, requires thus a 'graph preparation process' in order to homogenize the graphs. The paper is organized as follows: in Sect. 2 presents a motivating example illustrating the issues and challenges of this study. Section 3 describes the related works. Section 4 explains the steps of our approach. Section 5 presents our experimental study where we extended GTFS-Madrid-Bench [8] benchmark which considers Madrid metro network data and defines RDF-based mappings. Finally, Sect. 6 presents the conclusion and the future lines of research.

2 Motivating Example

To illustrate the strong heterogeneity of mapping languages, we have chosen three languages of three different families: **RML** [3] (**an RDF-based language**), which extends the W3C standard R2RML [2], **SPARQL-Generate** [5] (**a query-based language**) and **SHEXML** [4], (**a constraint-based language**). The question of comparing mappings in different languages poses different challenges. Data is in different formats (eg. csv, json and xml), which makes the comparison difficult, even between mappings of the same language, because the syntax may differ when manipulating data from one format to another (eg. JSON format requires defining iterator for jsonpath, etc.).

Furthermore, the mappings are in different languages which increases the complexity of comparing mappings due to the strong variety of the syntaxes and notations of mappings languages, as pointed out by different studies [4,9]. For example, in Fig. 1, the four mappings A, B, C and D have relations of equivalence, inclusion, containment and disjointness, which appear in their generated graph fragments (A′, B′, C′ and D′), but the detection of these relationships directly from the mappings, remains a very complex task for designers.

As an alternative solution, we have investigated the possibility of choosing a pivot language, and translating all the mapping languages to this pivot, to facilitate the comparison between mappings. Different translation scenarios are proposed in [10], where a common interchange language is cited among the translation possibilities. For example, RML can be chosen as a pivot language because it is an extension of R2RML (a W3C standard language) or the meta-language [9] proposed for describing different mapping languages. Different translation works are proposed, eg. from Shexml to RML [11], from YARRRML to RML[1], etc. However, this solution presents many limitations in our context: (i) the translation between mapping languages is still an open issue [10], where different translations are still not available (eg. from Sparql-generate to RML [11]), (ii) the lack of a common model shared by all languages especially for very heterogeneous languages [10], for example, the meta-language proposed in [9] does not cover all the mappings languages because of its limitation to represent SPARQL-based languages. Moreover, the mappings languages tend to evolve and new languages may appear. The survey of Van Assche et al. [12] revealed many mappings languages within a few years. Depending on the availability of translation algorithms between mapping languages for comparing the mappings is not a suitable solution for our context. Even if we have chosen three mapping

[1] https://github.com/RMLio/yarrrml-parser.

languages as representatives, our aim is to propose a solution that can cover any set of input mapping languages.

This pushed us to consider the solution of comparing the generated RDF graphs to highlight the relationships between defined mappings, through a reverse ETL vision. This solution is motivated by the fact that the defined mappings are supposed to be executed by the designer (at least in order to check their correctness). Our solution propose to extend this execution by an automatic comparison between the mappings; either using an incremental scenario, or by comparing a block of defined mappings simultaneously.

However, using the KG to detect these relationships also poses complex challenges. In the literature, several systems are proposed for existing mapping languages such as RMLMapper [13], SDM-RDFizer [14] and Morph-KGC [15] for RML, Sparql-Generate[2] and Shexml[3]. The mastery, the test of all these tools and the analysis of the graphs obtained by each of them was challenging, since they generate graphs following different formats, even when the same mapping is defined. For example, some systems generate duplicate triples that are not duplicated in the input dataset. As cited in [16], the issues related to the heterogeneity of data format and systems are challenging for RML language, in our case, such issues are intensified by the heterogeneity of the mapping languages chosen from different families.

Additionally, the generated RDF graphs may differ from one language to another. For example, in Fig. 1, even if the mappings (defined in Sparql-generate (A) and Shexml (B)) have the same semantics (i.e. are equivalent) and use the same input dataset, their respective RDF graphs (B′ and A′) are not equivalent, because the graph generated using Shexml contains the datatype of data used since it uses Shape Expressions (ShEx) to define the desired structure of the output [4]. We encountered different additional issues related to the encoding, serialization formats, special characters usage, etc. which required the preparation of the graphs generated by the mappings, in order to achieve a reliable comparison between them.

3 Related Works

In order to have an overview of studies related to declarative mappings management for RDF KGs, we conducted a light-weight SLR (systematic literature reviews) following the steps described in [17]. We focused on works that meet these requirements: articles written in English, working on heterogeneous data sources, using declarative mappings for RDF KG, following an approach that materializes the KG generated, that are published between 2015 and 2023. We have used different search engines: Google Scholar, ACM Digital Library, IEEE Xplore Digital Library, Springer Link and Science Direct.

We focused on journals cited in surveys [12,17] and the common conferences cited. We can classify the selected studies according to the life-cycle of mapping management, as follows:

[2] https://ci.mines-stetienne.fr/sparql-generate/playground.html.
[3] http://shexml.herminiogarcia.com/editor/.

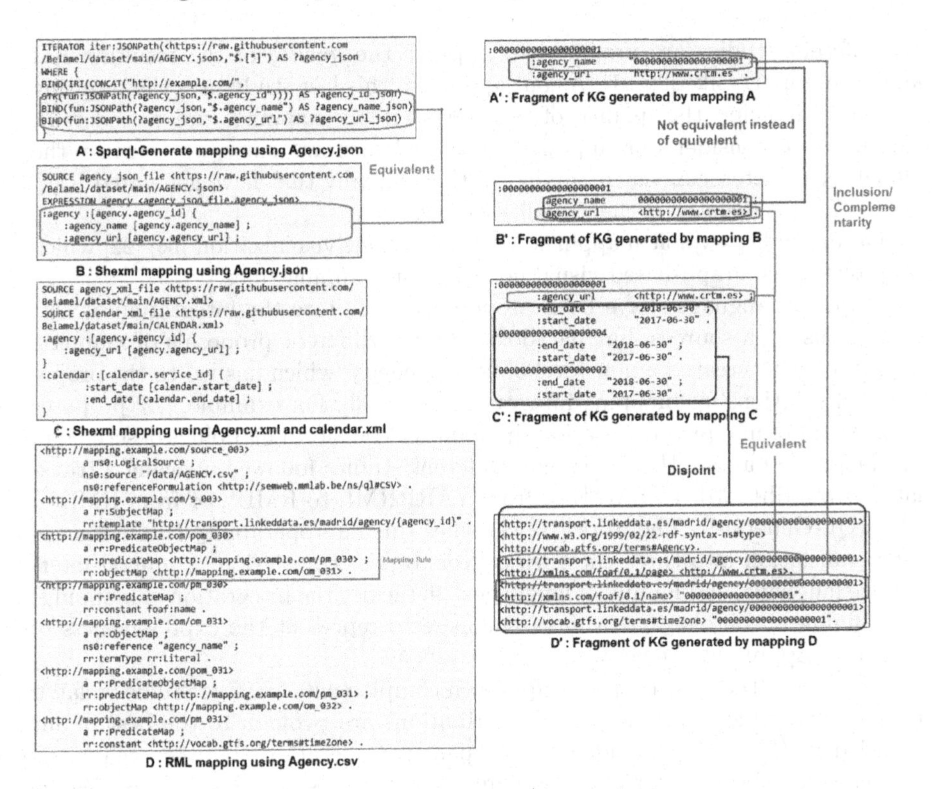

Fig. 1. Fragments of KGs generated by mappings RML, Shexml and Sparql-Generate LanguagesReal caption (The RML mapping in part D is provided in the benchmark "GTFS-Madrid-Bench" available in https://github.com/oeg-upm/kgc-eval/tree/master/mappings/rmlmapper).

(1) Mapping definition. Among the mapping languages, we can cite: RML [3] extending the standard W3C R2RML, SPARQL-GENERATE [5], ShExML [4], Helio [18], etc. Different materialization systems are also proposed for executing these mappings such as: RMLMapper [13], SDM-RDFizer [1,14], Morph-KGC [15] and Chimera [16] proposed to generate KG using RML language. Van Assche et al. [12] provide an informative overview of solutions and mapping languages for KG generation from heterogeneous (semi)-structured sources.
(2) Mapping optimization. Different studies manage the optimization of mappings execution. For example, [19] propose task-oriented tests to evaluate the performance of SDM-RDFizer [1,14]. [15] proposes to reduce execution time and memory when generating KGs using mapping partitions.
(3) Mapping Exploitation. By "exploitation" ***Exploitation***, we mean the use of the defined mappings in order to analyze or to enhance the KG and related mappings.

Different studies are proposed to improve the *quality* of the resulting KG and/or mappings like [20,21]. As our approach, this last study uses the resulting KG for enhancing the quality of mappings (RML mappings are considered). However, we consider our approach as complementary since we consider the global set of heterogeneous mappings of the company that have generated RDF graphs, independently of their quality management.

Other studies exploit mappings to enhance their visualization like [22] which proposes a rich graph-based visual notation for mapping rules.

Mapping languages have been proposed to facilitate the integration of heterogeneous data sources, but paradoxically, the different proposals of mapping languages have created a new level of heterogeneity, which has led to the proposition of solutions for mapping integration [9–11,23]. For example, [23] proposes the concept of mapping translation that aims to transform mappings described in one language into another language. Different studies followed allowing a translation from ShExML to RM [11], from YARRRML to RML[4], or from RML to SPARQL-Generate[5]. The work [10] classifies this interoperability through mapping translation into three categories: Peer-to-peer translation, Common interchange language, and Family of languages. [9] tackles the integration of mappings by defining an ontology as a meta-language to represent the expressiveness of existing mapping languages.

We believe that our proposed approach complements existing studies related to mapping exploitation, its main contributions are twofold, to the best of our knowledge: **(i)** it aims to identify comparative operations between mappings in a context of very heterogeneous mapping languages and data sources. Some studies [1,14] manage duplicates and overlap mappings, and [15] detects disjoint mappings, but these studies work within (RML based on R2RML) mapping language, and for optimization purposes. We note that the issue of identifying comparative operations has been investigated in other contexts like OLAP cube comparison [24] or ETL process [25]. However the idiosyncrasy of KG declarative mappings and the challenges related to their heterogeneity pushed us to investigate this issue in this particular context. **(ii)** The approach is managed within "reverse ETL" vision which tries to takes insights from the resulting KGs back to the mapping sources, for discovering comparative relationships between them, as a first step towards this vision. Finally, our approach can be used all along the KG life-cycle, following an incremental scenario each time a mapping is defined, or using a global set of defined executed mappings.

4 Our Approach

Our approach considers the following inputs and outputs:

(1) Inputs. We consider as inputs: (i) the enterprise KG and (ii) the mapping sources defined in three different mapping languages RML, Sparql-Generate and Shexml.

[4] https://github.com/RMLio/yarrrml-parser.

[5] https://github.com/sparql-generate/rml-to-sparql-generate.

(i) Knowledge Graph. We define a Knowledge Graph as an RDF graph. An RDF graph consists of a set of RDF triples where each RDF triple (s, p, o) is an ordered set of the following RDF terms: a subject $s \in U \cup B$, a predicate $p \in U$, and an object $U \cup B \cup L$. An RDF term is either a URI $u \in U$, a blank node $b \in B$, or a literal $l \in L$ [26].
(ii) The mapping source is formalized as follows: <MFile, <Map, Rules>, T, System, Team> where each mapping source is one file *MFile* having an extension (according to the used language) which contains mappings *Map* defining a set of mapping Rules m_1, m_2, ..., m_n that generates the triple set T (using correspondences with the input dataset), the mapping source is defined by a design *Team* and executed using a mapping *System*. Each map will define a correspondence between elements from the source to T.

Mapping rule. It typically refers to a set of guidelines or specifications that define how a dataset from one format is transformed into RDF triples (see Fig. 1).

RML Mapping. An RML mapping is defined as an RDF graph and is composed of one or more TripleMaps which define how the triples will be generated. A TripleMap is composed of a SubjectMap and zero or more PredicateObjectMaps [3], as shown in Fig. 1 (part D).

Sparql-Generate Mapping. SPARQL-Generate is based on a query language. It allows the generation of RDF from a set of RDF data and a set of documents in arbitrary formats [5] (see Fig. 1, part A). SPARQL-Generate is designed as an extension of SPARQL 1.1 [5].

Shexml Mapping. The Shexml mapping is an heterogeneous data mapping language based on Shape Expressions (ShEx)[6] (see Fig. 1, part B and C).

(2) Outputs. Our approach considers 04 relations between the mappings: Equivalence, Inclusion, Containment and Disjointness. As explained in Sect. 2, the comparison between mappings is reflected by the RDF graphs generated by these mappings. We note that we only consider ground RDF graphs, ie. that do not contain blank nodes, which make the comparison issue intractable [26].

(i) Equivalence. As defined in[7], two graphs G and G′ are considered equivalent or isomorphic if there is a bijection M between the sets of nodes of the two graphs, such that : (1) M(***lit***) = ***lit*** for all RDF literals ***lit*** which are nodes of G. (2) M(***uri***) = ***uri*** for all RDF URI references ***uri*** which are nodes of G. (3) The triple (s, p, o) is in G if and only if the triple (M(s), p, M(o)) is in G′.
(ii) Inclusion. It reflects a subgraph relation. A subgraph of an RDF graph is a subset of the triples in the graph[8].

[6] https://shexml.herminiogarcia.com/spec/#abstract.
[7] https://www.w3.org/TR/rdf11-concepts/.
[8] https://www.w3.org/TR/rdf11-mt/.

(iii) Containment. It is the inverse relation of inclusion. A graph G contains graph G′ when graph G′ is included in graph G. i.e. The graph G has some vertices or edges more than the graph G′.
(iv) Disjointness. A Disjointness between two graphs is defined when they share no nodes and no vertices: G = (N, E) and H = (N′, E′) are disjoint i.e. $N \cap N' = \emptyset$ and $E \cap E' = \emptyset$.

As cited in Sect. 2, our approach requires a first step of "graph preparation" before the second step of "mapping comparison".

4.1 Graph Preparation

Data preparation is defined as "the set of preprocessing operations performed in early stages of a data processing pipeline" [27]. Similarly to data preparation, the "graph preparation" step in our approach aims to identify the set of operations that preprocess the generated graphs related to mappings for their homogenization and comparison.

(1) Discovery: this step allows the analysis of the input mapping to identify the mapping language used, to check if the file extension is compatible with the mapping language used or simply to detect badly introduced files.
(2) Datatype unification: this issue occurs when the datatypes of a same resource in the generated graphs are different. For example, in Fig. 1, graph A' and B' generated using Sparql-Generated and ShExML respectively, are defined from the same input (string attributes "agency_name" and "agency_url" of object Agency), but their generated graphs indicate different datatypes. In Sparpql-Generate, both attributes are defined as Strings, and in ShExML they are defined as: Integer (for "agency_name") and URL (for "agency_url"), because ShExML uses Shape Expressions (ShEx) to define the desired structure of the output[9].
We treated this issue by unifying the datatypes in the generated graphs, by choosing the most appropriate datatype as a pivot datatype for these particular cases. In our context, both RML and Sparql-Generate preserve the datatypes of the input dataset (for most cases), we followed the same vision, and aligned the datatypes of the generated graphs from ShExML.
(3) Mapping enrichment: this step is related to the previous case. In certain mapping languages such as Shexml and RML, it is possible to directly adjust the mapping definition to achieve the intended datatype. In our future work, we aim to develop a tool that assists designers in selecting the most suitable datatype in the graphs, for their specific needs.
(4) Serialization unification: The output produced by KG generation systems for each mapping language exists in one of several serializations, such as turtle, N-triplets, etc. In particular, we notice that the Shexml is the only one that provides all the serializations. After an analysis of the resulting RDF

[9] https://github.com/kg-construct/mapping-challenges/tree/main/challenges/datatype-map.

graphs, we opted for turtle serialization as the pivot format, mainly because of its improved readability and ease of parsing.
(5) Duplicate data elimination: from the analysis of RDF graphs generated by the mapping languages with the same mapping and the same dataset as input, we noticed that the KG generated by the RML has more nodes compared to the other resulting KGs. This is explained by the fact that RMLMapper system used for executing RML mappings generates duplicate triples. We cleaned up these graphs by removing duplicated data.
(6) Encoding modification: only the graphs obtained via the RML mapping contains the encoding for the UTF-8 URL and the U+251C encoding for the "é". Consequently, in this scenario, we substituted each encoding with its corresponding character representation.
(7) Special characters elimination: The graphs generated by Sparql-Generate contains unnecessary special characters that do not exist in the data like "/" and "]". We detected and eliminated them.

We show the order of the graph preparation steps in the workflow of Fig. 2.

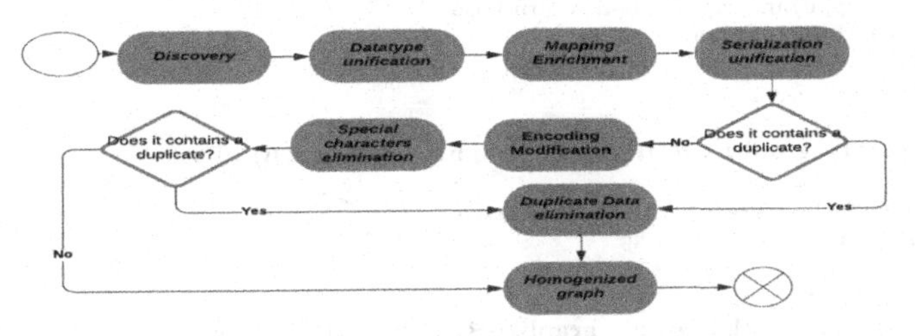

Fig. 2. Workflow of graph preparation.

We classify our graph preparation operations into the categories proposed by [27] as: **(1) Data discovery:**Discovery, **(2) Data structuring:** Encoding modification and Serialization unification, **(3) Data enrichment:** mapping enrichment, **(4) Data cleaning:** Duplicate data elimination and Special characters elimination and **(5) Data validation:** Datatype unification.

4.2 Mappings Comparison

In order to detect the relationships between the mappings, we used the ground graphs generated by the mappings, and returned these relationships back to the mappings. First, we take all the homogenized_graphs (scripted in Algorithm 1 as Hom_graphs) obtained from the previous phase, then we generate pairs of homogenized_graphs and push them into a list, then we launch a thread for each pair of graphs in order to start the comparison in parallel between the pairs. The expected total number of threads is the number of pairs which is equal to A_k^n, such that n is the number of mapping files and k=2 (graph pair): $A_k^n = \frac{(n)!}{k!(n-k)!}$. Finally, each pair of graphs are the parameters of the comparison function which first tests if these two graphs are equivalent. In this case we seek if they are isomorphic or not; in the negative case we test if one of the graphs is included in the other; for checking a Inclusion/Containment relationship between graphs. The last test checks the disjointness between the graphs.

Algorithmus 1 Compare mappings Algorithm

```
Inputs:List of Hom_graphs Lhomg
Outputs: Relationships between mappings
Function isEquivalent(graph1 : RDF graph, graph2: RDF graph) : Boolean
Begin
 1: if length(graph1) ≠ length(graph2) then return false
 2: end if
 3: for each s, p, o ∈ graph1 do
 4:     if not(s, p, o) in graph2 then return false
 5:     end if
 6: end for
 7: for each s, p, o ∈ graph2 do
 8:     if (not(s, p, o) in graph1) then return false
 9:     end if
10: end for
End
isInclude(graph1 : RDF graph, graph2: RDF graph) : Boolean
Begin
 1: for each s, p, o ∈ graph1 do
 2:     if (not(s, p, o) in graph2) then return false
 3:     else if r theneturn true
 4:     end if
 5: end for
End
isContainmentTo(graph1 : RDF graph, graph2: RDF graph) : Boolean
Begin
 1: if (graph2.isInclude(graph1)) then return true
 2: else if r theneturn false
 3: end if
End
AreDisjoint(graph1 : RDF graph, graph2: RDF graph) : Boolean
Begin
 1: for each s, p, o ∈ graph1 do
 2:     subjects_g1.add(s); predicates_g1.add(p); objects_g1.add(o) //subjects_g1, predicates_g1
    and objects_g1 are lists
 3: end for
 4: for each s, p, o ∈ graph2 do
 5:     subjects_g2.add(s); predicates_g2.add(p); objects_g2.add(o)
 6: end for
 7: if ((subjects_g1 ∩ subjects_g2) = ∅ & (predicates_g1 ∩ predicates_g2) = ∅ & (objects_g1 ∩
    objects_g2) = ∅) then return true else return false
 8: end if
End
Function CompareGraphs(graph1 : RDF graph, graph2: RDF graph) : Relationships
Begin
 1: if G1.isEquivalent(G2) then Equivalence
 2: else if G1.isIncludeIn(G2) then Inclusion
 3: else if G1.isContainmentTo(G2) then Containment
 4: else if G1.AreDisjoint(G2) then Disjointness
 5: end if
End
BeginAlgorithm
 1: for each homg ∈ Lhomg do
 2:     Add(homg, next homg) to liste_pairs
 3:     return liste_pairs
 4: end for
 5: for each pair ∈ liste_pairs do
 6:     CreateThread(target= CompareGraphs, arg=(G1,G2))
 7:     StartThread()
 8:     return Relations
 9: end for
End of Algorithm
```

Algorithm 1 formalizes these steps. This comparison between mappings can be performed on all mappings defined by the designer teams or it can also be performed incrementally, where each new mapping file is compared against existing defined mappings.

5 Experimentation

This part aims to illustrate the performance of our approach. We first show the results obtained in the graph preparation phase. Then we test our comparison algorithm. In both experiments, we extended the "GTFS-Madrid-Bench" benchmark [28] (which contains only 03 RML mapping files), with new mappings to obtain: 06 RML mapping files of all GTFS data (csv, json and xml), 18 Shexml mapping files and 43 sparql-generate mapping files. The mappings are executed using RMLMapper, and the online systems of Sparql-Generate and Shexml (see Sect. 2).

The experiments are performed using in an 11th Gen Intel(R) Core(TM) i7-1165G7 @ 2.80 GHz 1.69 GHz, 16 GB memory, and with the O.S. Windows 11 version 22H2. The proposed algorithms are implemented in Python and use RDFlib library.

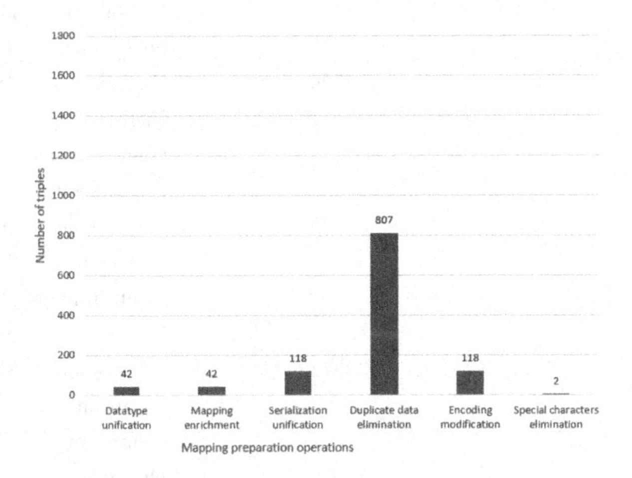

Fig. 3. Number of triples involved in a each graph preparation step.

The first test concerns the graph preparation process of our approach. Figure 3 shows the number of triples concerned by each step of this process. We note that the Discovery step concerns all the triples generated by the mappings. We notice that duplicate triples are numerous. When analysing the graphs, we noticed that this is due to the usage of RMLMapper system which generates duplicate data that are not present, neither in the input dataset nor in the mapping. The task of "special characters elimination" concerns few triples, because only Sparql-Generate generates such characters. Even if few triples are concerned, their presence may lead to different erroneous results when comparing the graphs. Additionally, looking for these problematic data (among the global set of RDF triples) is not an easy task for the designer. The second test concerns the mapping comparison process of our approach. Table 1 shows the results. The table includes the following columns: the number of comparison between pairs of homogenized graphs (each graph corresponds to a mapping file), the number of corresponding rules of mappings, number of triples concerned, and the number of relationships discovered between the mappings.

Table 1. Number of relationships in each test.

Number of pairs	Number of rules	Number of triples	Number of relationships	
3		63	Equivalence	1
			Inclusion	2
			Containment	2
			Disjoitness	0
6	33	69	Equivalence	1
			Inclusion	2
			Containment	2
			Disjoitness	3
120	1041	395294	Equivalence	78
			Inclusion	13
			Containment	13
			Disjoitness	29
496	1137	395390	Equivalence	406
			Inclusion	29
			Containment	29
			Disjoitness	61
1378	1230	395742	Equivalence	877
			Inclusion	253
			Containment	253
			Disjoitness	248
2211	5147	3574450	Equivalence	1154
			Inclusion	299
			Containment	299
			Disjoitness	758

Figure 4 (part A) illustrates this last column using a histogram to facilitate its reading. We started with 3 comparisons of pairs of graphs (i.e. comparison of 3 files of mappings, pair by pair) and we increased the number of mapping files at each iteration. The precision of the resulting relationships is proven by comparing the expected relationships (that we defined between the mappings) and the resulting relationships identified by the algorithm. The results obtained show that important relationships are discovered, which are difficult to identify by the designers, especially from different teams. These teams may discover equivalent mappings defined in different languages, know how other teams use the company dataset, and learn common or disjoint design patterns for defining the KG.

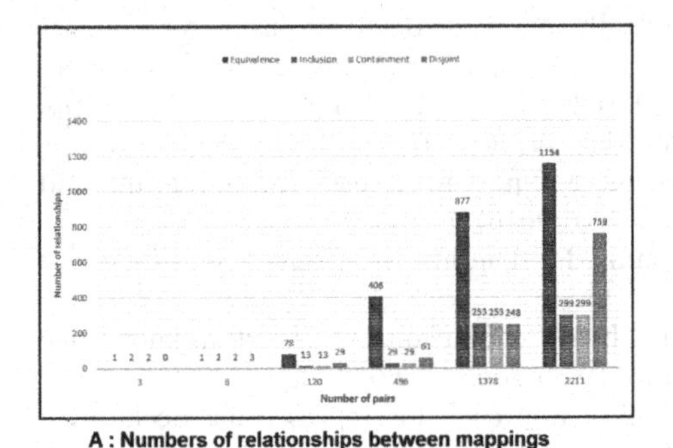

A : Numbers of relationships between mappings

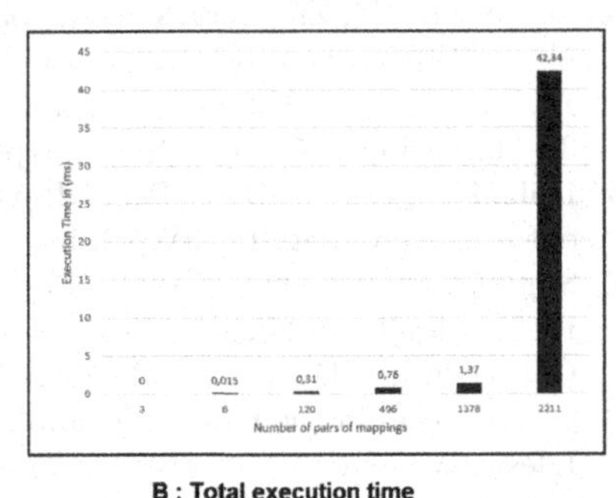

B : Total execution time

Fig. 4. Total number of relationships between mappings and execution time of experiments.

In Fig. 4 (part B), we have calculated the execution time (reported in milliseconds (ms)) in order to illustrate the efficiency of running the comparison of the graphs in parallel by the use of threads.

6 Conclusion

This article develops the concept of heterogeneous mappings comparison using a reverse ETL approach, i.e. based on the analysis of the generated KG. In a context of a company involving different designer teams in the enterprise KG project, the analysis of the different mappings may provide useful insights in the defined tasks, datasets and patterns defined by different teams, and consequently contribute to have a global picture of the KG life-cycle. As future works, we are implementing a modular system that automatically supports different mapping languages, uses graph matching and mining techniques to monitor the mappings and to identify trends, patterns and recommendations each time a new mapping attempt is defined. Also, the possibility of identifying comparative operations by comparing the syntax of input mapping languages, without executing the mappings, should be investigated.

References

1. Iglesias, E., et al.: Empowering the SDM-RDFizer tool for scaling up to complex knowledge graph creation pipelines. SWJ J. (2022)
2. Das, S., Sundara, S., Cyganiak, R.: R2RML: RDB to RDF mapping language. W3C recommendation. W3C, vol. 9 (2012)
3. Dimou, A., et al.: RML: a generic language for integrated RDF mappings of heterogeneous data. Ldow, vol. 1184 (2014)
4. García-González, H., et al.: ShExML: improving the usability of heterogeneous data mapping languages for first-time users. PeerJ Comput. Sci. **6**, e318 (2020)
5. Lefrançois, M., Zimmermann, A., Bakerally, N.: A SPARQL extension for generating RDF from heterogeneous formats. In: Blomqvist, E., Maynard, D., Gangemi, A., Hoekstra, R., Hitzler, P., Hartig, O. (eds.) ESWC 2017. LNCS, vol. 10249, pp. 35–50. Springer, Cham (2017). https://doi.org/10.1007/978-3-319-58068-5_3
6. Djilani, Z., et al.: MURGROOM: multi-site requirement reuse through graph and ontology matching. In: iiWAS, pp. 160–169 (2016)
7. Simitsis, A., Skiadopoulos, S., Vassiliadis, P.: The history, present, and future of ETL technology (2023)
8. Chaves-Fraga, D., et al.: GTFS-madrid-bench: a benchmark for virtual knowledge graph access in the transport domain. J. Web Semant. **65**, 100596 (2020)
9. Iglesias-Molina, A., et al.: An ontological approach for representing declarative mapping languages. Semant. Web (Preprint) 1–31 (2022)
10. Iglesias-Molina, A., Cimmino, A., Corcho, O.: Devising mapping interoperability with mapping translation. In: KGCW (2022)
11. García-González, H., Dimou, A.: Why to tie to a single data mapping language? Enabling a transformation from ShExML to RML. In: SEMANTiCS 2022, vol. 3235, p. 11 (2022)
12. Van Assche, D., et al.: Declarative RDF graph generation from heterogeneous (semi-) structured data: a systematic literature review. J. Web Semant. **75**, 100753 (2022)
13. Dimou, A., et al.: Automated metadata generation for linked data generation and publishing workflows. In: LDOW2016, pp. 1–10. CEUR-WS.org (2016)
14. Iglesias, E., et al.: SDM-RDFizer: an RML interpreter for the efficient creation of RDF knowledge graphs. In: CIKM, pp. 3039–3046 (2020)
15. Arenas-Guerrero, J., et al.: Morph-KGC: scalable knowledge graph materialization with mapping partitions. Semant. Web (Preprint), 1–20 (2022)
16. Composable semantic data transformation pipelines with chimera (2023)
17. Zaveri, A., et al.: Quality assessment for linked data: a survey. Semant. Web **7**(1), 63–93 (2016)
18. Cimmino, A., García-Castro, R.: Helio: a framework for implementing the life cycle of knowledge graphs. Semant. Web (Preprint), 1–27 (2022)
19. Iglesias, E.A., Vidal, M.-E.: Knowledge graph creation challenge: results for SDM-RDFizer (2023)
20. Randles, A., Crotti Junior, A., O'Sullivan, D.: A framework for assessing and refining the quality of r2rml mappings. In: iiWAS, pp. 347–351 (2020)
21. Heyvaert, P., et al.: Rule-driven inconsistency resolution for knowledge graph generation rules. Semant. Web **10**(6), 1071–1086 (2019)
22. Heyvaert, P., et al.: Specification and implementation of mapping rule visualization and editing: MapVOWL and the RMLEditor. J. Web Semant. **49**, 31–50 (2018)

23. Corcho, O., Priyatna, F., Chaves-Fraga, D.: Towards a new generation of ontology based data access. Semant. Web **11**(1), 153–160 (2020)
24. Vassiliadis, P.: A cube algebra with comparative operations: containment, overlap, distance and usability. arXiv preprint arXiv:2203.09390 (2022)
25. Berkani, N., Bellatreche, L., Khouri, S.: Towards a conceptualization of ETL and physical storage of semantic data warehouses as a service. Clust. Comput. **16**(4), 915–931 (2013)
26. Hogan, A.: Skolemising blank nodes while preserving isomorphism. In: WWW, pp. 430–440 (2015)
27. Hameed, M., Naumann, F.: Data preparation: a survey of commercial tools. ACM SIGMOD Rec. **49**(3), 18–29 (2020)
28. Harris, J.M.: Combinatorics and Graph Theory. Springer, Heidelberg (2008). https://doi.org/10.1007/978-0-387-79711-3

Machine Learning and Optimization

Exploring Synthetic Noise Algorithms for Real-World Similar Data Generation: A Case Study on Digitally Twining Hybrid Turbo-Shaft Engines in UAV/UAS Applications

Ali Aghazadeh Ardebili[1,2(✉)], Antonella Longo[1,5(✉)], Antonio Ficarella[3,5], Adem Khalil[4], and Sabri Khalil[4]

[1] Data Lab, Dept. of Engineering for Innovation, University of Salento, 7100 Lecce, LE, Italy
{ali.a.ardebili,antonella.longo}@unisalento.it
[2] CRISR Research Center, University of Salento, 7100 Lecce, LE, Italy
[3] Green Engine Lab, Dept. of Engineering for Innovation, University of Salento, 7100 Lecce, LE, Italy
antonio.ficarella@unisalento.it
[4] Engineering and Architecture Faculty, Istanbul Gelisim University, Istanbul, Turkey
[5] Italian Research Center on High Performance Computing, Big Data and Quantum Computing (ICSC), Bologna, Italy
https://sydalab.unisalento.it/, https://crisr.unisalento.it/, https://greenengine.unisalento.it/en/, https://www.supercomputing-icsc.it/

Abstract. An emerging technology for automating Unmanned aircraft is digitally twining the system, and employing AI-based data-driven solutions. Digital Twin (DT) enables real-time information flow between physical assets and a virtual model, creating a fully autonomous and resilient transport system. A key challenge in DT as a Service (DTaaS) is the lack of Real-world data for training algorithms and verifying DT functionality. This article focuses on data augmentation using Real-world Similar Synthetic Data Generation (RSSDG) to facilitate DT development in the absence of training data for Machine Learning (ML) algorithms. The main focus is on the noise generation step of the RSSDG for a common Hybrid turbo-shaft engine because there is a significant gap in transforming synthetic data to Real-world similar data. Therefore we generate noise through 6 different noise generation algorithms before Rolling Linear Regression and Filtering the noisy predictions through Kalman Filter. The primary objective is to investigate the sensitivity of the RSSDG process concerning the algorithm that is used for noise generation. The study's results support the potential capacity of RSSDG for digitally twining the engine in a Real-world operational lifecycle. However, noise generation through Weibull and Von Mises distribution showed low efficiency in general. In the case of Normal Distribution, for both thermal and hybrid models, the corresponding DT model has shown high efficiency in noise filtration and a certain amount

M. Mosbah et al. (Eds.): MEDI 2023, LNCS 14396, pp. 87–101, 2024.
https://doi.org/10.1007/978-3-031-49333-1_7

of predictions with a lower error rate on all engine parameters, except the engine torque; however, Students-T, Laplace, and log-normal show better performance for engine torque RSSDG.

Keywords: Noise Generation · Digital Twins · Unmanned Aircraft Systems · Synthetic data Generation · Data for Resilience · Realistic Synthetic Data

1 Introduction

Urban Air Mobility (UAM) has become a rising trend that garners significant attention from scholars and practitioners, aiming to establish a sustainable and resilient transport infrastructure. However, UAM encounters numerous technological and legislative challenges, including air traffic control, cybersecurity concerns, noise pollution [23], and ecological considerations.

On a positive note, electrified propulsion systems have been identified as a promising solution for fuel savings and emission reduction [10] and as a solution for automating unmanned aircraft, the utilization of Digital Twin (DT) technology has proven effective in enhancing system and entity performance, enabling predictive maintenance, and increasing safety standards [1].

Digital Twining technology is rapidly emerging as a means to enhance system/entity performance and improve predictive maintenance practices with a high level of safety [11]. Nevertheless, DT strongly relies on data-oriented solutions and operates with machine learning (ML) algorithms. The primary challenge in this approach lies in obtaining Real-world datasets to train the ML algorithms effectively. Especially in the aerospace domain, fabricating complex and costly entities to generate Real-world measurements poses a significant obstacle. Recognizing this challenge, the main goal of this research is to explore Real-world Similar Synthetic Data Generation (RSSDG) approaches. Given the constraints of time and cost involved in setting up physical test beds and collecting data from Real-world entities, the proposed solution is to leverage data augmentation through RSSDG techniques.

The main goal of this study is to investigate the generation of synthetic data using simulated Hybrid Turbo-shaft Engine data to predict engine behavior in various flight scenarios. The main contribution of this research is the development of a Digital Twin for the Hybrid Turbo-shaft Engine based on Augmented Synthetic Data. This approach can be a facilitator in the development of DT in case the developers do not have enough data to train the Machine Learning (ML) algorithm. On the other hand, the current twining approach provides a prospective ideal state of the engine used for the proactive monitoring of engine health in DT as an anomaly detection service. In brief, the study aims to fill the significant gap in Real-world similar RSSDG in the UAV domain. This approach begins by constructing a simulation model of the Hybrid Turbo-shaft Engine. The model is then linearized to increase the understanding of the relationships between the engine's parameters. Next, noise is added to the simulated dataset to replicate Real-world noise patterns. The final step involves validating the performance of the Digital Twin.

2 Background Review

This section presents a brief background of the Urban Air Mobility concept, RSSDG, and the significant role of Digital Twin technology in increasing the sustainability and safety aspects of the vision.

Urban Air Mobility (UAM). Researchers and practitioners in developing new infrastructures are actively exploring innovative solutions to address traffic congestion and provide faster, safer, and more efficient transportation systems. Initiatives such as Hyperloop and Urban Air Mobility (UAM) have gained attention in this regard. It's important to highlight the concept of "Vertical Takeoff and Landing" Vehicles (VTOL), which are considered the most common solution for infrastructure problems and traffic management [16].

One of the most studied electrified propulsion systems for UAM are Hybrid Turbo-shaft engines. This type of engine offers several advantages in terms of energy consumption, performance, and safety.

Digital Twining of UAS/UAVs. Like many new technologies, ensuring a high level of safety in complex systems requires advanced performance analyses. However, simulating such complex systems can be computationally expensive, and it is crucial to align these analyses with the Real-world performance of the system being studied. In this context, the use of Synthetic Data and encompassing different correlated parameters can be beneficial in replicating the behavior of the system under investigation. Utilizing Data-Driven simulations and Digital Twin (DT) technology supports building predictive models that enable real-time simulations that help with preventing undesirable scenarios. As a result, DT affords us a clear image of the system from a physical and operational point of view [9,17].

DT is widely used for real-time modeling of complex systems. DT models have demonstrated their effectiveness in handling complex systems by utilizing simplified models [22]; however, the state-of-the-art review reveals a significant gap in the studies of implementing DT in the UAV/UAS domain. The main reason is the lack of Real-world data to simulate the UAS and training ML algorithms for digitally twining the system.

UAS/UAV Real-World Similar Synthetic Data Generation (RSSDG) for Digital Twining. Most of the research belongs last 5 years. The majority of the studies have been done on Security [14] and navigation [19], and there are a handful of articles that are experimental with a test bed to collect Real-world data [18]. This fact unveils that there is a lack of physical testbeds and experimental data in this domain. correspondingly, developing/testing/employing AI algorithms for DT is not possible if there is no data for training the algorithm. this was the main motive to trigger the research on RSSDG for UAS/UAVs.

However, due to the lack of Real-world measurements/physical entities, researchers and academics have demonstrated that mathematical models, implemented using software like MATLAB for study purposes, can efficiently generate data and validate their approaches [12]. However, this process generates a simulation that produces data without noise, unlike Real-world data which is typically affected by various sources of noise. This noise can arise from transmission constraints, faults in sensor devices, and irregularities in sensing and transcription, etc. Building upon this motivation, current investigation aims to assess the performance of a predictive digital twin model based on Kalman Filtering and ML. The model aims to predict the behavior of a hybrid turboshaft engine using synthetic data with added noise.

3 RSSDG Process with Various Noise Generation Methods

In this section, the workflow of the employed RSSDG and DT approach will be detailed. The pipeline architecture of the workflow is shown in (see Fig. 1a). In the current article, The proposed DT model is founded on the recursive algorithms, Kalman Filter and Rolling Linear Regression which follows the workflow of the 6 main steps. The outline of the steps in the order is: [step1: Data Linearisation and referencing, step2: Adding noise, step3: Implementation of Linear regression, step4: Implementation of Kalman filter step5: Implementation of Rolling linear regression, step6: Printing the results in a function of mean squared error\mean].

In order to investigate the performance of the proposed DT model we adopted the simulation model which is previously published by Donateo, T. et al. [9] that provides a generated DATA set of the propulsion system (Hybrid Turboshaft Engine (see Fig. 1, where A, B and C respectively in Fig. 1b are The turboshaft engine, Electric machines and drivers, Li-on Battery, Hybrid Turboshaft Engine system, and Digital Twin and supervisory control system).

The information on the simulated mission of the hybrid powertrain with compressor degradation is described in the following Table 1.

Table 1. Simulated mission.

	TIMES (S)	SPEED (m*backslash*s)	ALTITUDE (m)	POWER (kW)
MISSION A START	0.1	30.6	0	48
MISSION A END	1650	30.6	0	48
MISSION B START	0.1	0	1150	172
MISSION B END	1246	1	1149	152
MISSION C START	0.1	0	7	168
MISSION C END	2079	1.59	6.22	147
MISSION D START	0.1	0	7	168
MISSION D END	935	1.54	6.41	151

The generated DATA is containing the input variables, state variables, and output variables of the engine for both thermal and electrical systems, after cleaning the data set the modeling of the DT will be described in the following subsections:

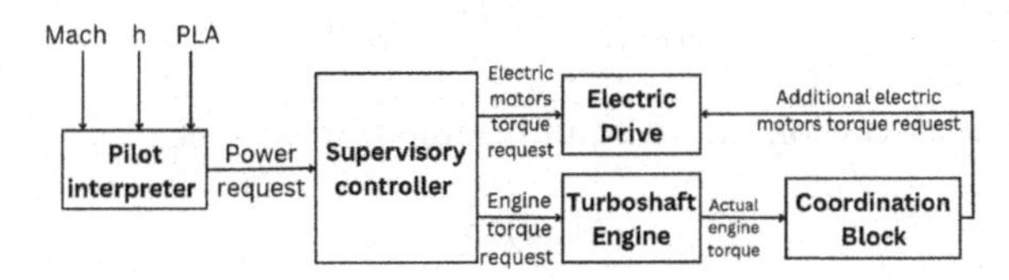

(a) Hybrid Electric Propulsion System model flow chart[8].

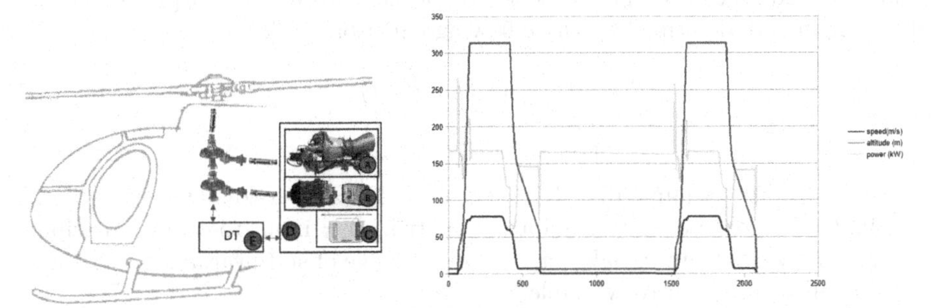

(b) Overview of the hybrid electric power systems [4].

(c) The behavior of the velocity, pressure, and power along the mission[4].

Fig. 1. Hybrid Turbo-shaft engine components and behavior illustration.

3.1 Data Linearisation

The linearisation of non-linear complex systems increase the understanding and the study of the behavior of the system without losing the relationship between the parameters. In the UAV domain, in particular, control systems engineering of UAVs, representing the nonlinear dynamical systems in the form of mathematical models through a set of state variables and I/O. If these variables are related by a first-order differential equation, the model is called state-space form [6,7]. The state space form of the system is formulated in the following way [3]:

$$\dot{x}(t) = f(x(t),\ u(t)) \tag{1}$$

$$y(t) = g(x(t),\ u(t)) \tag{2}$$

considering x is the state vector and U is the input, there are two continuous and differential functions in Eqs. 1 and 2 (f and g). Expanding differential functions in series around $\overline{x}$ and $\overline{y}$, the following equations are defined where the x and u are in the vicinity of a point $(\bar{x},\ \bar{u})$.

$$f(x,\ u) = f(\overline{x},\ \overline{u}) + A(x - \overline{x}) + B(u - \overline{u}) \tag{3}$$

$$g(x,\ u) = g(\overline{x},\ \overline{u}) + C(x - \overline{x}) + D(u - \overline{u}) \tag{4}$$

$$A = \left.\frac{\partial f}{\partial x}\right|_{\substack{x=\bar{x}\\ u=\bar{u}}},\quad B = \left.\frac{\partial f}{\partial u}\right|_{\substack{x=\bar{x}\\ u=\bar{u}}} \tag{5}$$

$$C = \left.\frac{\partial g}{\partial x}\right|_{\substack{x=\bar{x}\\ u=\bar{u}}},\quad D = \left.\frac{\partial g}{\partial u}\right|_{\substack{x=\bar{x}\\ u=\bar{u}}} \tag{6}$$

We define the deviation on an equilibrium state where:

$$f(\overline{x}, \overline{u}) = 0 \tag{7}$$

After defining deviations we can model our engine as a Linear-Time- Invariant (LTI) [3, 13, 20], the general non-linear form of the state and the output equations of the engine are described by the following equations [15]:

$$\dot{x}(t) = Ax(t) + Bu(t) \tag{8}$$

$$y(t) = Cx(t) + Du(t) \tag{9}$$

where A = State dynamic distribution matrix, B = input-to-state distribution matrix, C = state-to-output distribution matrix, D = input-to-output distribution matrix, y = output variable, $\dot{x}$ = rate of change of state variables, x = state variable, u = input control variable.

3.2 Referencing Data to Take Off Condition

Since the features of the data from an engine have varying scales which decreases significantly its readability.

By a consequence the patterns between features will be hard to study in this situation data normalization is an efficient solution, so we referenced all the features to the Take off condition.

Therefore, considering the reference equations (Eqs. 7 and 8), our model is described as follows where Table 2 defines the parameters:

$$\begin{bmatrix} Nc\_cref_{t+1} \\ PT3ref_{t+1} \\ TT4ref_{t+1} \\ Wfref_{t+1} \end{bmatrix} = [A] \begin{bmatrix} Nc\_cref_t \\ PT3ref_t \\ TT4ref_t \\ Wfref_t \end{bmatrix} + [B]\, PLA\_goverf_t \tag{10}$$

$$TQ\_pt_t = [C] \begin{bmatrix} Nc\_cref_t \\ PT3ref_t \\ TT4ref_t \\ Wfref_t \end{bmatrix} + [D]\, PLA\_goverf_t \tag{11}$$

Table 2. Engine Parameters (R.P.V. = Referenced parameter in take-off).

Parameter	Explanation	Parameter	Explanation
Nc_c	High pressure spool speed	$TT4$	Turbine inlet total temperature
Nc_cref	R.P.V.	$TT4ref$	R.P.V.
Nc_cref_t	Measured value	$TT4ref_t$	Measured value
Nc_cref_{t+t}	Predicted value	$TT4ref_{t+1}$	Predicted value
$PT3$	Compressor outlet total pressure	Wf	Fuel flow rate
$PT3ref$	R.P.V.	$Wfref$	R.P.V.
$PT3ref_t$	Measured value	$Wfref_t$	Measured value
$PT3ref_{t+1}$	Predicted value	$Wfref_{t+1}$	Predicted value

3.3 Noise/Error Generation

In order to simulate Real-world conditions for our Digital Twin model we will be comparing the Kalman filter's Performance with varying noise by drawing samples from a different distribution. Next, for each set of noisy data, we will proceed with the following steps outlined in Subsect. 3.3 and Subsect. 3.4. Finally, by comparing the results obtained, we will determine if the RSSDG is a noise generation-sensitive process. In the following subsections, 6 distributions that are used for noise generation are elaborated.

Gaussian Distribution. The Normal Distribution is a continuous probability distribution that is widely used. When generating noise, it is assumed that the real-valued random variable follows this distribution, resulting in an equal number of measurements above and below the mean value. (Fig. 2a). In the following equation, μ is mean and σ is the standard deviation.

$$f(x) = \frac{1}{\sigma\sqrt{2\pi}} e^{-\frac{1}{2}\left(\frac{x-\mu}{\sigma}\right)^2} \tag{12}$$

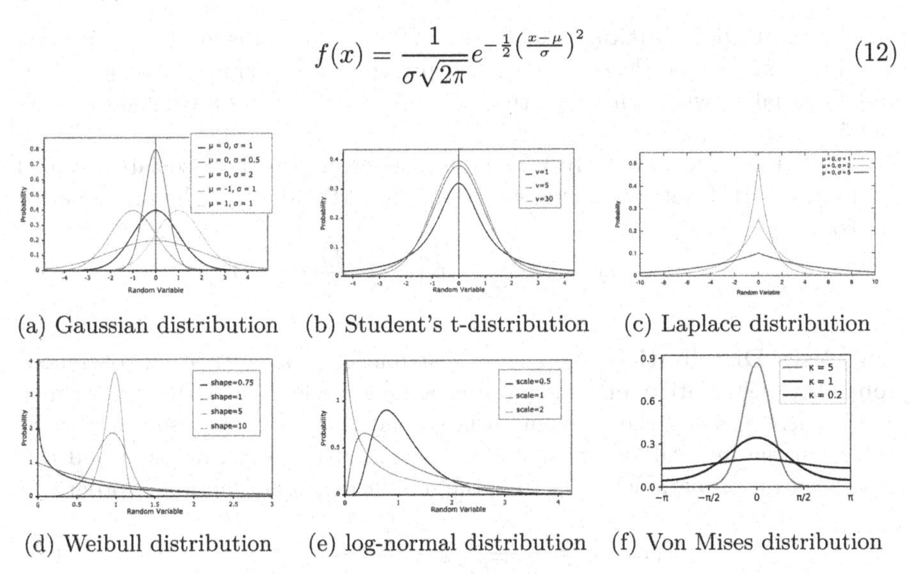

(a) Gaussian distribution (b) Student's t-distribution (c) Laplace distribution

(d) Weibull distribution (e) log-normal distribution (f) Von Mises distribution

Fig. 2. Distribution Curves.

Student's t -Distribution. The Student's t distribution is a family of curves characterized by a single parameter. This distribution finds extensive application in hypothesis testing, particularly when the population's standard deviation is unknown, and we rely on the mean value (Fig. 2b). The Probability density function (PDF) is described as follows where x is the random variable and ν is the degrees of freedom, Γ shows the gamma function:

$$f(x|\nu) = \frac{\Gamma\left(\frac{\nu+1}{2}\right)}{\sqrt{\nu\pi}\Gamma\left(\frac{\nu}{2}\right)}\left(1+\frac{x^2}{\nu}\right)^{-\frac{\nu+1}{2}} \tag{13}$$

Laplace Distribution. The Laplace distribution is a continuous distribution that can be defined as the difference between two independent variables with identical exponential distributions [2]. Below, we present the PDF in Eq. 14 for the Laplace Distribution With x being the random variable, μ being the location parameter (mean), b being the scale parameter:

$$f(x|\mu, b) = \frac{1}{2b}\exp\left(-\frac{|x-\mu|}{b}\right) \tag{14}$$

Weibull Distribution. The Weibull distribution is a versatile distribution that can generate various types of distributions depending on the shape parameter. [21]. In the following, the two-parameter Weibull distribution PDF is shown where λ is the scale parameter, and k is the shape parameter.

$$f(x|\lambda, k) = \begin{cases} \frac{k}{\lambda}\left(\frac{x}{\lambda}\right)^{k-1}\exp\left[-\left(\frac{x}{\lambda}\right)^{k}\right] & \text{if } x \geq 0 \\ 0 & \text{if } x < 0 \end{cases} \tag{15}$$

Log-Normal Distribution. The log-normal distribution is an alternative distribution that can be derived using the principle of maximum entropy for a random variable, where the logarithm of the variable follows a normal distribution [5].

The PDF is shown in Eq. 16 Where μ is the mean, σ is the associated normal distribution, erf denotes the error function. and erf^{-1} denotes the inverse error function.

$$f(x|\mu, \sigma) = \frac{1}{x\sigma\sqrt{2\pi}}\exp\left(-\frac{(\ln(x)-\mu)^2}{2\sigma^2}\right) \tag{16}$$

Von Mises Distribution. Von Mises distribution is a captivating continuous probability distribution mapped around a unit circle. The PDF is shown in Eq. 17 where x being the random variable and the equation assumes x to be defined on the interval [-π, π] or [0, 2π], μ being the mean direction, and κ is the concentration parameter. $I_0(\kappa)$ is the modified Bessel function of order 0.

$$f(x|\mu, \kappa) = \frac{1}{2\pi I_0(\kappa)}\exp\left(\kappa\cos(x-\mu)\right) \tag{17}$$

3.4 Prediction and Filtering

Rolling Linear Regression. To predict or reduce error rates, linear regression is an excellent statistical tool for creating a predictive model using a captured set of values for the response and explanatory variables. Rolling regression, also known as "moving period regression" or "rolling window regression," evaluates the changing relationships among variables over time, specifically measuring the outputs such as correlation and standard error from linear regression. This visualization allows for adjustments to the dataset as time progresses, whereas traditional linear regression models assume that parameters remain constant over time. The general procedure of linear regression is illustrated in Fig. 3. The results from the rolling regression will serve as input for implementing the Kalman Filter and investigating its advantages and efficiency in filtering noise and making more accurate predictions.

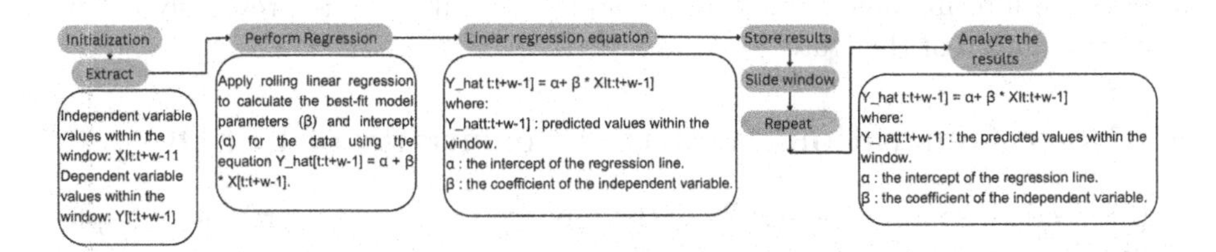

Fig. 3. Outline of the steps for Rolling Linear Regression.

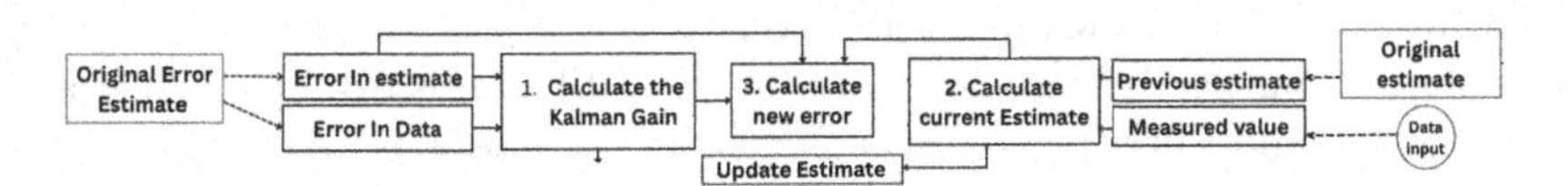

Fig. 4. The block diagram of the Kalman filter.

Implementation of Kalman Filter. Kalman filter is a recursive algorithm that adjusts the process state using real-time measurements (see Fig. 4). The reason that filtering algorithms are used is that in Real-world modeling scenarios, the measurements are not in all respect accurate. Therefore, when data-driven methods are implemented, we need to deal with the noise that is present in the collected data from IoT sensors. In the current study, instead of the noisy measurement, we will use data that are produced by a simulation model with the noise that is added to the simulation data (Subsect. 3.3). The Kalman Filter in this study process has two steps predict and update as follows (Eq. 19 formulates the predict step and Eq. 19 formulates the Update step):

$$\begin{bmatrix} Nc\_cref_{t+1} \\ PT3ref_{t+1} \\ TT4ref_{t+1} \\ Wfref_{t+1} \end{bmatrix} = [F_t] \begin{bmatrix} Nc\_cref_t \\ PT3ref_t \\ TT4ref_t \\ Wfref_t \end{bmatrix} + [B_t]\, PLA\_goverf_t + W_t \tag{18}$$

$$\begin{bmatrix} Nc\_crefMEA_{t+1} \\ PT3refMEA_{t+1} \\ TT4refMEA_{t+1} \\ WfrefMEA_{t+1} \end{bmatrix} = [H_t] \begin{bmatrix} Nc\_cref_t \\ PT3ref_t \\ TT4ref_t \\ Wfref_t \end{bmatrix} + V_t \tag{19}$$

where F_t= State transition model, B_t= Control-input model, H_t= Observation model, W_t= Process noise, V_t= Observation noise.

As mentioned earlier, the relationship between the inputs and outputs of the hybrid turboshaft engine is described using a state space model to represent the system's state. The difference between the actual behavior and the simulation data is attributed to uncertainty in the dynamic model, known as process noise. To address errors or uncertainties in the synthetic data generated by the simulation, the noise (already introduced in Subsect. 2.3) will be filtered to achieve behavior similar to Real-world conditions. Therefore, to calculate the next state, a recursive filtering algorithm will be employed, utilizing the previously calculated estimation of the state.

4 Noise Generation Algorithm Sensitivity of RSSDG

In this section, some of the results of the proposed method are presented graphically (Since in the current study, there are several parameters, just some examples will be illustrated). As it is mentioned in the Methodology section, after configuring the take-off condition, the noise is added to the data.

For instance, Fig. 5a) is presenting the noise generation on shaft speed using Gaussian distribution. The standard deviation is 0.1 in this example. Then Linear Regression and Filtering algorithms are used. Figure 5b is an example of the result for rolling linear regression for shaft speed. In Fig. 5b the blue line is referenced shaft speed (the data), and the line in orange is the result of rolling linear regression (prediction). Already with Fig. 5b predictions show accuracy on the data without noise.

(See Fig. 5c) shows the results of the rolling linear regression on noisy data. Time window 5000-time steps (500 s). The linear regression parameters change with time.

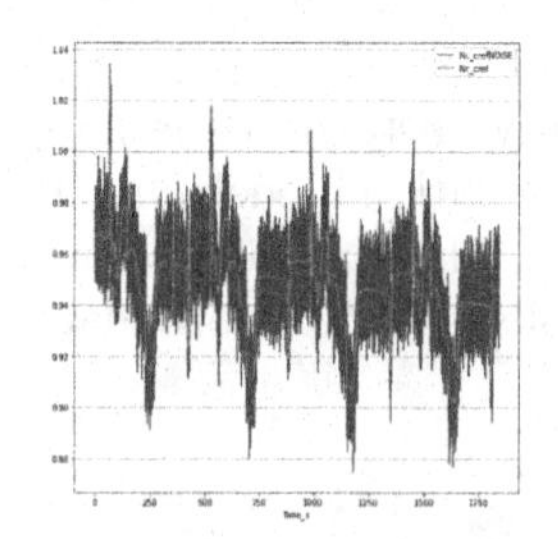

(a) Shaft speed with added noise.

(b) Rolling linear regression for shaft speed.

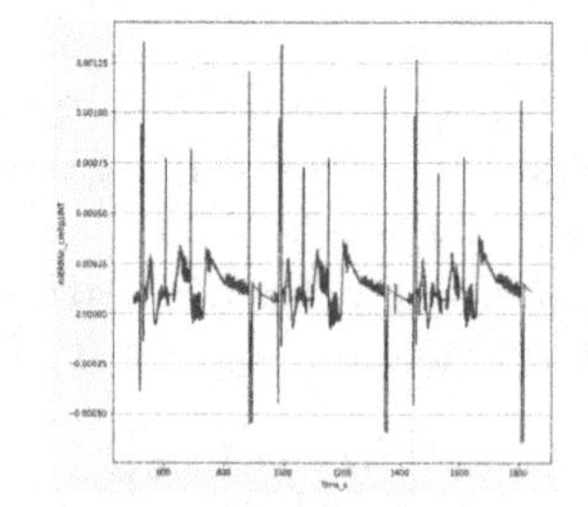

(c) Rolling linear regression on noisy data.

Fig. 5. RSSDG with Gaussian distribution for noise generation [4].

In the next section, the DT model performance will be discussed. Regarding the results, it performed effectively across the majority of the parameters as the results show that the combined algorithms have a significantly lower (mean squared error/ mean value) Index.

5 Validation

The results of this study have provided valuable insights into the performance of the suggested Digital Twin approach for a Hybrid Turboshaft Engine. Since the reference data (simulated data) are created by the simulation approach, validation through comparison with Real-world data generated by a physical Hybrid Turboshaft Engine is not possible. Therefore, the suggested model was analyzed based on its overall efficiency, accuracy in prediction, and noise filtration capabilities by mean squared error/mean value index.

After the application of the suggested DT model to different parameters of the Hybrid Turboshaft engine and in order to study its performance the function of Mean squared error/Mean value is used. In the first phase of current research project, the result from the Thermal Engine with noise generation by Gaussian function (Fig. 6a, 6b) and With the same simulation conditions, the results from the Hybrid model are reported (Fig. 6c) by Aghazadeh Ardebili et al. [4].

While the rolling linear regression alone has shown a weaker performance as a consequence using combined ML recursive algorithms is an efficient solution to get a higher accuracy rate. This approach is showing a weak performance with the engine torque therefore the linear model isn't optimal for the parameter. the results also show a high (mean squared error/mean value) index for the engine torque which means a weak performance in prediction and noise filtration, indicating that the used approach was not optimal for the torque.

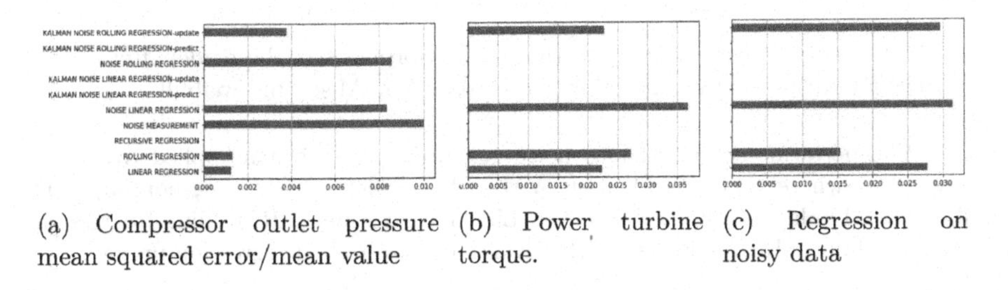

(a) Compressor outlet pressure mean squared error/mean value

(b) Power turbine torque.

(c) Regression on noisy data

Fig. 6. Power turbine torque with noise generation through Gaussian distribution.

In this article, we tried to use 6 different noise-generating algorithms within the same twining approach for data generation. We witnessed similar results of validation while using other distributions in Fig. 2. For both Thermal and Hybrid models, the DT model (Rolling Linear Regression + Klaman filter together) has shown efficiency in noise filtration and predictions with a lower error rate on all the parameters except with the engine torque. The low error rate indicates the great potential use of this approach in a variety of applications in the Urban Air Mobility field.

Figure 7a provides an overview of the model's performance when subjected to various noise generation algorithms: Students-t distribution, Von Mises distribution, Log-normal distribution, Weibull distribution, Laplace distribution, and Gaussian distribution, all applied to the engine torque. The model's performance is evaluated using the Mean Squared Error (MSE) divided by the Mean Value, which serves as a measure of accuracy for the model's predictions on the noisy data and the impact of noise on its performance. Specifically, the analysis focuses on two bars in the chart: the orange bar, representing the noisy results from linear regression (input for the Kalman filter), and the green bar, indicating the results after applying the Kalman filter.

Upon examination, we observe notably high Mean Squared Error rates for both the Von Mises distribution and the Weibull distribution, with values of 7 and 15, respectively. Consequently, we can draw the conclusion that these distributions contain numerous outliers, and the noise introduced does not conform to the underlying patterns of the data. As a result, we choose to eliminate these distributions from the analysis as they significantly hinder the readability and reliability of the results.

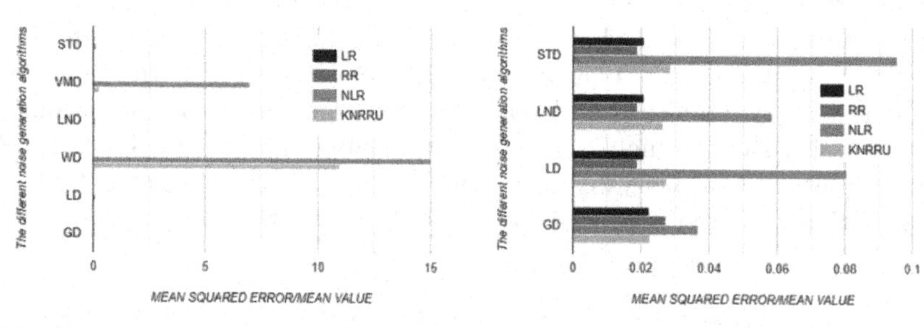

(a) The overall performance of the models for various noise generation

(b) Performance of the model after excluding Von Mises and weibull

Fig. 7. The performance for various noise generation Algorithms applied to Engine torque data; where STD: Students-t, VMD: Von Mises, LND: Log-normal, WD: Weibull, LD: Laplace, GD: Gaussian, LR: Linear Regression, RR: Rolling Regression, NLR: Noise Linear Regression, KNRRU: Kalman noise rolling regression-update.

Figure 7b provides a comprehensive analysis of the model's performance while excluding the Von Mises distribution and Weibull distribution noise generation methods. The chart offers valuable insights into the model's performance for different noise generation algorithms. For the Students-t distribution, the Mean Squared Error divided by the Mean Value (MSE/Mean) starts at 0.095 for the linear regression (orange bar) and significantly improves to 0.029 after applying the Kalman filter (green bar). Similarly, the Log-normal distribution shows a drop in MSE/Mean from 0.058 to 0.027 after Kalman filtering. The Laplace distribution also demonstrates improved performance, with MSE reducing from 0.08 to 0.023, while the Gaussian distribution sees a decrease from 0.027 to 0.023.

In conclusion, combining both rolling linear regression and Kalman filter consistently leads to improved performance across various noise generation methods. However, it is worth noting that the model exhibits weaknesses in handling noise related to the Engine torque parameter, particularly when compared to its performance on other engine parameters. This observation underscores the need for further investigation and optimization to enhance the model's overall effectiveness in dealing with noise variations in the Engine torque parameter.

6 Conclusions and Future Research Lines

The DT model performs well in terms of its overall efficiency; predictions, and noise filtration. These results have provided valuable data and knowledge related to the DT topic for Hybrid Turboshaft engines for further studies and investigations. The Low error rate of the DT model indicates that this type of DT holds excellent potential for use in a variety of applications in the Urban Air Mobility and Advanced Air Mobility field. Additionally, this study has provided a DT modeling method that could help designers, manufacturers, and academics to work on Hybrid propulsion systems and to drive further research and development.

The study's results support the potential capacity of RSSDG for digitally twining the engine in a Real-world operational lifecycle. However, noise generation through Weibull and Von Mises distribution showed low efficiency in general. In the case of Normal Distribution, for both thermal and hybrid models, the corresponding DT model has shown high efficiency in noise filtration and a certain amount of predictions with a lower error rate on all engine parameters, except the engine torque; however, Students-T, Laplace, and log-normal show better performance for engine torque RSSDG.

The main limitation of this study is the absence of Real-world data from a physical entity. At the moment, there is no openly accessible data for Hybrid-turbo shaft engines in this domain, mainly due to its status as a cutting-edge technology in the realm of UAM (Urban Air Mobility).

A key future research line is employing synthetic data for anomaly detection by bringing DT as a Service (DTaaS) into action. Also, We intend to explore alternative research directions in the future This involves collecting data from various Vertical Take-Off and Landing (VTOL) platforms and comparing the results to gain insights into Real-world behavior of noise and noise patterns. Subsequently, the potential future study is to train a machine learning (ML) algorithm to incorporate Real-world noise characteristics into the simulated data, thereby generating engine data that closely resembles Real-world scenarios.

Acknowledgements. The research was partially supported by the Ph.D. school of the University of Salento, Dep. of the complex systems engineering - XXXVI cycle, the RIPARTI regional project - dataEnrichment for Resilient UAS (assegni di

RIcerca per riPARTire con le Imprese)-POC PUGLIA FESRTFSE 2014/2020, CUP F87G22000270002 and the Italian Research Center on High Performance Computing, Big Data and Quantum Computing (ICSC) funded by EU - NextGenerationEU (PNRR-HPC, CUP:C83C22000560007).

References

1. Definition of a digital twin. https://www.digitaltwinconsortium.org
2. Abramowitz, M., Stegun, I.A., Romer, R.H.: Handbook of mathematical functions with formulas, graphs, and mathematical tables (1988)
3. Abughali, A., Habash, O., Elshamy, A., Alansari, M., Alhammadi, K.: Design and analysis of a linear controller for parrot AR drone 2.0. In: 2022 International Conference on Electrical and Computing Technologies and Applications (ICECTA). IEEE (2022). https://doi.org/10.1109/icecta57148.2022.9990127
4. Aghazadeh Ardebili, A., Ficarella, A., Longo, A., Khalil, A., Khalil, S.: Hybrid turbo-shaft engine digital twining for autonomous air-crafts via AI and synthetic data generation. Preprints.org (2023). https://doi.org/10.20944/preprints202307.0981.v1
5. Antoniou, I., Ivanov, V.V., Ivanov, V.V., Zrelov, P.: On the log-normal distribution of stock market data. Phys. A **331**(3–4), 617–638 (2004)
6. Aoki, M.: State Space Modeling of Time Series. Springer, Berlin Germany (1990). https://doi.org/10.1007/978-3-642-75883-6
7. Bay, J.S.: Fundamentals of linear state space systems. Irwin Professional Publishing, Maidenhead(Sep, McGraw-Hill International Editions Series) (1998)
8. Chiodo, L.S., Donateo, T., Ficarella, A.: Effect of coordination on transient response of a hybrid electric propulsion system. Int. J. Aviat. Sci. Technol. **3**(01), 4–12 (2022)
9. Donateo, T., De Pascalis, C.L., Strafella, L., Ficarella, A.: Off-line and on-line optimization of the energy management strategy in a hybrid electric helicopter for urban air-mobility. Aerosp. Sci. Technol. **113**, 106677 (2021)
10. Duo, W., Zhou, M., Abusorrah, A.: A survey of cyber attacks on cyber physical systems: recent advances and challenges. IEEE/CAA J. Automatica Sinica **9**(5), 784–800 (2022)
11. Erkoyuncu, J.A., del Amo, I.F., Ariansyah, D., Bulka, D., Roy, R., et al.: A design framework for adaptive digital twins. CIRP Ann. **69**(1), 145–148 (2020)
12. Esposito, A., Lo Iacono, F., Orlando, C., Navarra, G., Alaimo, A.: Whole body vibration during simulated flight via uncertain models and interval analysis. Mech. Adv. Mater. Struct. **30**, 4397–4406 (2022)
13. Gehrig, D., Göttgens, M., Paden, B., Frazzoli, E.: Scale-corrected monocular-slam for the ar.drone 2.0 (2017). https://doi.org/10.3929/ETHZ-A-010897472
14. Iqbal, D., Buhnova, B.: Model-based approach for building trust in autonomous drones through digital twins. In: 2022 IEEE International Conference on Systems, Man, and Cybernetics (SMC), pp. 656–662 (2022). ISSN: 2577-1655
15. Jaw, L., Mattingly, J.: Aircraft engine controls. American Institute of Aeronautics and Astronautics New York, NY, USA (2009)
16. Kulkarni, S., Panicker, R., Kadeppagari, M., Elahi, I.: Next-gen maintenance framework for urban air mobility vehicles. Technical report, SAE Technical Paper (2022)

17. Li, Z., Ma, Y., Wei, Z., Ruan, S.: Structured neural-network-based modeling of a hybrid-electric turboshaft engine's startup process. Aerosp. Sci. Technol. **128**, 107740 (2022)
18. Madni, A.M., Erwin, D., Madni, C.C.: Digital twin-enabled MBSE testbed for prototyping and evaluating aerospace systems: lessons learned. In: 2021 IEEE Aerospace Conference (50100), pp. 1–8 (2021). ISSN: 1095-323X
19. Miao, J., Zhang, P.: UAV visual navigation system based on digital twin. In: 2022 18th International Conference on Mobility, Sensing and Networking (MSN), pp. 865–870 (2022). https://doi.org/10.1109/MSN57253.2022.00140
20. Nejad, H.H., Sauter, D., Aberkane, S.: On-line scheduling and fault detection in NCS with communication constraints in drone application. In: 2010 Conference on Control and Fault-Tolerant Systems (SysTol). IEEE (2010). https://doi.org/10.1109/systol.2010.5675989
21. Shakhatreh, M.K., Lemonte, A.J., Moreno-Arenas, G.: The log-normal modified Weibull distribution and its reliability implications. Reliab. Eng. Syst. Saf. **188**, 6–22 (2019)
22. Wang, J., Moreira, J., Cao, Y., Gopaluni, B.: Time-variant digital twin modeling through the Kalman-generalized sparse identification of nonlinear dynamics. In: 2022 American Control Conference (ACC), pp. 5217–5222. IEEE (2022)
23. Wei, L., Justin, C.Y., Briceno, S.I., Mavris, D.N.: Door-to-door travel time comparative assessment for conventional transportation methods and short takeoff and landing on demand mobility concepts. In: 2018 Aviation Technology, Integration, and Operations Conference, p. 3055 (2018)

Understanding Mobile Game Reviews Through Sentiment Analysis: A Case Study of PUBGm

Yang Yu[1(✉)], Tai Dinh[2(✉)], Fangyu Yu[1], and Van-Nam Huynh[1]

[1] School of Knowledge Science, Japan Advanced Institute of Science and Technology (JAIST), Nomi 923-1211, Japan
{yangyu,s2010185,huynh}@jaist.ac.jp

[2] The Kyoto College of Graduate Studies for Informatics (KCGI), 7 Tanaka Monzencho, Sakyo Ward, Kyoto 606-8225, Japan
t_dinh@kcg.ac.jp

Abstract. Mobile games are increasingly gaining popularity within the gaming industry due to their unique features and experiences, distinct from conventional platforms like desktops and consoles. Understanding player feedback and reviews is crucial for enhancing game services and optimizing user experiences. This paper focuses on analyzing player reviews within the realm of mobile esports. We introduce a comprehensive framework that employs topic modeling and sentiment analysis to extract insightful keywords from a vast collection of reviews. Utilizing the Latent Dirichlet Allocation algorithm, we uncover diverse topics within the reviews. Furthermore, we exploit Bidirectional Encoder Representations from Transformers (BERT) combined with a Transformer (TFM) downstream layer for precise sentiment analysis, capturing players' sentiments towards various topics. The experiment was conducted on a dataset containing six million English reviews collected up to March 2023 for the mobile game PUBGm from Google Play. The experimental results demonstrate the framework's proficiency in efficiently identifying player concerns and revealing significant keywords embedded in their reviews, thereby supporting mobile esports game operators to refine services and elevate the gaming experience for all players.

Keywords: Mobile Esports · Topic Modeling · Sentiment Analysis

1 Introduction

Sentiment analysis [8] aims to systematically analyze opinions and emotions in text. This analytical technique has found diverse applications across various sectors, such as marketing, business, and hospitality. These fields are crucial in assessing customer sentiments, refining brand strategies, and elevating overall customer experiences. Aspect-based sentiment analysis (ABSA) is a more detailed form of sentiment analysis that divides text data and defines sentiment based on its aspects. This approach enables understanding the sentiment associated with individual components of a product, service, or topic, providing a more granular and insightful perspective on users' opinions and feelings.

M. Mosbah et al. (Eds.): MEDI 2023, LNCS 14396, pp. 102–115, 2024.
https://doi.org/10.1007/978-3-031-49333-1_8

Esports is a rapidly growing industry, with over two billion players and spectators worldwide. This sector generates billions of dollars in revenue annually, showcasing its significant economic impact. For the last decade, sentiment analysis has gained significant traction within the game industry [1,3,5,7,9,11,13–16]. By analyzing player discussions, feedback, reviews, and in-game interactions, sentiment analysis provides invaluable insights into player satisfaction, areas of improvement, and emerging trends. This support aids developers in identifying strengths and weaknesses within games and guiding the design process, allowing them to tailor experiences that align with player preferences. However, most research on game reviews generally focuses on game rating and selection. It considers esports games as common as other games but lacks an independent analysis of esports game review contents. In addition, they focused only on games designed for personal computers (PCs) and consoles but not for mobiles. Mobile games hold a unique appeal due to their accessibility and convenience. Their mobility, compact size, and easy installation make them attractive to users of different ages, genders, and countries.

A prominent entity in the realm of esports is the widely recognized game PlayerUnknown's Battlegrounds (PUBG), along with its mobile iteration, PUBG Mobile (PUBGm). Developed by PUBG Corporation, PUBG, and PUBGm brought the Battle Royale genre into the mainstream, reshaping competitive gaming. The community surrounding both PUBG and PUBGm is evolving rapidly, with numerous platforms and forums emerging where players can provide evaluations, share comments, express opinions, and offer feedback about the games. Sentiment analysis can offer a comprehensive lens through which the multifaceted dimensions of PUBG can be understood. By examining player sentiments, opinions, and feedback, developers can gain insights into various critical aspects of the game. This includes gameplay mechanics, visual and audio elements, map designs, loot distribution, game modes, updates, community interaction, and the competitive scene. These insights empower developers to make informed decisions that refine and elevate individual elements of PUBG and foster an enriched and engaging gaming experience for players.

These above observations motivate us to focus on understanding the feedback from the enormous number of players on PUBG and PUBGm. The contributions of this study are highlighted as follows:

- We have collected six million reviews for PUBG Mobile (PUBGm) from Google Play. This extensive collection of reviews has allowed us to construct a comprehensive dataset that captures a wide range of player perspectives. This dataset will be publicly available to the research community, facilitating further exploration and analysis.
- This work represents the first attempt to mine insights from mobile game reviews, building upon the framework proposed in [13]. This framework consists of two key components: topic modeling and sentiment analysis. Leveraging the Latent Dirichlet Allocation (LDA) algorithm, the framework identifies diverse topics within reviews. These identified topics were then employed in a prevalence analysis to reveal the connections between players' concerns

and various esports games. Furthermore, using Bidirectional Encoder Representations from Transformers (BERT) combined with a Transformer (TFM) downstream layer enables accurate detection of players' sentiments toward different topics.
- We identify key aspects that significantly impact PUBG and PUBGm and translate the results into easily understandable visualizations. These visualizations shed light on the diverse landscape of player sentiments, facilitating a clear understanding of the range of opinions and experiences within the PUBG and PUBGm communities. The key findings can support game providers to enhance their services and offer improved feedback for future development.

The rest of this paper is organized as follows. Section 2 gives a brief overview of the related work. Section 3 introduces the analysis framework. Section 4 shows data information and experimental results. Finally, Sect. 5 draws a summary and outlines directions for future work.

2 Related Work

Several pioneering works in game reviews date back to 2009, when Bond et al. [3] conducted a study where they identified the attributes of a high-quality game through the analysis of game reviews. Zagal et al. [15] considered the significance of game reviews as a key aspect of video game journalism and a common medium for discussing games. They focused on game reviews from popular online platforms to gain insights into their structure and influence. Livingston et al. [9] pointed out that game reviews and ratings affect commercial success. Zagal et al. [16] demonstrated the relationship between game rating and sentiment words chosen by players. Gifford [5] analyzed the differences in reviews between video games and films. Lin et al. [7] pointed out that game reviews differ from mobile app reviews in several aspects, and both positive and negative reviews could be useful to game operators. Baowaly et al. [1] designed a gradient-boosting algorithm to identify helpful and negative reviews within Steam game reviews. They also constructed a regression-based model to predict review scores. In another study, Ruseti et al. [11] employed support vector machines, multinomial Naive-Bayes, and deep neural networks to categorize reviews into positive, neutral, and negative sentiments. Recently, Yu et al. [14] delved into the appeal of the Dark Souls series through an analysis of player-uploaded Steam reviews using a topic modeling approach.

More recently, Yu et al. [13] proposed a hybrid approach of topic modeling and sentiment analysis to analyze the vast number of game reviews from four esports games on Steam, including TEKKEN7, Dota2, PUBG, and CS:GO. The framework comprises several steps. After preprocessing the dataset, the main step is to use Latent Dirichlet Allocation (LDA) for topic modeling. The LDA technique, initially introduced by Blei et al. [2], has played a crucial role in uncovering underlying themes within extensive collections of text data. LDA serves as a generative probabilistic model, assuming that each document within the dataset is composed of a chance combination of potential topics. Each topic is characterized by its word distribution across a vocabulary. LDA employs a sequential process of analyzing topics and word distributions to identify the most suitable number of topics. It calculates coherence scores for various topic numbers and compares their respective values. This step facilitated automatically identifying topics and associated keywords for the esports game. The subsequent step utilized these topics for prevalence analysis and guidance during annotation. The framework then trains a BERT-based model with a Transformer layer as the downstream component, tailored to excel in sentiment analysis tasks. An annotated dataset fine-tunes the model based on the identified topics and keywords detected from the previous step. The sentiment polarity of each topic is generated through the trained BERT model.

This paper expands upon the model proposed in [13] to address mobile game reviews. As demonstrated in [13], this hybrid approach leverages both unsupervised and supervised learning to enhance the precision of sentiment analysis. The experimental results provide in-depth insights and valuable customer feedback to esports game operators, enabling them to refine their services and offer an improved gaming experience for all players. The detail of the analysis framework is introduced in the next section.

3 The Analysis Framework

Figure 1 illustrates the framework for conducting aspect-based sentiment analysis on mobile esports game reviews. As previously mentioned, this framework is an extension of the model proposed in [13] specifically tailored for mobile game reviews. The core components of topic modeling and aspect-based sentiment analysis have been retained. The initial phase involves collecting raw data comprising reviews for mobile games, which are subsequently subjected to preprocessing procedures employing NLP techniques. These techniques encompass key information extraction, noise removal, spelling correction, stemming, lemmatization, and tokenization. Subsequently, the preprocessed dataset is fed into Latent Dirichlet Allocation (LDA) for conducting topic modeling. This step facilitates automatically identifying topics and corresponding keywords associated with mobile esports games. In the next step, the BERT-based model is trained with a Transformer layer as the downstream layer to perform sentiment analysis tasks. The sentiment polarity of each topic is accessed through the trained BERT model.

BERT [4] leverages masked language models to pre-train deep bidirectional representations. It achieves high performance on sentence and token-level tasks. In the analysis framework, we use the pre-trained "bert-base-uncased model"[1] with default parameter settings, such as 768 for the number of hidden layers in the Transformer encoder and 512 for the maximum sequence length.

Given a sentence $S = (s_1, s_2, \ldots, s_t)$, where $1 \leq t \leq 512$ represents the sentence length, the embedding space is characterized by vectors that encapsulate

[1] https://github.com/huggingface/transformers.

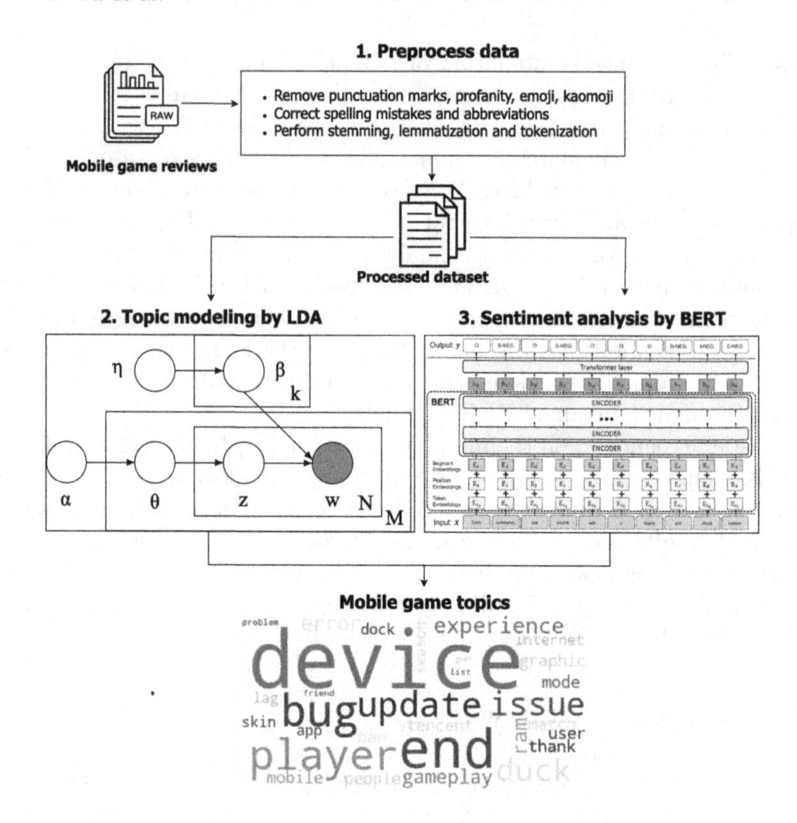

Fig. 1. The workflow of the analysis framework.

the semantic meaning of individual words. These vectors ensure that words with similar meanings exhibit closely aligned values. BERT's input embeddings are assembled as $E = (e_1, e_2, \ldots, e_t)$, which encompasses the summation of token embeddings, segmentation embeddings, and position embeddings. The subsequent steps of the implementation closely mirror the original BERT architecture. The transformer layer is employed without delving into an extensive model structure description. To adhere to BERT's constraints on the input size, we assess the token count within each review. For reviews surpassing the permissible length, we partition them into multiple paragraphs.

The Transformer architecture, originally introduced by Vaswani et al. in 2017 [12], provides a robust feature extraction approach while departing from traditional RNN architectures in NLP tasks. Notably, since BERT is composed of the Transformer model's encoder, we employ a Transformer layer with an identical architecture to the BERT encoder. The computational procedure of the Transformer model can be disassembled into several sequential steps, outlined below:

$$Q^{(i)}, K^{(i)}, V^{(i)} = \left(W^{(Q)}{}_i, W^{(K)}{}_i, W^{(V)}{}_i\right) \cdot A \tag{1}$$

$$\mathrm{Att}^{(i)}(Q, K, V) = \mathrm{softmax}\left(\frac{Q^{(i)}(K^{(i)})^T}{\sqrt{d_k}}\right) \cdot V^{(i)} \tag{2}$$

$$\text{Self-Att}_{MH}(Q, K, V) = \mathrm{Concat}\left(\mathrm{Att}^{(1)}, \mathrm{Att}^{(2)}, \ldots, \mathrm{Att}^{(i)}\right) \cdot W^{(O)} \tag{3}$$

$$\hat{A} = \mathrm{LN}\left(A + \text{Self-Att}_{MH}(Q, K, V)\right) \tag{4}$$

$$y = \mathrm{LN}\left(\hat{A} + \mathrm{FFN}\left(\hat{A}\right)\right) \tag{5}$$

In the above equations, $Q^{(i)}$, $K^{(i)}$, and $V^{(i)}$ represent the Query, Key, and Value matrices generated from the input matrix A. The feed-forward network (FFN) consists of a simple fully-connected neural network employing ReLU as its activation function [12]. Equations 1 through 3 describe the computational process of the multi-head self-attention mechanism with i heads. Meanwhile, Eqs. 4 and 5 illustrate the residual connection step [6] and the utilization of layer normalization. Finally, a linear layer followed by softmax activation is added to the output of the Transformer model's layer, resulting in the final prediction.

4 Experimental Results and Analysis

4.1 Experimental Dataset

For data collection, we used an API[2] to collect around six million English reviews up until March 2023, sourced from Google Play. We then refined this dataset by excluding specific columns that were considered non-essential. The detailed information regarding each retained column is provided in Table 1.

Table 1. Description of Google Play Reviews DataFrame.

Column Name	Description
reviewId	Unique identifier for each review
userName	Username of the reviewer
content	Actual text of the review
score	Star rating given by the user (1–5)
thumbsUpCount	Number of 'likes' the review received
at	Date and time when the review was written
replyContent	Text of the developer's response, if any
repliedAt	Date and time when the reply was posted

[2] https://pypi.org/project/google-play-scraper/.

Following the previous steps, we observed that some comments related to esports were either in languages other than English, consisted of emoticons, or contained characters of ambiguous significance. Consequently, we decided to conduct preliminary text processing, removing non-English terms and inappropriate language. The detailed distribution for PUBGm is illustrated in Table 2.

Table 2. Detailed Properties for PUBGm dataset.

Rating	Count	Percentage
1 star	925,769	14.87%
2 stars	151,392	2.43%
3 stars	225,403	3.62%
4 stars	373,452	6.00%
5 stars	4,549,245	73.08%
Total	6,225,261	100.00%

4.2 Topic Modeling

We utilized the LDA algorithm to identify the prevalent topics and associated keywords within the PUBGm reviews. To ascertain the ideal number of topics, we relied on the coherence value [10]. As depicted in Fig. 2, the coherence values were examined for a range of topics from 3 to 50. After analysis, we settled on 16 topics for PUBGm because the coherence values reached a local minimum, indicating these were the optimal counts.

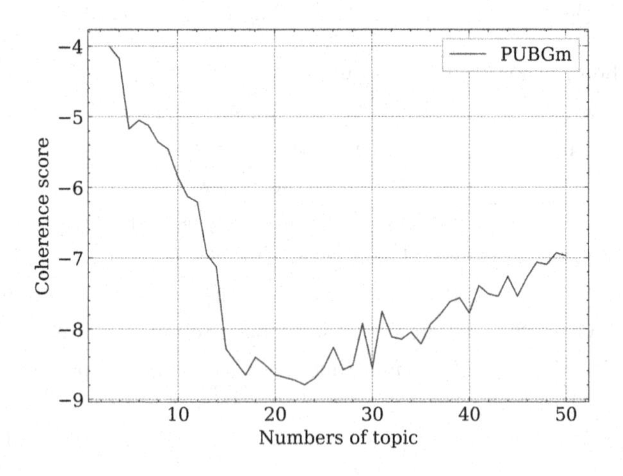

Fig. 2. Coherence scores of PUBGm.

The 16 topics, along with exemplar keywords, are detailed in Table 3. Building upon insights from earlier investigations [13], we proceeded to classify these topics into two distinct categories, grounded in their respective attributes: Game-related Topics (GRT) and Player-related Topics (PRT).

Table 3. Keyword Examples and Inferred Topics of PUBGm.

ID	Inferred Topics	Keyword Examples
1	grapics	graphic*0.033, fps*0.004
2	character	skin * 0.007, outfit* 0.007, zombie* 0.018
3	map	map * 0.022, world * 0.002
4	optimization	bug*0.031, issue*0.010, glitch*0.010
5	update	update*0.096, season*0.024
6	gameplay	gameplay * 0.005, mode*0.003, metro * 0.005
7	community	friend * 0.012, people * 0.005, winner* 0.012
8	server	server*0.033, ping*0.013, lag *0.005
9	region	bangladesh* 0.003
10	platform	money * 0.001, app * 0.080, application*0.007
11	teamwork	team * 0.003, mate *0.001
12	cheating	hacker * 0.038, cheater * 0.002
13	skill	weapon * 0.002, gun * 0.002, battle* 0.013
14	learning curve	noice * 0.004, level* 0.005
15	ranking	rank*0.002, ban*0.001
16	device	phone*0.023, ram*0.016, mouse* 0.002

4.3 Sentiment Analysis

Sentiment Distribution Across All Ratings
Initially, we explored how sentiment fluctuates across different aspects of the game. Figure 3 illustrates the sentiment distribution, offering insight into players' sentiments concerning various aspects across all ratings.

The positive sentiment towards **graphics** emphasizes the game's success in visual appeal, suggesting a well-executed combination of high-quality textures, effective lighting, and compelling artistic direction. It's a testament to the game developers' dedication to delivering a visually immersive experience. The mixed opinions on **character** and **map** highlight the challenges of catering to diverse player preferences. While character design might involve issues of balance or customization options, feedback on maps could touch on aspects like terrain diversity and navigational challenges. It underscores the importance of iterative design and community engagement. The dissatisfaction with **optimization**

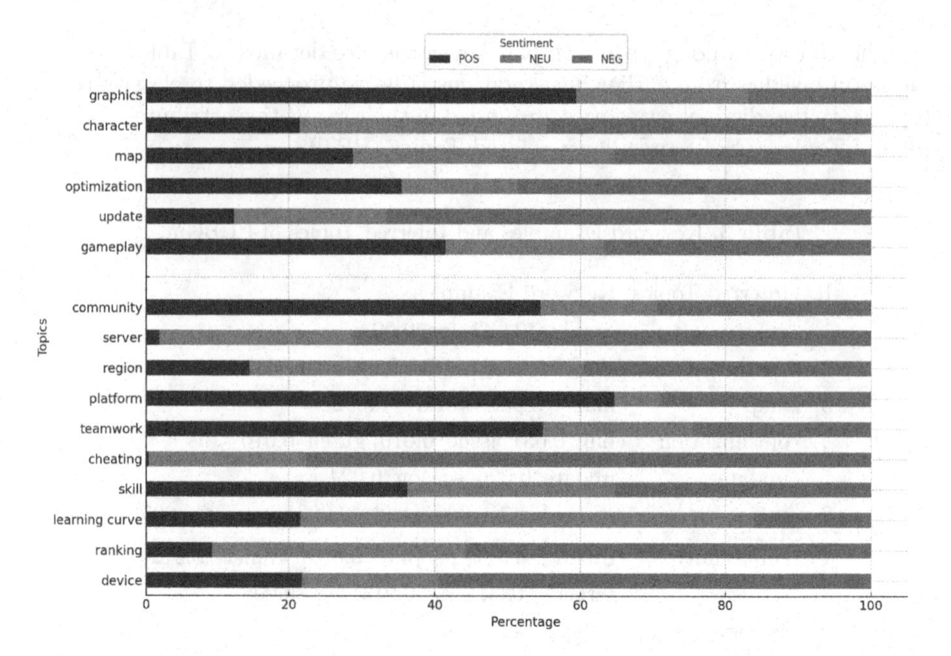

Fig. 3. Overall Topics Sentiments for PUBGm.

suggests potential technical issues that might hinder gaming. This could range from frame rate drops to glitches, indicating areas where technical refinement is crucial for a seamless gaming experience. Negative sentiments around **update** might point to issues with how new content or changes are rolled out. This could involve gameplay alterations or bug introductions, emphasizing the importance of comprehensive testing and clear communication with the player base. General appreciation for **gameplay** indicates that the core mechanics and mission designs are well-received, but there's always room for further refinement to enhance player engagement.

Sentiment Distribution Across Different Ratings

To gain a more detailed insight into player feedback, in this section, we explored the sentiment distribution across various ratings. Figures 4 and 5 showcases this comprehensive sentiment breakdown, illuminating how player sentiments evolve with their overall rating of the game. At higher ratings, the sentiments predominantly lean positive, reflecting satisfaction with various game aspects. However, as we navigate towards lower ratings, critical feedback becomes more pronounced, offering insights into specific areas that might have led to player dissatisfaction. This granular view serves a dual purpose: celebrating the game's strengths and pinpointing avenues for enhancement to cater to a broader player base.

GRT Group Sentiments. User feedback from PUBGm provides important insights into both the game's standout strengths and areas that might need

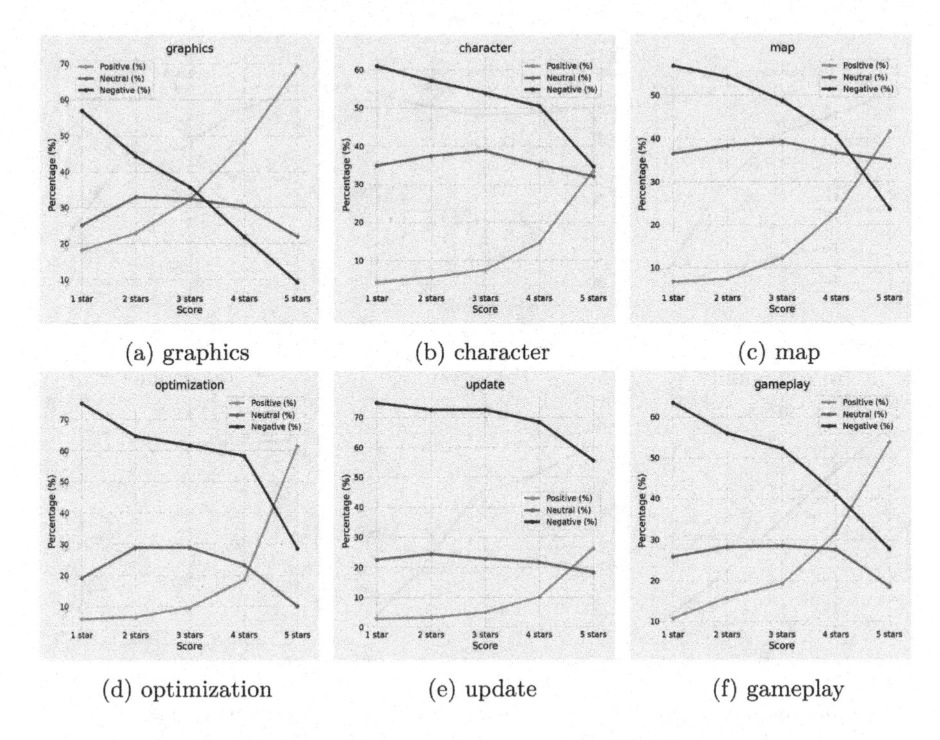

(a) graphics (b) character (c) map

(d) optimization (e) update (f) gameplay

Fig. 4. GRT group Sentiments for PUBGm.

further attention within the GRT category. Two aspects, **update** and **optimization**, stand out as the most polarizing. While a significant portion of users express satisfaction, especially evident in the positive feedback for higher ratings, there's an unmistakable section of the user base voicing concerns. This dichotomy becomes clearer when we examine Fig. 4, which visually captures the sentiment distribution of the GRT group. **Graphics** in PUBGm are generally applauded, showcasing it as one of the game's strengths. However, it's worth noting a subset of users who might be experiencing graphical issues or have particularly high expectations, leading to some negative feedback. On the other hand, aspects like **gameplay**, **map**, and **character** don't garner strong reactions from the majority. The feedback suggests that these elements, while integral, might neither be standout features nor major concerns for most players. For PUBGm to sustain its momentum and remain a beloved title, it's imperative for developers to delve deeper into areas causing a rift in user sentiments. By understanding the root causes of concerns and addressing them head-on, they can ensure a consistently enjoyable experience for their expansive community.

PRT Group Sentiments. The vast array of feedback from the PUBGm player base offers invaluable insights into the game's strengths and weaknesses. An analysis of sentiment distribution is presented in Fig. 4, elucidating the variances in player sentiments across different aspects of the PRT group. From **server** issues

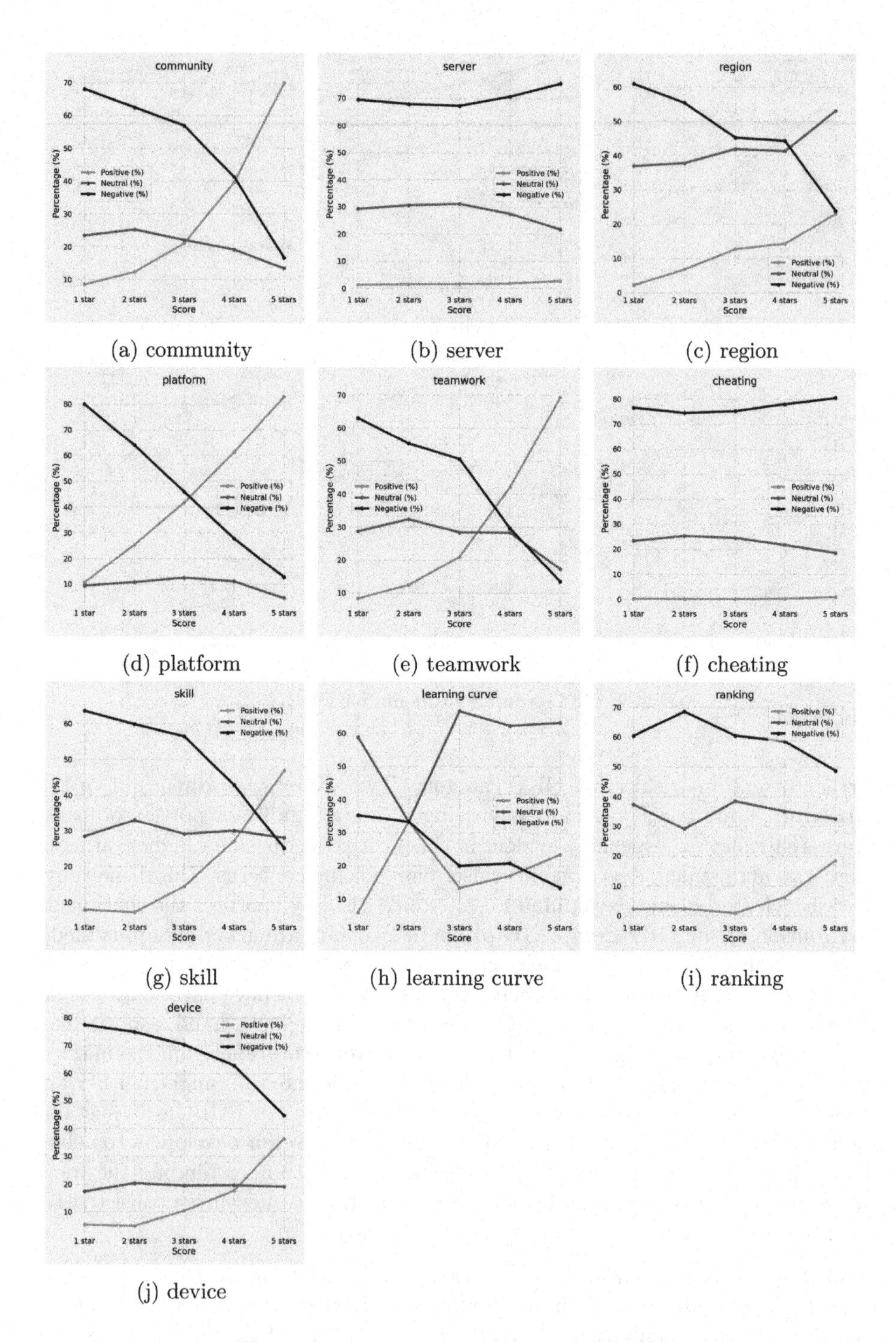

(a) community (b) server (c) region

(d) platform (e) teamwork (f) cheating

(g) skill (h) learning curve (i) ranking

(j) device

Fig. 5. PRT group Sentiments for PUBGm.

to platform adaptability, players have voiced their experiences, and a deeper dive into this data unveils some crucial findings.

- **Server Issues:** A significant portion of the feedback revolves around the game's servers. Players often associate their gaming experience with *server* performance. The substantial negative feedback, especially in the lower star ratings, paints a clear picture: server-related issues are a predominant concern. The onus is on the game operators to enhance server stability, as it stands paramount in retaining and satisfying the vast player base.
- **Cheating Concerns:** Another major area of concern for the players is *cheating*. The pronounced negative feedback in this category suggests that cheating or unfair gameplay mechanics might be prevalent, affecting user experience. Ensuring fair gameplay by implementing stringent anti-cheat measures can enhance trust and satisfaction among its players.
- **Consistency in Certain Game Aspects:** Some game facets, including *community*, *teamwork*, *region*, *learning curve*, and *ranking*, demonstrate consistent feedback. These areas neither stand out as exceptional nor as problematic. They offer a stable experience, lacking standout features that might elicit strong reactions but also avoid significant pitfalls.
- **Device Compatibility Issues:** Lastly, feedback pertaining to device compatibility or performance paints a concerning picture. Higher negative feedback, especially among lower star ratings, underscores the need for *optimization*. Ensuring the game runs smoothly across a wide range of devices can broaden its appeal and enhance user experience.

In summation, while PUBGm excels in certain areas like platform support, there are evident concerns regarding *server*, *cheating*, and *device* performance. Addressing these pressing issues can bolster the game's reputation and ensure sustained user satisfaction.

5 Conclusion

In the realm of mobile gaming, PUBGm stands as a testament to the potential of mobile platforms. Yet, it seems to grapple with issues that have persisted since its transition from a PC-based game. The similarities in challenges faced by both PUBGm and its PC counterpart, PUBG, are evident. Despite analyzing a vast collection of 6 million English reviews for PUBGm, our topic modeling indicates a striking resemblance between the two, save for an additional theme related to "device" for PUBGm. This suggests that the developers merely transitioned the game to mobile without significant modifications or updates - a trend that appears prevalent among many game developers aiming for cross-platform offerings.

The feedback from PUBGm players reveals a meaningful result. The game's graphical excellence showcases the dedication of its developers to provide an immersive experience, a fact further underscored by the widespread acclaim for its visuals. Conversely, diverse opinions on character and map designs highlight the ever-present challenge in game development: catering to the varied tastes of a global audience. A closer examination of sentiments, especially when segmented by ratings, offers richer insights. Positive feedback is more associated with higher ratings, whereas constructive criticism becomes evident in lower ratings, guiding the direction for possible future enhancements. However, the analysis also brings to light certain challenges. Issues like technical glitches and the need for optimization, as well as concerns about updates, underscore the necessity for not only introducing new features but ensuring their flawless execution. The issue of cheating is also evident, necessitating robust anti-cheat systems. In comparing PUBGm to its PC predecessor, PUBG, the mobile version faces unique challenges. Server performance emerges as a primary concern, emphasizing its pivotal role in defining a game's success. Additionally, with the diverse range of mobile devices in use today, the game's compatibility across these devices becomes not just desirable but essential. In essence, while PUBGm boasts significant strengths that have endeared it to many, it also faces critical challenges. For PUBGm to continue its reign and resonate with its expansive community, it is essential for the developers to address the highlighted concerns. By heeding this feedback and continually refining the game, PUBGm can fortify its position as a premier gaming title, delighting both seasoned players and newcomers alike.

There are certain limitations in this current work. First, we didn't delve into mobile MOBAs, another dominant mobile e-sports genre, exemplified by titles like *League of Legends: Wild Rift.* Additionally, our analysis was constrained to the Android platform. Despite iOS users being fewer in number compared to Android, their average expenditure on apps tends to be higher. This may discrepancy influence players' overall satisfaction. Lastly, we were unable to secure substantial data from rapidly growing mobile gaming markets such as China and Southeast Asia. Addressing this gap should be a priority for upcoming research endeavors.

Dataset and Source Code. The experimental datasets and source code can be found at this repository: https://github.com/YYdeeplearning/MEDI2023.

References

1. Baowaly, M.K., Tu, Y.P., Chen, K.T.: Predicting the helpfulness of game reviews: a case study on the steam store. J. Intell. Fuzzy Syst. **36**(5), 4731–4742 (2019)
2. Blei, D.M., Ng, A.Y., Jordan, M.I.: Latent dirichlet allocation. J. Mach. Learn. Res. **3**(Jan), 993–1022 (2003)
3. Bond, M., Beale, R.: What makes a good game? using reviews to inform design. People Comput. XXIII Celebrating People Technol., 418–422 (2009)
4. Devlin, J., Chang, M.W., Lee, K., Toutanova, K.: BERT: pre-training of deep bidirectional transformers for language understanding. In: Proceedings of the 2019 Conference of the North American Chapter of the Association for Computational Linguistics: Human Language Technologies, Volume 1 (Long and Short Papers), pp. 4171–4186. Association for Computational Linguistics (2019)

5. Gifford, B.: Reviewing the critics: examining popular video game reviews through a comparative content analysis. Ph.D. thesis, School of Communication, Cleveland State University (2013)
6. He, K., Zhang, X., Ren, S., Sun, J.: Deep residual learning for image recognition. In: Proceedings of the IEEE Conference on Computer Vision and Pattern Recognition, pp. 770–778 (2016)
7. Lin, D., Bezemer, C.P., Zou, Y., Hassan, A.E.: An empirical study of game reviews on the steam platform. Empirical Softw. Eng. **24**(1), 170–207 (2019)
8. Liu, B.: Sentiment Analysis and Opinion Mining. Morgan & Claypool Publishers, San Rafael, California (2012)
9. Livingston, I.J., Nacke, L.E., Mandryk, R.L.: The impact of negative game reviews and user comments on player experience. In: Proceedings of the 2011 ACM SIGGRAPH Symposium on Video Games, pp. 25–29 (2011)
10. Newman, D., Lau, J.H., Grieser, K., Baldwin, T.: Automatic evaluation of topic coherence. In: Human language technologies: The 2010 Annual Conference of the North American Chapter of the Association for Computational Linguistics, pp. 100–108 (2010)
11. Ruseti, S., Sirbu, M.-D., Calin, M.A., Dascalu, M., Trausan-Matu, S., Militaru, G.: Comprehensive exploration of game reviews extraction and opinion mining using NLP techniques. In: Yang, X.-S., Sherratt, S., Dey, N., Joshi, A. (eds.) Fourth International Congress on Information and Communication Technology. AISC, vol. 1041, pp. 323–331. Springer, Singapore (2020). https://doi.org/10.1007/978-981-15-0637-6_27
12. Vaswani, A., et al.: Attention is all you need. In: Advances in Neural Information Processing Systems, vol. 30 (2017)
13. Yu, Y., Dinh, D.T., Nguyen, B.H., Yu, F., Huynh, V.N.: Mining insights from esports game reviews with an aspect-based sentiment analysis framework. IEEE Access **11** (2023)
14. Yu, Y., Nguyen, B.H., Dinh, D.T., Yu, F., Fujinami, T., Huynh, V.N.: A topic modeling approach for exploring attraction of dark souls series reviews on steam. In: Intelligent Human Systems Integration (IHSI 2022): Integrating People and Intelligent Systems, vol. 22. AHFE Open Access (2022)
15. Zagal, J.P., Ladd, A., Johnson, T.: Characterizing and understanding game reviews. In: Proceedings of the 4th International Conference on Foundations of Digital Games, pp. 215–222 (2009)
16. Zagal, J.P., Tomuro, N., Shepitsen, A.: Natural language processing in game studies research: an overview. Simul. Gaming **43**(3), 356–373 (2012)

Data-Driven and Model-Driven Approaches in Predictive Modelling for Operational Efficiency: Mining Industry Use Case

Oussama Hasidi[1,2](✉), El Hassan Abdelwahed[1], My Abdellah El Alaoui-Chrifi[3], Aimad Qazdar[4], François Bourzeix[2], Intissar Benzakour[3], Ahmed Bendaouia[1,2], and Charifa Dahhassi[5]

[1] LISI Laboratory, Faculty of Sciences Semlalia, Cadi Ayyad University, Marrakesh, Morocco
Oussama.hasidi@ced.uca.ma

[2] AI Department, Moroccan Foundation for Advanced Science, Innovation and Research (MAScIR), Rabat, Morocco

[3] R&D and Engineering Center, Reminex, Managem Group, Marrakes, Morocco

[4] ESTIDMA Laboratory, ENSA, Ibn Zohr University, Agadir, Morocco

[5] Math and Computer Science Department, ENSAM, Moulay Ismail University, Meknes, Morocco

Abstract. In this study, we explore the effectiveness of a hybrid modelling approach that seamlessly integrates data-driven techniques, specifically Machine Learning (ML), with physics-based equations in Simulation. In cases where real-world data for industrial processes is insufficient, a simulation tool is employed to generate an extensive dataset of process variables under varying operating conditions. Subsequently, this dataset is utilized for training the Machine Learning model. The paper showcases a practical use case of this hybrid modelling approach, revealing a model that consistently demonstrates strong predictive accuracy and reliability within the specific industrial context we investigate. By merging the insights derived from physics-based understanding with the adaptability of data-driven Machine Learning, the hybrid model offers a comprehensive solution for precise and accurate predictions.

Keywords: Predictive Modelling · Machine Learning · Simulation · Hybrid approach · Process Monitoring

1 Introduction

In today's world, modelling has become incredibly important for understanding complex systems, making predictions, and making smart choices. With everything being so interconnected and data-driven, modelling is like a guiding light that helps us make sense of complicated things. Researchers, industries, and

M. Mosbah et al. (Eds.): MEDI 2023, LNCS 14396, pp. 116–127, 2024.
https://doi.org/10.1007/978-3-031-49333-1_9

practitioners are investing a lot in modelling to better understand how systems work and to find solutions that can make a real impact.

In the realm of modelling, two primary approaches are employed. One involves using extensive data and computer learning, while the other relies on well-established scientific principles. These are known as data-driven modelling and physics-based modelling. Data-driven modelling employs computers to uncover patterns within large datasets. It's particularly valuable in fields such as healthcare and finance. Conversely, physics-based modelling applies scientific laws to construct simulated versions of real-world scenarios. This aids in comprehending phenomena like fluid motion or material behavior.

Machine learning and simulation both share a common objective: the anticipation of system behavior through data analysis and mathematical representation [1]. In the context of our study, these approaches hold intrinsic significance. Machine learning, as evident in domains like image classification, linguistic analysis, and socio-economic exploration, has exhibited exceptional achievements. These triumphs are especially pronounced when dealing with scenarios characterized by limited causal insights but extensive datasets. Conversely, simulation finds its historical roots in disciplines such as natural sciences and engineering. Notably, computational fluid dynamics relies on simulation to understand intricate causal inter-plays, while areas like structural mechanics employ it for evaluating structural performance encompassing reactions, stresses, and displacements.

This paper introduces a hybrid approach for developing predictive models and discusses a use case of model development for industrial froth flotation process using Machine Learning and Simulation. By leveraging a simulator to generate data and Artificial Neural Networks to construct the core virtual model, this hybrid approach harnesses the strengths of both model-driven and data-driven methods. This strategic combination overcomes the limitations posed by each approach - the scarcity of real data in model-driven methods and the complexities faced by data-driven methods in handling intricate systems.

The remainder of this paper is organized as follows: Sect. 2 further discusses the data-driven and physics-driven modelling approaches, the advantages of each one and their challenges. Section 3 elucidates the hybrid approach where the both approach could assist each other and mitigate the challenges faced by each one. Section 4 outlines a use case of model development of froth flotation process using hybrid approach for minerals processing advanced monitoring. Section 5 presents the results and analysis of the study, including the evaluation of the predictive model for a functioning flotation cell. Finally, Sect. 6 concludes the paper with a summary of key findings and recommendations for future research.

2 Modelling Approaches

In this section, we elucidate the two modelling methodologies through a conceptual framework designed to enhance their transparency and facilitate meaningful comparisons between their respective components.

2.1 Data-Driven Approach (i.e Machine Learning)

Data-driven modelling with machine learning has gained significant traction in recent years due to the explosion of available data and advancements in algorithms. This approach involves training models on large datasets to learn complex patterns and relationships, enabling them to make accurate predictions and inform decision-making processes.

The benefits of data-driven modelling are numerous and impactful. By harnessing the power of big data, organizations can make data-informed decisions, enhancing efficiency and accuracy. Moreover, these models excel at handling nonlinear relationships and can uncover hidden insights within data. For example, in genomics, machine learning has facilitated the identification of novel disease-associated genes and pathways, revolutionizing our understanding of complex genetic disorders. Furthermore, data-driven modelling's adaptability to diverse data types, from text to images, has paved the way for innovations in natural language processing, image recognition, and autonomous vehicles.

Nevertheless, several challenges accompany this approach. One key challenge is the issue of data quality and bias. Biased or incomplete datasets can lead to skewed model outcomes and reinforce existing prejudices. Another challenge is model interpretability, where complex algorithms like deep neural networks can be difficult to explain, raising questions about transparency and accountability in decision-making. Addressing these challenges necessitates a comprehensive understanding of data collection practices, algorithmic transparency, and adherence to regulatory guidelines.

2.2 Physics-Driven Approach (i.e Simulation)

Physics-driven modelling through simulation is a potent tool for understanding and predicting complex real-world phenomena. However, this approach is not without challenges that need to be carefully considered. The goal of a simulation is to predict the behavior of a system or process based on the underlying physical principles. To achieve this, a mathematical model is formulated using differential equations that capture the causal relationships between different variables. These models are often developed through extensive research, starting with the derivation of equations from theoretical physics principles, followed by validation through experiments.

One major challenge is the dependence on accurate input parameters and assumptions in the mathematical models. Small inaccuracies in these inputs can lead to significant discrepancies in the simulation results. In complex systems with numerous interacting components, it can be difficult to precisely estimate all necessary parameters, introducing uncertainties in the predictions. Moreover, assumptions made during model formulation might not hold true in all scenarios, further affecting the reliability of simulations. Addressing these challenges requires rigorous sensitivity analysis to identify the impact of input variations and a thorough understanding of the assumptions' validity.

Computational intensity is another challenge in physics-driven simulations, particularly when dealing with complex behaviors such as quantum mechanics or highly nonlinear interactions. Simulating such phenomena demands substantial computational resources and advanced numerical techniques. As the complexity of the system increases, the computational cost can become prohibitive, requiring researchers to strike a balance between accuracy and computational feasibility.

Furthermore, while physics-driven models are adept at capturing well-defined causal relationships, they can struggle to represent emergent behaviors and complex interactions that arise in systems with many interdependent components. For instance, in agent-based simulations, where individual agents interact based on predefined rules, it can be challenging to predict emergent patterns that arise from the collective behavior of the agents. This limitation can be addressed by integrating data-driven approaches that leverage machine learning techniques to capture these complex interactions.

Table 1. Summary of advantages and limitations of data driven and model driven approaches.

	Data-driven approach	Model-driven approach
Advantages	Simplicity	Better understanding of complex systems
	Faster decision-making	Generalized to similar problems
	Increased accuracy with more data	Accuracy: limited errors
Limitations	Quality and quantity of data	Lack of flexibility
	High cost of data collection, storage, and analysis	Technical expertise required
	Technical expertise required	Time-consuming

In summary, as represented in Table 1 the data-driven approach may be advantageous in situations where the physical system is complex and difficult to model or where there is a large amount of data available. However, this approach can be limited by the quality and availability of data, and may not always be suitable for systems with complex nonlinear behaviors [2]. On the other hand, the model-driven approach may be advantageous in situations where the physical system is well-understood and there is a good understanding of the underlying physics. However, this approach can be limited by the accuracy of the mathematical model and the assumptions made in developing the model.

3 Combination of Data-Driven and Model-Driven Approach

Considering the advantages and limitations of both approaches, we propose a hybrid modelling approach for the development of accurate and reliable predictive models for leveraging the advantages of the both methodologies. The combination of data-based and knowledge-based modelling is motivated by applications that are partly based on causal relationships, while other effects result from hidden dependencies that are represented in huge amounts of data [1] (Fig. 1).

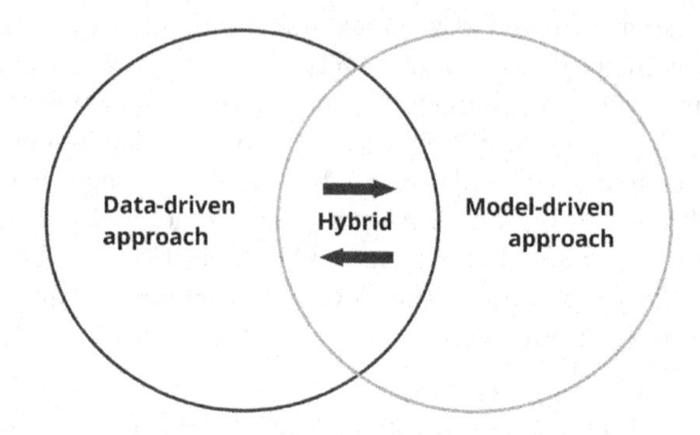

Fig. 1. Data-driven and Model-driven modelling approaches combination.

The data-driven approach could assist the model-driven approach in various ways. Tiny Machine Learning-based algorithms could be advantageous for the inference of complex measurements that are hard to measure or that are not bound to be directly measured. Statistical methods could also play a major role in studying and investigating hidden parameters dependencies. Moreover, Data assimilation and Reinforcement Learning play an important role in guarantying the Predictive Layer accurate performance after deployment.

Inversely, the model-driven approach could as well assist the data-driven approach. In the modelling phase for instance, additional training data could be generated based on physical and phenomenological equations to be fed to Machine Learning models. Furthermore, testing Machine Learning models is less risky if firstly validated in a simulation-based environment that is built with physics and phenomenological equations.

Overall, the hybrid approach has the potential to improve the accuracy and robustness of predictive models. Diverse are the applications that include the combination of the two methods in different modelling areas [3–5]. In the next section, we will showcase one way of combining data-driven and model-driven approaches in predictive modelling for operational efficient in the mining industry using Simulation and Machine Learning.

4 Use Case: Froth Flotation Predictive Model Using Hybrid Approach

4.1 Froth Flotation Overview

Froth flotation is the most common way to separate minerals in the mining industry. Since it was first used in factories in 1905, many researchers have contributed to improving how we understand the process [6]. This method is the most widely used way to separate valuable minerals from useless rock. It works by taking advantage of the different ways minerals react with water. Some minerals don't like water (hydrophobic), while others do (hydrophilic). When we stir the mixture and add air, minerals that don't like water stick to tiny bubbles and rise to the top as foam. Minerals that like water stay in the liquid. This helps us get the valuable minerals out for further use (Fig. 2).

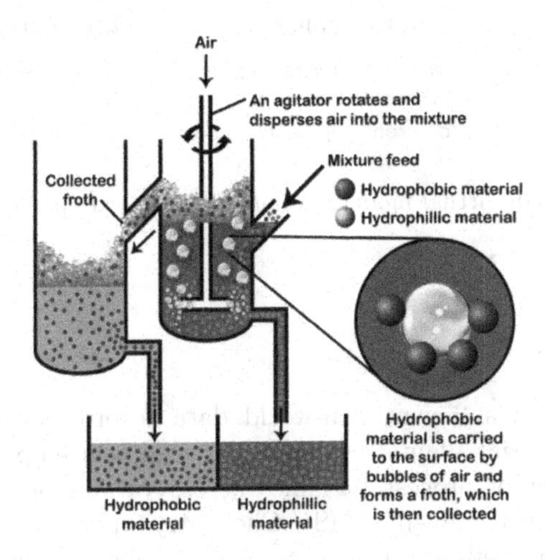

Fig. 2. A schematic representation of flotation process.

Advances in control and optimisation of the froth flotation process are of great relevance since even very small increases in recovery lead to large economic benefits [6]. However, the implementation of advanced control and optimisation strategies has not been completely successful in flotation since ever. This is because flotation performance is affected by a great number of variables that interact with each other, while unmeasurable disturbances in the process further complicate the implementation of efficient strategies [7]. Besides the multiple chemicals reagents (Collectors, Modifiers and Frothers), froth flotation is influenced by several operating factors. Many parameters involved in the process and have a crucial impact on the quality of the froth floated and therefore the concentration of the minerals resulted: the pH within the liquid, the rate of oxidation of the ore, the grain size of the feed, The viscosity of the pulp, air flows, agitators speed,...etc.

Acquiring a specific knowledge of the operational behavior of the process and trying to ensure a certain stability of the flotation yield (in terms of concentration/grade) is considered a complicated and tough challenge. Hence the necessity of the development of an accurate and reliable predictive models that comprehends the underlying phenomenon of the process and that incorporates these affecting parameters as an advanced operations monitoring system [8]. In the further section we will discuss the development of Machine Learning based predictive model of the froth flotation cell using phenomenological, kinetic and physics-based simulation data.

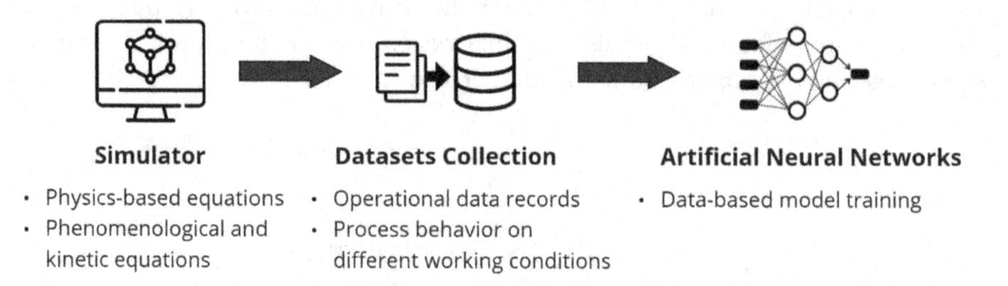

Fig. 3. Froth flotation virtual model development: Hybrid approach methodology.

4.2 Methodology

Given the scarcity of sufficient real-world data records for essential process parameters in industrial plants, a simulation tool was employed to generate an extensive dataset of process variables across various operating conditions (Fig. 3). This simulator, known as HSC Chemistry, operates on a model-driven approach, integrating well-established physical and chemical principles that govern the froth flotation process. The simulator incorporates essential equations, such as the calculation of the overall flotation rate k (1/min) of a cell, which is derived from the overall recovery (mass pull) R (%) and cell residence time W (min) using the formula:

$$k = \frac{R}{W(100 - R)} \tag{1}$$

The Eq. 1 play a crucial role in accurately simulating the flotation process and provide valuable insights into the system's behavior under different operating conditions.

By running simulations with HSC Chemistry, we obtained a wealth of data that enabled the training of the data-driven component of our hybrid approach.

4.3 Operational Data Generation and Collection

In this study, we conducted data collection using HSC Chemistry [9], a minerals processing simulator, as illustrated in Fig. 4. Given the limited availability of operational data from the industrial flotation plant, we used a small amount of industrial data to pre-configure the simulations in the software. These data comprised crucial variables such as feed rate, water flow rate, pulp density, flotation rate and mineral grades. Through multiple simulations, we generated a diverse dataset encompassing a wide range of operating conditions. HSC Chemistry was chosen as the simulator due to its remarkable accuracy and flexibility in simulating various mineral processing scenarios. The simulator allowed us to obtain an extensive amount of data that would have been challenging and costly to acquire through traditional experimental methods in real industrial settings. Following data generation, we carefully curated the dataset to ensure it captured the most pertinent and critical process parameters. This comprehensive dataset served as the foundation for developing the predictive model. This approach not only optimized data collection efforts but also ensured a high level of accuracy and efficiency in the development of predictive models for the froth flotation process.

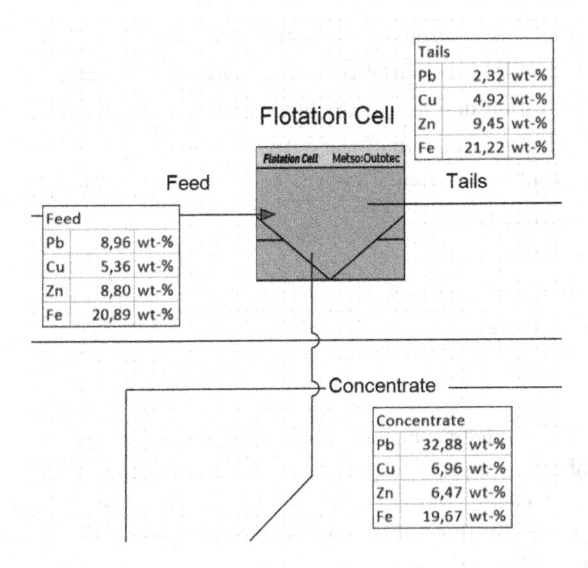

Fig. 4. Single flotation cell simulation using HSC Chemistry Sim.

4.4 Artificial Neural Networks Training

Following the data collection, the gathered dataset was employed to train a Neural Network model, enabling predictions of the froth flotation process's behavior under new operating conditions. This Neural Network model was built on a data-driven approach, learning the intricate nonlinear relationships between the input and output variables. Through training the ANN model with the dataset generated from the simulator, we established a predictive model for the froth flotation process.

The predictive model facilitated the estimation of mineral grades before and after entering a specific flotation cell, based on feed characteristics and control variables. For this study, a feed-forward backpropagation network structure was adopted, with the appropriate configuration determined by adjusting the number of neurons, hidden layers, transfer functions, and optimization methods.

Satisfactory outcomes were achieved through careful selection of the number of layers and neurons in each layer. The final architecture network is presented in Table 2.

Table 2. Architecture of the trained Feed Forward Neural Network.

Structure's parameters	ANN
Number of Layers	3
Number of neurons in the input layer	64
Number of neurons in the hidden layer	64
Number of neurons in the output layer	4
Activation function of the input hidden layer	ReLu
Activation function of the hidden layer	ReLu
Activation function of the output layer	Linear
Optimizer function	Adegrad
Learning rate	0.01
Loss function	MSE
Number of epochs	500
Batch size	32

To optimize the model's accuracy, the training and testing sets were allocated 80% and 20% of the samples, respectively. While maintaining a fixed structure for the Artificial Neural Network (ANN), the learning hyperparameters were fine-tuned using the training set. This iterative process helped enhance the performance of the model and improve its predictive capabilities.

5 Results and Discussion

The integration of operational records from the industrial plant into the HSC Chemistry software allowed us to establish simulations for the froth flotation process. Utilizing these simulated datasets, we trained a Neural Network to estimate mineral grades in the concentrates based on feed and control variables. The Neural Network exhibited exceptional accuracy, achieving a predictive capability of 95% with an MSE of less than 2. This confirms the model's capability to accurately estimate mineral grades in the concentrates using input variables.

Figures 5, 6, 7, and 8 visually demonstrate the comparison between actual and predicted values of minerals-of-concern grades, showcasing the robust predictive capability of the Neural Network model. Combining the simulator with the Neural Network model resulted in a powerful hybrid model. The simulator provided a physics-based foundation that encompassed the process's underlying principles, while the Neural Network model incorporated data-driven insights to capture complex nonlinear relationships between input and output variables.

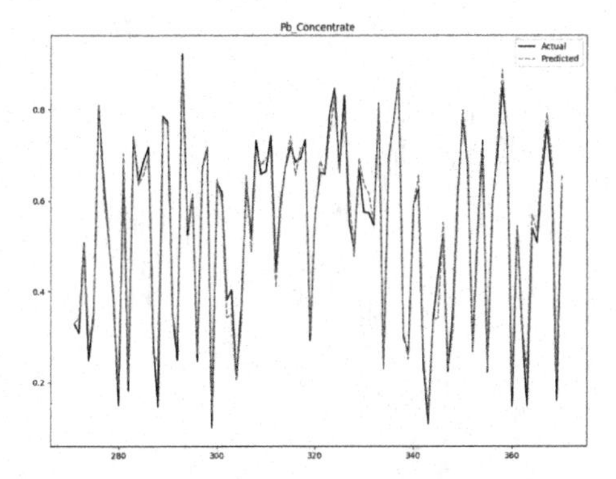

Fig. 5. Grades of Lead Pb% in the flotation cell's concentrates: Actual values are represented in blue and the Neural Networks predictions values are represented in orange. (Color figure online)

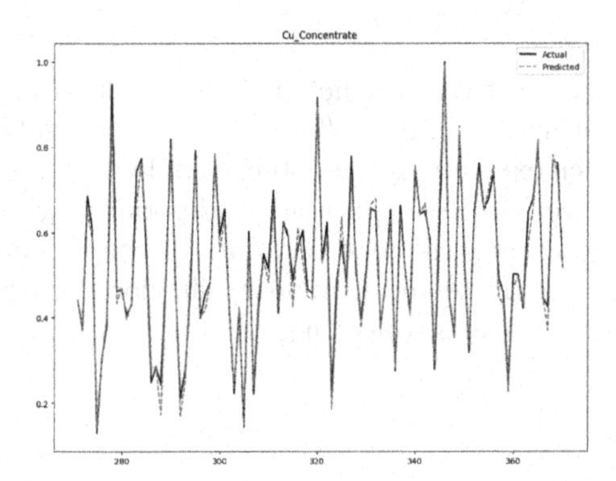

Fig. 6. Grades of Copper Cu% in the flotation cell's concentrates: Actual values are represented in blue and the Neural Networks predictions values are represented in orange. (Color figure online)

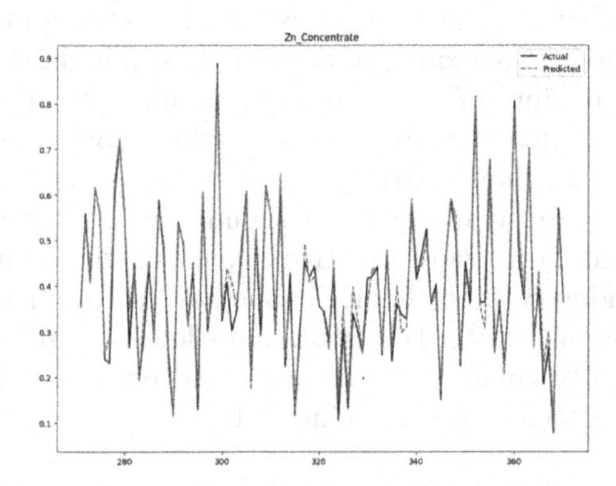

Fig. 7. Grades of Zinc Zn% in the flotation cell's concentrates: Actual values are represented in blue and the Neural Networks predictions values are represented in orange. (Color figure online)

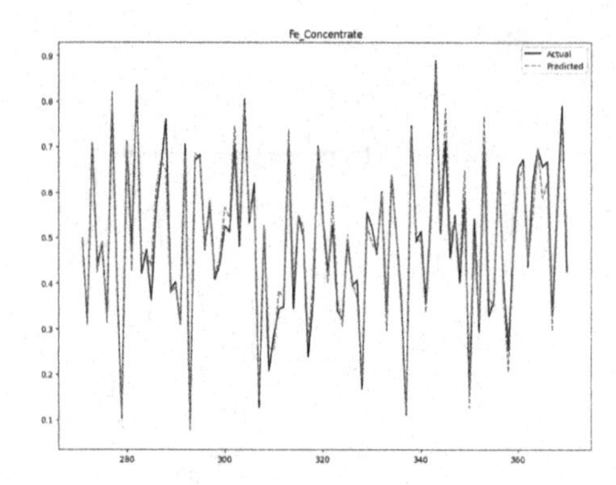

Fig. 8. Grades of Iron Fe% in the flotation cell's concentrates: Actual values are represented in blue and the Neural Networks predictions values are represented in orange. (Color figure online)

This hybrid model effectively predicted the froth flotation process's behavior under novel operating conditions, offering valuable insights for process optimization. Our approach exemplifies the potential of hybrid modelling techniques in developing predictive models for intricate industrial processes, such as froth flotation. By synergizing physics-based and data-driven models, we successfully captured the process's intricate and nonlinear behavior, empowering us to make precise predictions under diverse operating conditions.

6 Conclusion

In conclusion, this study successfully showcased the creation of a predictive model for the froth flotation process through a hybrid approach, combining simulation and Machine Learning. The simulator generated crucial operational data, which was then utilized to train a Neural Network, resulting in the creation of a robust virtual model. The outcomes of the model exhibited impressive accuracy, with a predictive capability of 95% and an MSE of less than 2. This promising result underscores the potential application of this hybrid modelling approach using simulators and Machine Learning for an advanced operations monitoring.

Moreover, this research emphasizes the significance of Digital Twins in industrial productions and their transformative impact on the mining sector [8,10]. By incorporating Industry 4.0 technologies such as the Internet of Things (IoT), Cyber-Physical Systems (CPS), Big Data, and Cloud Computing, mining operations can significantly enhance productivity and efficiency, fostering positive economic and environmental sustainability [11].

Acknowledgements. This work is supported by the Ministry of Higher Education, Scientific Research and Innovation, the Digital Development Agency (DDA) and the CNRST of Morocco through Al-Khawarizmi program. This publication is part of the work undertaken by the consortium of partners which is composed of MAScIR (Moroccan Foundation for Advanced Science, Innovation and Research), Reminex; the R&D and Engineering subsidiary of Managem group, UCA, ENSIAS and ENSMR. We would like to thank the Managem Group and its subsidiary CMG for allowing the conduction of this research on its operational site as an industrial partner of this project.

References

1. Rueden, L., Mayer, S., Sifa, R., Bauckhage, C., Garcke, J.: Combining machine learning and simulation to a hybrid modelling approach: current and future directions. In: Advances In Intelligent Data Analysis XVIII, pp. 548–560 (2020)
2. Liao, L., Köttig, F.: A hybrid framework combining data-driven and model-based methods for system remaining useful life prediction. Appl. Soft Comput. **44**, 191–199 (2016)
3. Erge, O., Oort, E.: Combining physics-based and data-driven modelling in well construction: hybrid fluid dynamics modelling. J. Nat. Gas Sci. Eng. **97**, 104348 (2022). https://www.sciencedirect.com/science/article/pii/S1875510021005436
4. Song, H., Liu, X., Song, M.: Comparative study of data-driven and model-driven approaches in prediction of nuclear power plants operating parameters. Appl. Energy **341**, 121077 (2023). https://www.sciencedirect.com/science/article/pii/S0306261923004415
5. Zhang, S., et al.: Combing data-driven and model-driven methods for high proportion renewable energy distribution network reliability evaluation. Int. J. Electr. Power Energy Syst. **149**, 108941 (2023). https://www.sciencedirect.com/science/article/pii/S0142061522009371
6. Michaud, L.: Froth Flotation: A Century of Innovation (2017). https://www.911metallurgist.com/blog/froth-flotation-century-innovation
7. Bendaouia, A., et al.: Digital transformation of the flotation monitoring towards an online analyzer. In: Hamlich, M., Bellatreche, L., Siadat, A., Ventura, S. (eds.) SADASC 2022. Communications in Computer and Information Science, vol. 1677, pp. 325–338. Springer, Cham (2022). https://doi.org/10.1007/978-3-031-20490-6_26
8. Hasidi, O., et al.: Digital Twins-Based Smart Monitoring and Optimisation of Mineral Processing Industry. In: Hamlich, M., Bellatreche, L., Siadat, A., Ventura, S. (eds.) SADASC 2022. Communications in Computer and Information Science, vol. 1677, pp. 411–424. Springer, Cham (2022). https://doi.org/10.1007/978-3-031-20490-6_33
9. Roine, A.: HSC Chemistry® [Software], Metso Outotec, Pori (2021). Software available at www.mogroup.com/hsc
10. Sircar, A., Nair, A., Bist, N., Yadav, K.: Digital Twin in hydrocarbon industry. Petrol. Res. (2022)
11. Qassimi, S., Abdelwahed, E.H.: Disruptive innovation in mining industry 4.0. Distrib. Sens. Intell. Syst. 313–325 (2022)

Approach Based on Bayesian Network and Ontology for Identifying Factors Impacting the States of People with Psychological Problems from Data on Social Media

Mourad Ellouze(✉) and Lamia Hadrich Belguith

ANLP Group MIRACL Laboratory, FSEGS, University of Sfax-Tunisia, Sfax, Tunisia
ellouzemourad@yahoo.fr, lamia.belguith@fsegs.usf.tn

Abstract. Nowadays, social networks provide relevant information that is used in many contexts for different objectives. However, the major challenges remain at the level of processing this data, which is generated in a specific way. In this context, we propose in this paper a hybrid approach based on Bayesian network and ontology techniques for formalizing textual data published on social media by people with personality disorders. The objective of this task is to identify the main factors that have a significant impact on the state of sick persons. Our proposed approach is composed of three major steps: data collection and preprocessing, the construction of a set of Bayesian networks, and finally the incorporation of semantic components into the constructed networks. Our proposed approach takes advantage of both statistic and linguistic techniques, which can provide explainable and enriched results at multiple hierarchical levels. In addition, our approach addresses language issues like the evolution of the lexicon over time, the ellipsis phenomenon, etc. For the evaluation of our proposed approach, we have used two different methods, and in general, we achieved an accuracy rate equal to 83% for correct links prediction.

Keywords: Bayesian Network · Ontology · Personality Disorder · Social Media · Natural Language Processing

1 Introduction

In 2022, the World Health Organization (WHO) affirms that the personality disorder disease (PD) affects 970 million persons in the world, representing approximately one of every eight persons[1]. According to the same source, this number is expected to increase as a result of social and economic difficulties. These persons have a negative impact on the growth of countries. In this context, the WHO estimates that depression and anxiety disorders cost the global economy one

[1] https://www.who.int/fr/news-room/fact-sheets/detail/mental-disorders.

M. Mosbah et al. (Eds.): MEDI 2023, LNCS 14396, pp. 128–141, 2024.
https://doi.org/10.1007/978-3-031-49333-1_10

trillion dollars annually[2]. However, the report of the recent cited study reveals that governments spend an average of 3% of their budgets on mental health. The diagnosis of a mental disorder is different compared to other types of disorders. In fact, this disorder affects a person's thinking, emotions, behavior and interpersonal relationships, in contrast to physical diseases, which have tangible symptoms that can be detected through X-rays, blood tests or radiological examinations. Another issue is that we can expect any reaction from the patient to be diagnosed; in reality, he may lie during a psychological diagnostic for fear of being judged or stigmatized, or in order to protect his own image. For this reason, psychologists and psychiatrists believe that the greatest time to diagnose a person is when he is relaxed and acting spontaneously in a free and uncontrolled environment.

Social networks are among the environments where everyone is able to share their thoughts and express their opinions, without fear of being oppressed or controlled. This prompted researchers to conduct a significant number of studies on identifying certain psychological information on social media (like isolation, aggressiveness, recurrent suspicions, etc., [5]). Hence, the relevance of automating information extraction tasks using advanced techniques of computer science such as artificial intelligence (AI) and Big data tools that are able to analyse a large amount of data, extract relevant knowledge, and ensure generalization.

In this paper, we aim to track the states of people with personality disorders by determining the major elements impacting their status. These elements include generally different encountered problems and needs of the sick person, which can be identified by analysing its textual production on Facebook. The choice of Facebook is made since it is the most popular type of social media that provides greater flexibility in producing large amounts of data. This work enables specialists and persons in charge of people's health around the world to better understand the state of people with PD. In addition, this work can assist in discovering improved methods to support them in order to avoid critical situations such as aggressiveness or extremism. The proposed approach's implementation is focused on the three key objectives mentioned below:

- Collecting appropriate data generated by sick people including details about their primary preoccupation. This data is then preprocessed in order to be treatable by the machine.
- Building a collection of coherent Bayesian networks that can present the primary causes of the illness using various linguistic resources.
- Incorporating the semantic aspect into each network in order to make them understandable and interpretable, as well as to take into account the semantic relationships between words.

The organization of our work in this paper is as follows: in the second section, we present the various research studies that have been conducted in this context. The third section is devoted to presenting our proposed approach and detailing the various resources employed. The assessment of the proposed approach is

[2] https://www.who.int/fr/news-room/fact-sheets/detail/mental-health-at-work.

presented in the fourth section. Finally, we close this paper with a conclusion and some perspectives presented in the fifth section.

2 Related Works

Several studies proposed in the literature are based on the analysis of the textual production of people with PD on social media. The majority of these studies have focused on detecting the presence of users with PD, predicting the disorder's type [6] and symptoms [5], detecting the behavior of sick users towards a specific phenomenon such as COVID-19 [8], and so on. However, being limited only by the detection of people with psychological problems is not enough. In fact, this operation cannot allow to assist these people or to acquire useful knowledge.

For this reason, several researchers have focused their research studies on formalizing textual data in a specific format allowing to manage the risk related to these people. In this context, Ellouze et al. [7] have proposed an ontology allowing to manage the risk related to the epidemic COVID-19. This ontology may provide a relevant opportunity to present crucial factors associated to this epidemic that lead individuals with PD to be in an unstable state. In 2022, Ellouze et al. [9] have proposed a comparative study about different models based on various formats (graph, tree, rules) allowing the extraction of dependency links demonstrating the interdependence between the different states of people having a negative degree of personality traits (not conscientious, not extroverted, etc.). In the same context, [2] aims to identify characteristics that may impact behavior disorders over time. This work is done by taking advantage of a behavioral graph (sbGraph) which is composed by nodes that express terms that represent various aspects, such as interest, work, etc. The different achieved nodes are connected through semantic arcs derived from existing linguistic resources. Finally, the created graph model is enhanced by language information in the form of keywords that might have an emotive and cognitive impact on the personality. In [11], authors presented an analytical study demonstrating the factors affecting the misuse of social networks. The various variables explored in this study include a variety of psychological elements such as self-esteem, fear of missing out scale, daily time expenditure rate, and so on. The result of this study is a path analysis model that illustrates the degree of fit between the different factors investigated (measured using different metrics such as $\chi 2$).

In a similar context, Alavijeh et al. in [1] have proposed an analytic method. The purpose of the method proposed is to study the relationship between users' psychological disorders and their musical preferences on Twitter, based on an analysis of their musical tastes shared. This study considers various aspects, including words used, linguistic style applied, sentiment and emotion patterns expressed, thematic interests, and underlying semantics illustrated.

User habits and interests can also provide insight into the mental state of individuals suffering from psychiatric issues. For this reason, Si et al. in 2019 [13] have focused on analyzing users' interests on social networks to discover useful dependencies between their interests. The purpose of this study was to determine the degree of dependence between various types of information gathered from social network info-boxes intended for expressing interests and hobbies, etc. The different dependence links were determined using multiple metrics, such as Confidence, Support, Lift, etc. According to our study, we distinguish four crucial criteria in our field of study: (i) the importance of data quality due to the sensitivity of our context related to the medical field: the different studies have generally relied on social network data obtained in a hazy manner, but in our work we have sought reliable data (data of people who admit their illness). (ii) treatment objective: we note that the various studies have generally been based on a hypothesis that needs to be validated or denied, such as the study of the impact of music on personality disorders. Whereas in our context our study is more generic since it is supposed to detect implicit links even for a human being. (iii) employed approach: several researchers have employed a single technique (statistical or linguistic). While in our work, we have based our approach on the hybridization of several techniques in order to make it more profound. (iv) dealing with specific linguistic issues: the proposed approaches for the textual study relied on superficial treatment and did not address specific issues related to natural language processing challenges (NLP), such as the evolution of lexicons over time.

3 Proposed Approach

In this study, we describe a method illustrated in Fig. 1 for developing a hybrid model that combines a probabilistic model based on Bayesian network with a semantic model based on ontology to take advantage of both statistical and semantic aspects. Our proposed model is composed of a set of nodes that represent a state, problem, phenomena, or requirement related to a sick person. These nodes are connected by dependency links that illustrate the causes and effects of a situation that could be either the beginning of a problem (inappropriate state) or the resolution of a problem.

The proposed model involves analyzing the textual production of users in a Facebook group "xxxxx" which allows users with PD and unstable status to discuss their preoccupations and share their experiences in order to obtain or provide a relevant recommendation. For that, our model may be an appropriate solution for identifying the various factors that lead users to become unstable. In addition, it may be beneficial for getting a sense of what might happen as a result of this state, also for determining whether there is a solution to this problem. It should be noted that our proposed model's architecture enables it to address linguistic issues such as lexical analysis (several words can express the same idea) through the semantic aspect provided by the ontology. We describe below in detail the process of our work.

3.1 Data Collection and Preprocessing

Data Collection. Our data applied in this study were obtained from a Facebook group that allows people with PD to discuss in French and change their

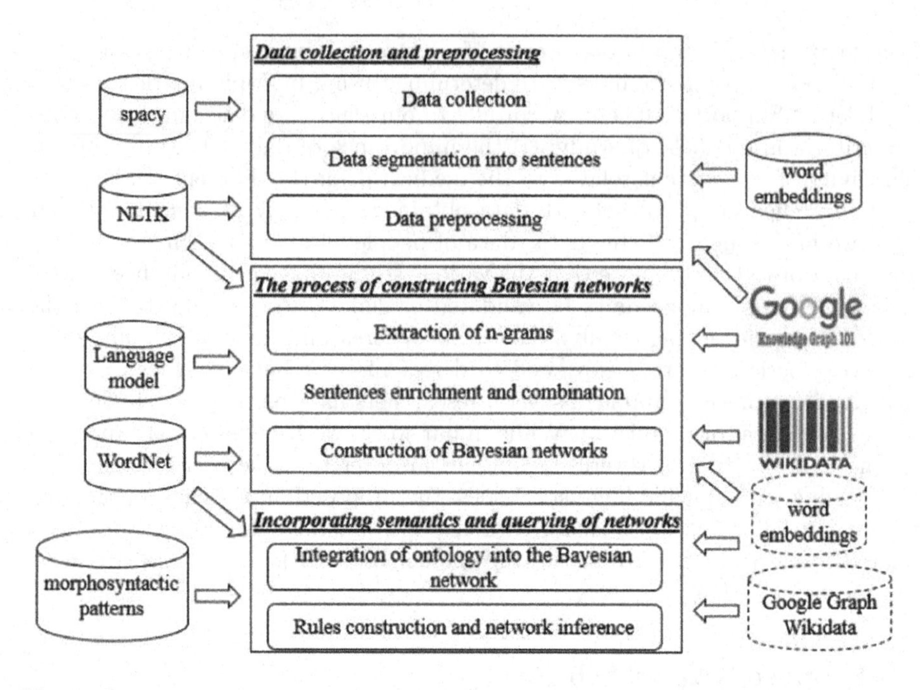

Fig. 1. Our proposed approach for formalizing a text published on social networks based on both Bayesian networks and ontology.

experiences about the difficulties, concerns and treatment of personality disorder disease. The members of this group have acknowledged their psychological problems, for this reason the annotation process is not necessary in this case. In addition, this data source ensures reliable information while avoiding particular phenomena such as irony, sarcasm, variety of fields, etc. We provide in Table 1 a detailed description of the size of the collected corpus. After collecting the data, we take specific steps to validate our data at the data value level to ensure that the data's content is relevant and meaningful. Moreover, we normalize our data in a machine-processable way, especially since it came from Web 2.0, where user interaction is a crucial operation for data generation.

Table 1. Overview of the size of the corpus collected from Facebook.

Description	Size
The number of posts	1419
The number of users related to the collected posts	304
The average posts per user	4.67
The number of users with more than 5 posts	87
The number of users with more than 10 posts	30
The number of posts with more than 280 characters	654

Data Segmentation into Sentences. In this phase, we aim to segment the text portion into sets of sentences by determining the end-of-sentence punctuation using the spacy resource[3]. This process was performed since we revealed that the size of posts is important and it provides a variety of information. This variety is due that the authors need to be more specific when explaining their circumstances in order to retrieve relevant replies later. In our case, segmenting text into independent sentences may cause a loss of the semantic aspect. In fact, our empirical study revealed an extensive use of the linguistic phenomena ellipsis, which is employed by users to prevent repetitions during the textual production process. For this reason, we have employed the following treatment process to address the issue of finding a sentence containing a pronoun, e.g. they, them, her, him, etc.: (i) extract all nouns that precede this sentence in the same post, (ii) identify the grammatical category of the unknown element and all nouns identified from (i) through a morpho-syntactic analysis, (iii) determine the correspondence between the unknown element and the other nouns listed, based on a comparison of gender (masculine/feminine) and forms (singular/plural), (iv) if there is an overlap, we use the word embeddings technique [12] to determine the relationship between each noun in the released list and all words with the grammatical category noun, verb, adjective in the sentence containing the unknown word. The following example demonstrates the processing applied to a specific sentence in this stage.

Before the Processing: *Je suis suivie par un psychiatre en HP. Je le vois peu, hélas, impossible d'avoir un rdv en urgence avec lui. (I'm being followed by a psychiatrist in a hospital. Unfortunately, I don't see him very often and it's impossible to get an emergency appointment with him.)*

After the Processing: *Je suis suivie par un psychiatre en HP. Je psychiatre vois peu, hélas, impossible d'avoir un rdv en urgence avec psychiatre. (I'm being followed by a psychiatrist in a hospital. Unfortunately, I don't see psychiatrist very often and it's impossible to get an emergency appointment with psychiatrist.)*

Data Preprocessing. The process of preprocessing data consists of removing unnecessary elements such as stop words, that are used by everyone independently of the context to ensure the articulation between ideas. Our current process involves also removing from our corpus: symbols, numbers, and other elements, since in our case we focus primarily on the linguistic aspect. Next, we proceed to normalize the data by converting capital letters at the beginning of sentences to lowercase letters and abbreviations to their raw format such as *OMS* to *Organisation mondiale de la santé, (World Health Organization (WHO))*, through the resource Google Knowledge Graph [10]. The task of normalization involves transforming the various words related to our corpus into a common root using the NLTK resource, in order to avoid morpho-syntactic distinction between words and focus only on the semantic aspect.

[3] https://spacy.io/.

3.2 Extraction of n-grams

In this phase, we intend to extract the various elements that will act as nodes in the Bayesian network. These elements can be considered as problems encountered or a treatment to be followed in the context of psychological disorders. For the detection of these elements, we first employed a statistical method based on the occurrence of frequent words for the case of simple words, as well as the detection of elements that frequently appear together for the case of compound words. This method is performed using the n-grams (1,2,3 g) technique of NLTK resource. The second step involved extracting key words, this task was performed using a linguistic method based on morpho-syntactic patterns e.g.: noun+noun: *urgence psychiatrique, médecin généraliste (psychiatric emergency, general practitioner)*. For the remainder of our process, we kept the common elements between the results of these two methods.

3.3 Sentences Enrichment and Combination

In this phase, we focused on enriching our corpus with other sentences to avoid being limited to the data of our corpus. In order to accomplish this task, we conducted a research on the Google Knowledge Graph database for the combination of two words among words extracted from the previous step. The combination of the two words to be searched is chosen using the degree of binding returned by the word embedding resource. This operation was done to strengthen the relevant connections between the dependent words. The result of each query is a sentence containing the two words. The example (1) shows the results returned by the Google Knowledge Graph resource for the combination of the two words: *trouble+sommeil (disorder+sleep)*.

(1) *Le trouble dissociatif lié au sommeil est un* ***trouble*** *du* ***sommeil*** *rare, compris dans les parasomnies impliquant des phénomènes psychopathologiques. (Dissociative sleep disorder is a rare sleep disorder, included in parasomnias involving psychopathological phenomena.)*

After that, we proceed to group sentences that express the same semantic ideas. Each group of sentences is designed to form a Bayesian network in the next steps. In fact, the fundamental idea of our work is to construct a set of simple and coherent Bayesian networks. In our work, we have decided to make the clustering based on sentences rather than on posts in order to be more precise (calculating similarity between sentences is more precise than calculating similarity between posts).

For performing the sentences clustering task, we have employed the following process: (i) check sentences that share a semantic relationship between their concepts, using the semantic relations synonym and hyponym (subtype) existing in wordnet, (ii) measure the semantic similarity between sentences of each group derived from (i) using the pretrained model of SBERT [14]. This process is due to the fact that we can find two sentences that mention bipolar disorder, but one of them focuses on the problem of sleep and the other on social integration. It should be noted that we have set a threshold (stop condition) of 20 different words among the list of n-grams in each sentences group. The choice of using a threshold is that we aim to build a simple network for each sentences group that does not require a huge computation time, also to prevent generating a lot of errors.

3.4 Construction Of Bayesian Networks

In this phase, we aim to build for each group of sentences extracted from the previous step a Bayesian network. To accomplish this task, we extracted the list of n-grams associated with each sentences group. Then, we calculated the occurrence of each of these elements (words) at each sentence level of the same group to determine which elements occurred together.

At this point, we recall that we have grouped semantically similar concepts together. For example, according to an empirical study, we have discovered that suicide and violence (both consequences of psychiatric problems) belong to the same category of danger. For that, we have considered these two words as the same node. This is due to the fact that the term suicide was not frequently used in the corpus, but it is nonetheless a relevant concept in our study context. To do that, we have used two methods, the first is based on the three main functions: synonym, subtype (hyponym), and supertype (hypernym) that are available in WordNet. The second method is based on words that have the same semantic description according to the Wikidata resource. The Bayesian network was built automatically using the pyArgum library [3], which is based on Bayes' theorem (1), allowing the estimation of the probability that an event (A) will happen (in our context: the appearance of the word "A") based on the occurrence of another event (B) (the appearance of the word "B").

$$P(A/B) = \frac{P(B/A) * P(A)}{P(B)} \tag{1}$$

The Fig. 2 presents two examples illustrating the results of the current step when it was applied to two collections of sentences.

Translation of the Components of Fig. 2 (a): {alcool: alcohol; phobie: phobia; bless=blessure: injury; pension: pension; urgence: emergency;}

Translation of the Components of Fig. 2 (b): {troubl=trouble: disorder; sensib=sensibilité: sensitivity; bless=blessure: injury; autodestruct: self-destruct; tristesse: sadness; solitud=solitude: loneliness; irrationnel: irrational; aide: help;}

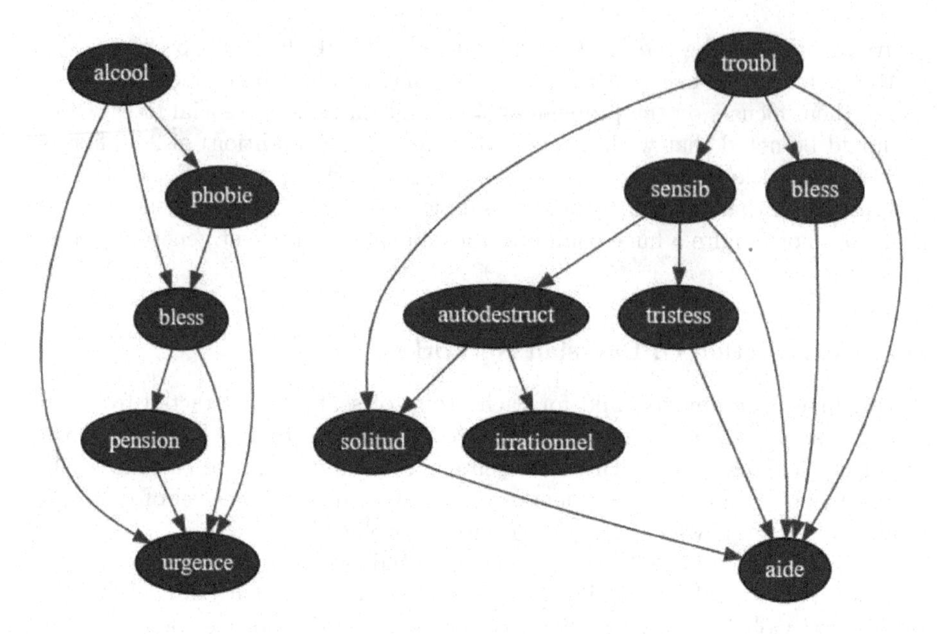

Fig. 2. Two examples of automatically generated Bayesian networks.

3.5 Integration of Ontology into Bayesian Networks

At this level, we have a collection of Bayesian networks that share an important degree of semantic similarity (concerns of people with PD). In this step, we focus on integrating the semantic aspect into all of our networks. The objective is to define a formal representation of vocabulary that is related to the anxieties of people with PD in order to increase the knowledge base of psychiatrists and psychologists by making our networks more understandable and interpretable. To achieve that, we aim to: (i) consider each node in the Bayesian network as a concept of ontology and to associate each dependency relationship with an element that describes the semantic aspect of this relation, which is done using: (a) Google knowledge Graph: by launching a request which contains the names of the two nodes, we receive a list of outputs grouping (result.name, result.description, resultScore, result.@type, result.detailedDescription), in this section we are interested in the value of the attribute result.description. For example, the value of this attribute for a query containing the two words *alcool+phobie (alcohol+phobia)* is *Anxiolytique (Anxiolytic)*.

(b) our corpus: by selecting words that appear in the sentence that contains the two nodes, but the selected words are neither adjectives, prepositions, or determinants. The word used to determine the name of the relationship is the one that is most dependent to the two nodes. The degree of dependency is measured using the two resources wordnet and word embeddings. If the two methods have provided multiple proposals, our final choice will be based on the word which belongs to the medical lexicon [4], this is to make our network more scientific. Figure 3 (a) depicts the improvement produced by the current step on an excerpt from Fig. 2 (a).

(ii) assign an ontology to each node (concept), in this way we ensure that our network contains the most possible number of terms and can answer to the greatest number of user queries. In reality, the user might specify the word psychiatrist in his query by using the word doctor or physician. For that, we assign to each node name four direct relations (synonym, hyponym, hypernym, sibling (under the same category)). The different relations of an ontology are detected through the resource wordnet. It should be noted that in order to determine a word's sibling relationship, we first apply the hypernym relation to determine its parent. Then, we apply the hyponym relation to this parent to identify their children. After that, we measure the degree of relationship between each of these words (children) and the main word through Google Knowledge Graph, word embeddings and wordnet resources. For example, to detect elements having a sibling relationship with the word *Psychiatre (Psychiatrist)*, we first identified many elements such as *radiologiste, dermatologue, pédiatre (radiologist, dermatologist, pediatrician, etc.)*. However, it is obvious that these elements do not have a strong relationship with the word psychiatrist. For this reason, we measured the degree of connection between each word in this list and the word psychiatrist, which enabled us to retain the word *neurologiste (neurologist)*. Figure 3 (b) demonstrates the automatic construction of an ontology for the node *tristesse (sadness)*.

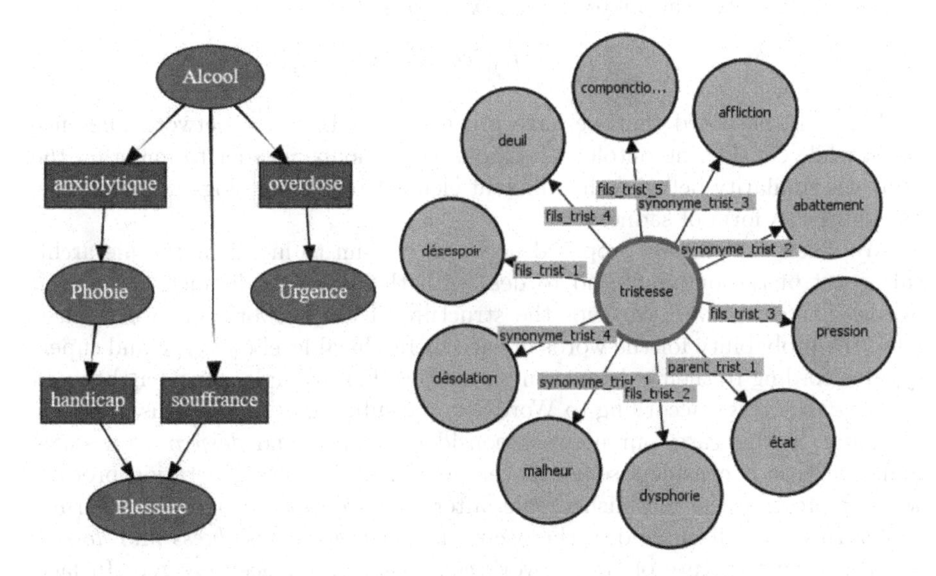

Fig. 3. Two examples illustrating the incorporation of the semantic aspect into our constructed models.

Translation of the Components of Fig. 3 **(a):** {alcool: alcohol; phobie: phobia; blessure: injury; anxiolytique: anxiolytic; overdose: overdose; handicap: disability; souffrance: suffering; urgence: emergency;}

Translation of the Components of Fig. 3 **(b):** {tristesse: sadness; affliction: affliction; abattement: depression; malheur: unhappiness; état: state; désespoir: despair; désolation: desolation; dysphorie: dysphoria; deuil: grief; componction: compunction; pression: pressure; synonyme_trist: synonym of sadness; parent_trist: parent of sadness; fils_tris: son of sadness;}

3.6 Rules Construction and Network Inference

In this step, we focus on processing rules allowing to perform the inference of our Bayesian networks. We recall that at this level, we have a set of networks that are enriched at both lexical and semantic levels. The recently cited networks occasionally share common nodes that were not grouped together during the step *2.3 Sentences enrichment and combination* because the similarity rate between sentences containing them did not exceed the threshold. These nodes may cause problems with our networks' inference because they may result an overlap at the level of rules generated. For this reason, we focus at this level on combining rules together. For example, if in a network we have identified that event C is triggered when events A and B occur and in another network we have identified that the same event C is triggered when events E and F occur. In this case, we transform these two rules into the following rule (2):

$$P(C/A, B)||P(C/E, F) \tag{2}$$

It should be noted that we have processed the Bayesian network inference rules, while considering ontology relations and axioms in order to maintain the semantic similarity between the different elements of the ontology. For example, mourning is a form of sadness.

We recall that in our proposed network, we aim to highlight the hierarchical aspect of the ontology and to deal with the semantic distinction between words. Therefore, we have kept the structure of the network and we recalculated the probability for the words of each hierarchical level (subtype and supertype) according to their existence in our corpus. For example, consider the word *tristesse (sadness)*; according to WordNet, the subtype of this word is *désespoir (despair)*; in this case, our network considers *sadness* and *despair* as a single node, because it considers semantic relations. And in rules generation process, we first preserve the established rule, after that we build two additional rules to determine the likelihood of the occurrence of the event *sadness* and *despair* based on the structure of the network that has already been created. In fact, we just focused in this step on learning parameters that can offer a hierarchical response from the general to the specific.

4 Evaluation

After experimenting our approach, we obtained 152 Bayesian networks (between 5 and 20 nodes) with 491 distinct concepts (simple and compound words). In this step, we focused on evaluating the results obtained. It should be noted that our work was applied to textual data that is published on social networks, therefore the lexicon obtained is wide and varied. This may make the evaluation based on probability difficult since the concept of probability is mainly based on the precision. In general, this concept is applied to structured and numerical data. For this reason, we have employed two methods to assess the various results achieved: human expert evaluation and Gruber criteria evaluation used in [7]. For the human-based evaluation, we evaluated the quality of 20 (arbitrary choice) Bayesian networks, given the difficulty of evaluating 152 networks. The first evaluation relates to the assessment of the task of grouping nodes that can be figured together in a network (the nodes that are connected in a dependency relationship). The second evaluation involves the assessment of the task of predicting elements that can trigger another element (event, state, phenomenon, etc.). To accomplish that, we first asked our expert to anticipate the elements that might trigger an element based on the nodes of our network. Then, we provided our results which included the dependencies in order to could compare the results and judge each relation of our network as (validate, maybe, false) according to their expertise. We have been tolerant in the evaluation of our results, in fact we have considered the answers "maybe" as a right answer (since the objective of this work is to detect hidden links). Furthermore, our lexicon is related to social networks, implying that the evaluation is done diligently (it is not a precise science). The evaluation metrics used to measure the performance of our results are both F-measure (3) and concept error rate (CER) (4). The values of the evaluation metrics achieved are detailed in Table 2.

$$F - measure = \frac{2 * recall * precision}{recall + precision} \tag{3}$$

$$CER = \frac{incorrect\ response\ predcition}{response\ reference} \tag{4}$$

Table 2. The evaluation of the task of formalizing data.

Description	Result
The number of evaluated networks	20
The number of nodes related to the networks evaluated	300
The average of correct nodes per network	86%
The F-measure related to the clustering task	78%
The CER involves evaluating the task of elaborated links	17%

For the second evaluation method, we asked an expert to measure the performance of our semantic Bayesian network according to Gruber's criteria. According to our expert's interpretation, our network has provided clarity because all their nodes are represented by French-language domain terms. Furthermore, our network is extensible due to its ability to accept other words and relationships, and it can ensure the evolution of the lexion through time. In terms of coherence, our expert believes that our network is generally coherent at nodes, links, and ontology levels. In addition, in our case, the encoding minimum deformation requirement is ensured by presenting many concepts linked to our study field.

In the end, we can affirm that our approach has ensured a significant social impact that enables to discover the causal links of several consequences related to the personality disorder. The objective of our proposed approach is to enrich the knowledge base of psychiatrists, psychologists and sociologists. This is in contrast to other works in the literature that have focused on just the detection of PD on social networks, which provides a superficial, rigid and not interpretable result. It should also be noted that the combination of linguistic and statistical aspects has made it possible to combine the advantages of probability and semantic techniques. The use of ontology has enabled us to overcome the issues of the lexical approach. In fact, our approach is focused on the semantic analysis rather than the existence of terms. As well, the use of ontology allows for providing answers at several levels, from the generic to the specific. In addition, our approach addresses issues of the evolution of the lexicon over time by using a variety of evolutionary linguistic resources, like wikidata, wordnet, etc. As perspectives, we aim to incorporate the concept of time into our network, as well as to work on the standard of network quality parameter (QoS). We also aim to enhance our study by examining the figurative linguistic factors involved in the textual production of these people.

5 Conclusion

In this paper, we presented a hybrid approach for the identification of factors affecting the state of people with PD by analysing their data published on social media in order to assist psychiatrists and sociologists to understand the specificities of their states. The proposed approach is mainly composed of two tasks: developing a collection of Bayesian networks and incorporating ontology into the built networks. By querying our semantic Bayesian network, we can identify the primary concerns of people with personality disorder disease, as well as the various causes and solutions of the disease, based on their analysis expressed in their textual production. Our proposed approach addresses a variety of linguistic challenges, including the lexical approach and the evolution of lexical over time. Moreover, it has the ability to provide an interpretable and enriched results. In our future works, we aim to incorporate to our networks the temporal factor and highlight the specifics related to the writing style of sick people. Thus, we aim to validate our constructed networks at a high standard of performance level.

References

1. Alavijeh, S.Z., Zarrinkalam, F., Noorian, Z., Mehrpour, A., Etminani, K.: What users' musical preference on twitter reveals about psychological disorders. Inf. Process. Manage. **60**(3), 103269 (2023)
2. Beheshti, A., Moraveji-Hashemi, V., Yakhchi, S., Motahari-Nezhad, H.R., Ghafari, S.M., Yang, J.: personality2vec: enabling the analysis of behavioral disorders in social networks. In: Proceedings of the 13th International Conference on Web Search and Data Mining, pp. 825–828 (2020)
3. Ducamp, G., Gonzales, C., Wuillemin, P.H.: aGrUM/pyAgrum: a toolbox to build models and algorithms for probabilistic graphical models in Python. In: International Conference on Probabilistic Graphical Models. PMLR (2020)
4. Duclos, C., et al.: Le vocabulaire médical, les ressources terminologiques, le codage de l'information en santé. In: Informatique médicale, e-Santé, pp. 11–41. Springer, Cham (2013). https://doi.org/10.1007/978-2-8178-0338-8_2
5. Ellouze, M., Hadrich Belguith, L.: A deep learning architecture based on advanced textual language models for detecting disease through its symptoms associated with a reinforcement learning algorithm. In: Fill, H.G., van Sinderen, M., Maciaszek, L.A. (eds.) ICSOFT 2022. Communications in Computer and Information Science, vol. 1859, pp. 207–229. Springer, Cham (2022). https://doi.org/10.1007/978-3-031-37231-5_10
6. Ellouze, M., Hadrich Belguith, L.: A hybrid approach for the detection and monitoring of people having personality disorders on social networks. Soc. Netw. Anal. Min. **12**(1), 67 (2022)
7. Ellouze, M., Mechti, S., Belguith, L.H.: Approach based on ontology and machine learning for identifying causes affecting personality disorder disease on Twitter. In: Qiu, H., Zhang, C., Fei, Z., Qiu, M., Kung, S.-Y. (eds.) KSEM 2021. LNCS (LNAI), vol. 12817, pp. 659–669. Springer, Cham (2021). https://doi.org/10.1007/978-3-030-82153-1_54
8. Ellouze, M., Mechti, S., Belguith, L.H.: A hybrid approach based on linguistic analysis and fuzzy logic to ensure the surveillance of people having paranoid personality disorder towards COVID-19 on social media. Int. J. Gen Syst **52**(3), 251–274 (2023)
9. Ellouze, M., Mechti, S., Hadrich Belguith, L.: A comparative study on the extraction of dependency links between different personality traits. SN Comput. Sci. **3**(6), 495 (2022)
10. Fensel, D.: Introduction: what is a knowledge graph? In: Knowledge Graphs, pp. 1–10. Springer, Cham (2020). https://doi.org/10.1007/978-3-030-37439-6_1
11. Gori, A., Topino, E., Griffiths, M.D.: The associations between attachment, self-esteem, fear of missing out, daily time expenditure, and problematic social media use: a path analysis model. Addict. Behav. **141**, 107633 (2023)
12. Levy, O., Goldberg, Y.: Dependency-based word embeddings. In: Proceedings of the 52nd Annual Meeting of the Association for Computational Linguistics (Volume 2: Short Papers), pp. 302–308 (2014)
13. Si, H., et al.: Association rules mining among interests and applications for users on social networks. IEEE Access **7**, 116014–116026 (2019)
14. Wang, B., Kuo, C.C.J.: SBERT-WK: a sentence embedding method by dissecting BERT-based word models. IEEE/ACM Trans. Audio Speech Lang. Process. **28**, 2146–2157 (2020)

Social Recommendation Using Deep Auto-encoder and Confidence Aware Sentiment Analysis

Lamia Berkani[1,2](✉), Abdelhakim Ghiles Hamiti[2], and Yasmine Zemmouri[2]

[1] Laboratory for Research in Artificial Intelligence (LRIA), USTHB University, 16111 Bab Ezzouar, Algiers, Algeria
lberkani@usthb.dz

[2] Department of Artificial Intelligence and Data Sciences, Faculty of Informatics, USTHB University, 16111 Bab Ezzouar, Algiers, Algeria

Abstract. The development of online social networks has attracted increasing interest in social recommendation. On the other hand, recommender systems based on deep learning and sentiment analysis techniques are currently widely used to solve the problem of data sparsity. However, only a few attempts have been made in social-based recommender systems. This article focuses on this issue and proposes a novel hybrid approach named CASA-SR (Confidence Aware Sentiment Analysis-based Deep Social Recommendation). Our approach exploits sentiment analysis by detecting fake reviews and combines predictions generated by collaborative and content-based filtering. A neural architecture has been adopted using an auto-encoder and a multilayer perceptron neural network. Moreover, our approach integrates social information, including users' trust (credibility and similarity degrees). Experimental results conducted on different datasets showed significant improvements in recommendation performance according to the state-of-the-art work.

Keywords: Social recommendation · hybrid sentiment analysis · fake reviews · deep learning · auto-encoder · MLP

1 Introduction

The development of online social media has attracted increasing interest in social recommendation. Previous work demonstrated that the integration of social information, as auxiliary information, can enhance the performance of traditional recommender systems. Several works have integrated social information with collaborative filtering-based methods [1–3].

On the other hand, deep learning techniques have been recently applied in recommender systems to solve the cold start and data sparsity problems and further enhance the recommendation accuracy and performance. Current models mainly use deep neural networks to learn user preferences on items for recommendations. However, only few initiatives have been conducted in social-based recommendation field [4–7].

M. Mosbah et al. (Eds.): MEDI 2023, LNCS 14396, pp. 142–155, 2024.
https://doi.org/10.1007/978-3-031-49333-1_11

To overcome the rating data sparseness, users' comments are being used for rating prediction. These reviews can express users' overall satisfaction on the items through their preferred, non-preferred or neutral opinions. Several research works are applying sentiment analysis (SA) in recommender systems [8]. Some works are based on hybrid deep-learning models. For instance Dang et al. [9] integrated a hybrid SA model into the collaborative filtering, by combining the CNN and LSTM models in different orders. This approach was proposed in the context of social networks, but did not take social information into account in the recommendation process. Berkani and Boudjenah [5], integrated a hybrid SA model to a deep neural network model including social information with friendship and trust features. However this work didn't take into consideration fake reviews. Recommendation systems are vulnerable to intrusions (due to a lack of security) or to the sharing of fake information (many malicious users add misleading information). Thus, recommendation systems based on SA can be manipulated or disrupted by the presence of fake reviews. Recently, some studies have proposed approaches for detecting fake reviews [10, 11]. However, to the best of our knowledge, no work proposed in a social context has combined a hybrid SA model with fake reviews detection along with social information formalization.

In this article, we propose a novel social-based recommender model using a confidence aware hybrid sentiment model to improve the user-item rating matrix by predicting missing ratings and correcting inconsistent values. By exploiting the updated matrix, our system generates predictions using a hybrid recommendation algorithm which combines social information with collaborative and content-based filtering algorithms using an auto-encoder and an MLP network, respectively. Extensive experiments conducted on two datasets demonstrated the effectiveness of our model compared to the state-of-the-art approaches and baselines.

The remainder of this article is organized as follows: Sect. 2 presents some related work. Section 3 and Sect. 4 present respectively the conception of our approach with the associated experiments. Section 5 highlights the most important contributions of this work and proposes some future perspectives.

2 Related Work

The development of online social media has favored the expansion of social recommendation, where several researchers are interested in proposing approaches that integrate social information. Different research works included social trust in recommender systems [2, 3].

With the latest achievements and the great potential for learning effective representations, DL models are being exploited recently in recommender systems. He et al. [12] proposed the widely used Neural Collaborative Filtering algorithm (NCF) using a Generalized Matrix Factorization (GMF) and an MLP to model the linear and non-linear relationship between users and items. Berkani et al. [13, 14] proposed the Neural Hybrid Filtering model (NHF) based on GMF and Hybrid MLP. To predict ratings in recommender systems, Rama et al. [15] proposed a discriminative model that integrates features from auto-encoders with embeddings in a deep neural network.

On the other hand, the flexibility and accessibility of social networks has enabled millions of people to subscribe and post their comments on these platforms, expressing their preferences and/or feelings about items. Given that people often choose comments rather than numerical ratings, researchers tried to get around the problem of rating data sparsity by leveraging textual reviews. These works are using sentiment analysis (SA) to predict their current preferences for given items [16–18]. Some of the proposed approaches are based on hybrid DL models [5, 9]. This combination has significantly improved the recommender system's performance.

However, despite much attention paid to DL and SA in recommender systems, the state of the art shows that only few works have used DL in social-based recommender systems. For instance, Bathla et al. [7] proposed AutoTrustRec, a recommender system with direct and indirect trust and DL using auto-encoder. However, this work did not exploit the users' reviews for improving the recommendation performance. Berkani et al. [4] developed the SNHF model, incorporating social information in the NHF. However this work has only considered friendship and trust degree that has been calculated with MoleTrust algorithm [19].

The review of related work demonstrates the significant results achieved by the application of DL techniques and sentiment models in recommender systems. Proposed work in a social context has considered only a few features to model social information, including friendship and trust (direct and/or indirect). While other important factors can be considered such as the credibility and the influence of users in social networks. Moreover, as reviews often contain fake good or fake bad information, some studies focused on fake reviews detection. Li et al. [10] exploited the interactivity of review information and used the confidence matrix to measure the relationship between rating outliers and misleading reviews. Birim et al. [11] focused on detecting fake reviews through topic modelling. Similarly, we will propose in this article a novel approach using DL and confident aware hybrid sentiment analysis by detecting fake reviews. We will also focus on modeling other features of social information.

3 Our Approach

We have proposed an approach called CASA-SR (Confidence Aware Sentiment Analysis-based Deep Social Recommendation) including four main modules, as illustrated in Fig. 1:

1. Hybrid sentiment analysis module with fake reviews detection (enabling the construction of a confidence matrix). This module updates the rating matrix.
2. Collaborative filtering based on neural architecture using an auto-encoder. This module uses the updated rating matrix considering user by user (line by line), and generates an output prediction.
3. Content-based filtering: This module uses an MLP network to generate a prediction based on user and item features.
4. Hybrid recommendation: a final neural layer combines users' social information with the results generated from the collaborative and content-based recommendations.

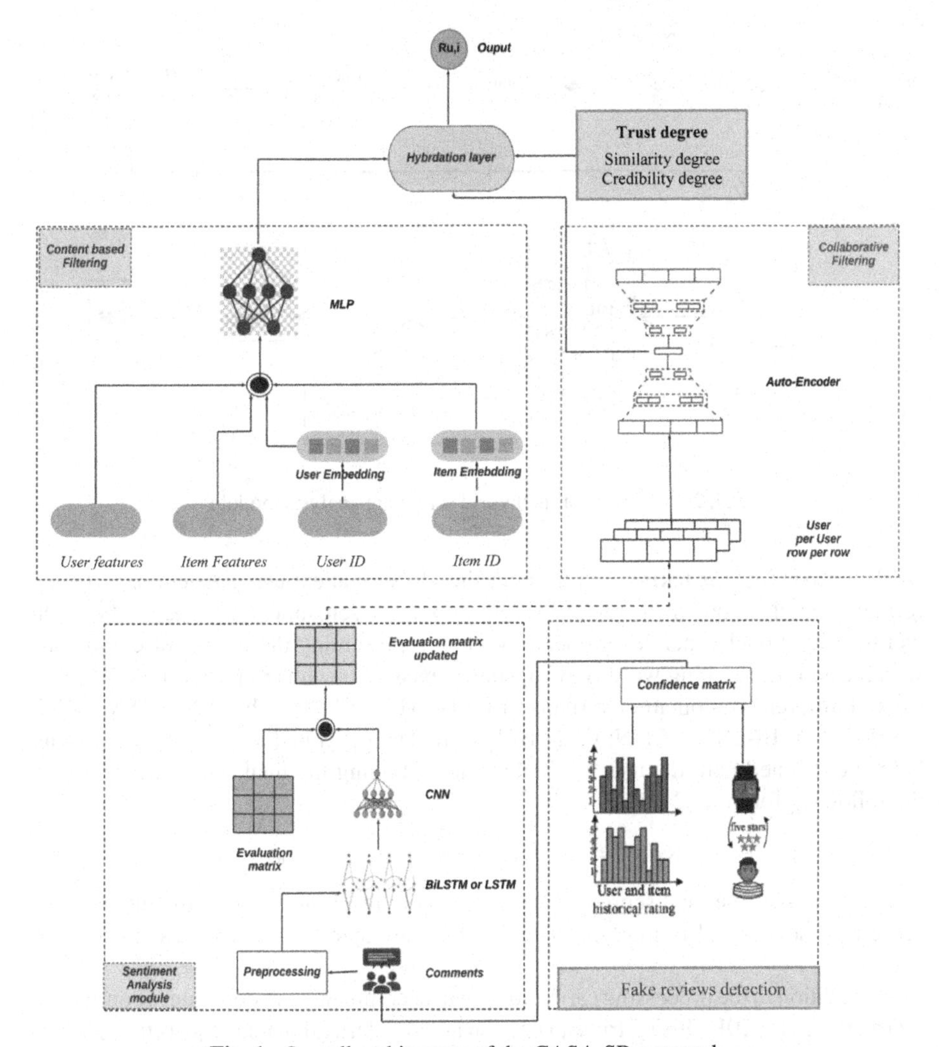

Fig. 1. Overall architecture of the CASA-SR approach.

3.1 Confidence Aware Sentiment Analysis

This module converts comments into a numerical score (from 1 to 5) using a sentiment analysis process based on a combination of an LSTM / Bi-LSTM recurrent neural network and a CNN convolutional neural network, then applies normalization to obtain values between 1 and 5. Figure 2 illustrates this process:

After extraction of the textual data (comments associated with each user's opinion about an item), a set of pre-processing operations is carried out, including: tokenization, used to fragment the comment into sub-words; cleaning (i.e. remove stop words, taking care to retain adjectives and remove all punctuation marks); encoding and padding to homogenize the lengths of the resulting sequences; and generation of embeddings using the BERT model (Bidirectional Encoder Representations from Transformers) [20].

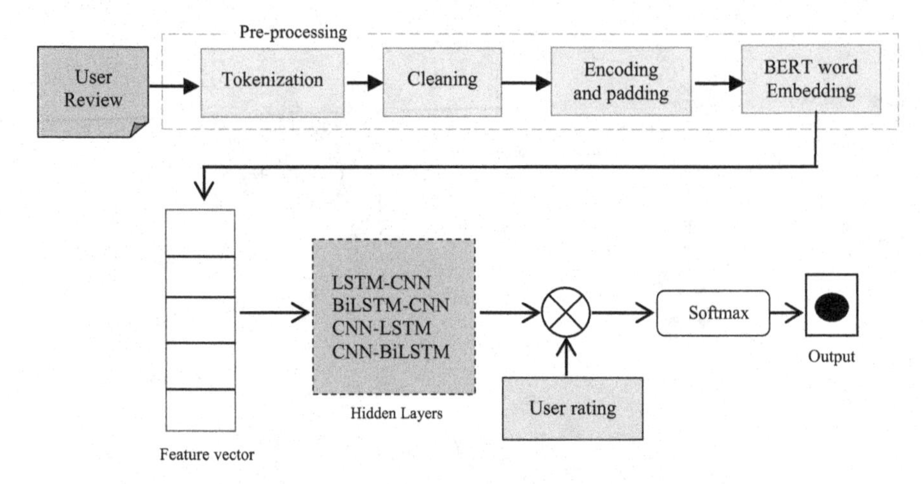

Fig. 2. Structure of the hybrid sentiment analysis module

BERT encodes words taking into account the global context using Attentional Transformers, which allows it to understand the complex semantic relationships between words. This model generates sequence vectors representing the tokens of comments, with the aim of matching words having similar meanings with similar vectors. We considered different combination variants of LSTM (Bi-LSTM) with CNN: LTSM-CNN; CNN-LSTM; Bi-LSTM-CNN; CNN-Bi-LSTM. The generated sentiment-based rating is then combined with the user's initial rating, obtaining the final rating, according to the following formula [21].

$$score_f = \alpha * score_s + (1 - \alpha) * score_R \tag{1}$$

where: score-f is the final score; score-s is the score resulting from sentiment analysis; score-r is the original rating; and α represents the balancing factor between the two values.

In addition to the processing carried out on user comments, we construct a confidence matrix (Q) as in [10]. This matrix can be seen as a regularization that adjusts the reviews. For convenience, the values of the matrix are between 0 and 1. The value of each element of the matrix is calculated by the following confidence degree function [10]:

$$Q_{ij} = F(R_{ij}, \beta) \begin{cases} e^{-Relu\left(\left|\sqrt{\sum_{p \neq j, R_{ip} \leq 3} R_{ip}} - R_{ij}\right| - \beta\right) - Relu\left(\left|\sqrt{\sum_{q \neq i, R_{ip} \leq 3} R_{ip}} - R_{ij}\right| - \beta\right)}, & R_{ij} \leq 3 \\ e^{e^{-Relu\left(\left|\sqrt{\sum_{p \neq j, R_{ip} > 3} R_{ip}} - R_{ij}\right| - \beta\right) - Relu\left(\left|\sqrt{\sum_{q \neq i, R_{ip} > 3} R_{ip}} - R_{ij}\right| - \beta\right)}}, & R_{ij} > 3 \end{cases} \tag{2}$$

where: β: is the deviation rate; Q: is the confidence matrix; R: is the rating matrix;

The process of calculating confidence values shows that when a user assigns a high (or low) rating to an item, this rating deviates considerably from the average of the user's previous ratings and the average of the item's previous ratings. Therefore, the corresponding comments are given a low weight, and a high probability of being considered as false comments.

3.2 Collaborative Filtering

This module takes as input the ratings matrix generated from the sentiment analysis module and processes it via an auto-encoder (AE) neural network. The objective of the AE is to learn a latent representation of the rating matrix by taking this matrix as input line by line and compressing it into a lower-dimensional latent representation through several layers of neurons, one smaller than the previous one. The size of a given layer is the size of the previous layer divided by 2.

Next, the decoder considers this latent representation and transforms it into a reconstruction of the input matrix line via the same number of layers as the encoder, but in this case, one larger than the previous one. This means that the size of a layer is the size of the previous layer multiplied by 2. During the training process, the auto-encoder attempts to minimize the difference between the rows of the original evaluation matrix and their reconstructions. Once the AE has been trained, the latent representation is used to generate personalized item recommendations.

3.3 Content-Based Filtering

For content-based filtering, we add the descriptive item and user information, then concatenate them with the embeddings of their respective identifiers. These vectors are then given as input to a multilayer MLP neural network structure, which will generate the prediction as output.

We have distributed the number of neurons per layer in decreasing order. This number is inversely proportional to the number of nodes per layer. We also set the number of nodes for the last layer, called 'predictive factors'.

3.4 Social Information Modeling

Social information between users will be modeled by considering the concept of trust. The degree of trust between two users will be used to select the closest people to a given user. It is necessary to filter the user's trusted contacts according to their proximity to the user (it is more likely that a user will seek advice from a person who is closer to him/her than from another person) as well as according to their credibility.

The trust degree between two users will take into consideration the degree of similarity between them and the degree of credibility of the second user:

$$D_{Trust}(u_1, u_2) = \alpha.D_{Similarity}(u_1, u_2) + (1 - \alpha).D_{Credibility}(u_2) \tag{3}$$

where: α is the importance weight between the degree of similarity and credibility.

Similarity Degree. We consider that two users u1 and u2 are similar if they interact with the same items. Similar assessment selection implies that users trust each other. To calculate this degree, we use the cosine similarity measure [5]:

$$D_{Similarity}(u1, u2) = \frac{\sum r_{u1,i}.r_{u2,i}}{\sqrt{\sum r_{u1,i}{}^2}.\sqrt{\sum r_{u2,i}{}^2}} \tag{4}$$

where: $r_{u1,i}$: is the evaluation of user u_1 on item i; and $r_{u2,i}$: is the evaluation of user u_2 on item i;

Credibility Degree. For the calculation of the degree of credibility we have considered the following formula which takes into account the calculation of the fake reviews rate, the competence degree and the participation degree:

$$D_{Credibility}(u) = \delta.D_{Fake}(u) + \lambda.dD_{Competence}(u) + \gamma.D_{Participation}(u) \tag{5}$$

where: γ, λ and δ: are weights expressing a priority, with: $\gamma + \lambda = 1$ and δ is negative.

Fake Reviews Rate. This rate is calculated according to the number of fake reviews posted by a given user *u*, based on the total number of fake reviews. The greater the number of fake reviews posted by a user, the less credible that user is.

$$D_{Fake}(u) = \frac{NbFakeReview(u)}{NbFakeReviewTotal} \tag{6}$$

Competence Degree. We consider a user to be competent if he/she has rated the items "correctly" compared to their average ratings, where the average rating of an item is calculated based on the ratings of all users of the system, according to the following formula [22]:

$$D_{Competence}\left(u, I_j\right) = \frac{\left|r_{u,j} - avg\left(I_j\right)\right|}{k} \tag{7}$$

where: I_j: is the item number j; $r_{u,j}$: represents the evaluation of the user *u* on the item I_j; and $avg\left(I_j\right)$: is the average evaluation of the item I_j relative to all users of the system.

The degree of competence when considering all the items is calculated as follows [22]:

$$D_{Competence}(u) = \frac{1}{n}.\sum\nolimits_{j=1}^{n} D_{Competence}(u, I_j) \tag{8}$$

where: *n* represents the number of items.

If a user's opinion is far from the average of other users' opinions, then he/she will lose out in credibility.

Participation Rate. This rate is calculated on the basis of the number of evaluations performed by a user u, according to all the evaluations in the system.

$$D_{Participation}(u) = \frac{\alpha.NbReview(u) + \beta.NbRating(u)}{NbTotalReview + NbTotalRating}$$

where β is the importance degree between Reviews and Ratings.

3.5 Hybrid Recommandation

The hybrid recommendation module combines the results obtained from the collaborative and content-based filtering modules, including social information. The last active layer of the MLP multi-layer neural network, related to the content-based module, will be concatenated to the common layer between the encoder and the decoder of the auto-encoder, related to the collaborative module, as well as to the value obtained from the average preference of the user's trusted persons. This concatenation will be fed as input to an active neural layer of 'SoftMax' function, providing a vector of normalized probabilities representing the probability distribution over the different classes (ranging from 1 to 5). The class with the highest probability will be the predicted score.

4 Experiments

We present in this section the experiments performed on two different datasets. We first performed some preliminary evaluations to set the parameter values. Then we evaluated the SA module by comparing the different hybrid models. Next, we evaluated the hybrid recommendation algorithm, by verifying the contribution of SA, social information and fake reviews detection. Finally, we compared our model with existing related works.

4.1 Datasets

For the training and evaluation of our method, we used two datasets from the Yelp[1] social network. We extracted two data samples related to the "Restaurant" and "Shopping" categories. Table 1 shows the corresponding statistics:

Table 1. Dataset statistics

Dataset	#Users	#Items	#Ratings	#Reviews	Density
Yelp-Shopping	2,935	9,637	61,967	61,967	0.2%
Yelp-Restaurant	4,346	15,588	391,924	391,924	0.57%

For the training of the sentiment module, we used the 'IMDB' dataset comprising 50,000 comments, based on the Internet Movie Database (IMDB). Each comment is associated with a sentiment label, which can be either "positive" or "negative". This dataset is balanced in terms of the number of positive and negative comments.

4.2 Evaluation Metrics

We used the Mean Absolute Error (MAE) and the Root Mean Square Error (RMSE) evaluation metrics. MAE and RMSE have been used as they are the most popular predictive metrics to measure the closeness of predictions relative to real scores:

$$MAE = \frac{\sum_{u,i \in \Omega} |r_{u,i} - p_{u,i}|}{|\Omega|} \tag{9}$$

$$RMSE = \sqrt{\frac{\sum_{u,i \in \Omega} (r_{u,i} - p_{u,i})^2}{|\Omega|}} \tag{10}$$

where:

Ω: set of test assessments and $|\Omega|$ indicates the cardinality of the set Ω;
$r_{u,i}$: is the rating given by the user u on the item i; and.
$p_{u,i}$: is the rating prediction of the user u on the item i.

[1] https://www.yelp.com.

4.3 Baselines

We have compared our model with the following related works:

- **PMF:** the probabilistic matrix factorization approach, widely used in CF.
- **SVD++:** the matrix factorization technique that exploits the concept of "Singular value decomposition" to improve the performance of the CF algorithm.
- **SocialMF:** an approach that enriches the PMF model by integrating social information [23].
- **DeepCoNN:** Deep Cooperative Neural Networks, a DL-based recommendation technique that exploits the reviews to generate recommendations [24].

4.4 Experimental Parameters

We used the Python language version 3.8.5, exploiting several libraries (e.g. Tensorflow, Keras, Scikit-learn, Pandas, NumPy). To train our models, we used the Google Colab platform, offering the following features: 2 GB RAM; 2 virtual cores CPU and a 12 GB GPU. For data distribution, 80% of the dataset was reserved for training of our models, the remaining 20% for the test. Inspired by He et al. [12], we trained the different models separately, then globally, in order to evaluate the results of the different architectures. The parameters considered during the training were: the number of iterations (epochs), the batch size, the optimization function and the cost function. We used the ADAM optimization function to adjust the weights and attributes of the neural architectures during the training stage. Moreover, we evaluated the AE and MLP models by varying the following parameters:

- AE: variation of the latent dimension of the AE core (LDim = 32, 64) and the number of neuronal layers (#Layers = 1–5),
- MLP: variation of embedding size (ES = 16, 32, 64), number of neuronal layers (#Layers = 1–5),

We obtained the following best values, which will be considered in the rest of our experiments: for AE, the LDim was equal to 64 with a single layer. For MLP, the best performance was obtained with 3 layers and an ES equal to 32.

4.5 Results and Analysis

Evaluation of Hybrid SA Models. By varying the hyper-parameters, we were able to set the following best values: number of epochs for training fixed at 5, with a batch_size of 32, the 'PMSProp' optimization function was chosen as it gave better performance than 'ADAM' and 'SGD'.

Then, we evaluated the four combinations of SA models, LSTM-CNN; BiLSTM-CNN; CNN-LSTM and CNN-BiLSTM, varying the number of LSTM / BiLSTM recurrent units (#RU = 20, 60, 12, 200) and the number of convolutional layers of CNN (#CL = 1, 3, 5).

We can see from Table 2 that each architecture offers better performance with different empirical parameters. The best results were obtained with the CNN-LSTM combination, with 120 recurrent units and 5 convolutional layers. Moreover, this combination is

Table 2. Evaluation summary of the different SA combinations

Models	MAE	RMSE	#CL	#RU	Time-Train (S)
LSTM-CNN	0.1570	0.3073	1	20	6687.19
BiLSTM-CNN	0.1543	0.3201	2	20	11795.25
CNN-LSTM	**0.1318**	**0.3028**	5	120	**2076.71**
CNN-BiLSTM	0.1401	0.3036	1	200	2191.02

characterized by the shortest training time. Accordingly, we'll consider the CNN-LSTM combination in the rest of our evaluation.

Contribution of SA on the Different Models We evaluated the contribution of SA on the different models: the collaborative filtering prediction model (AE), the content-based filtering prediction model (MLP) and the hybrid model (AE-MLP). All the results obtained have shown the contribution of integrating the SA in terms of MAE and RMSE evaluation metrics. Table 3 illustrates the MAE evaluations of the hybrid model with and without SA using the Yelp-Restaurant dataset. The best performance is obtained with LDIM = 64, 1 layer for the AE, 5 layers for the MLP and an embedding size equal to 16.

Table 3. MAE performance of AE-MLP hybrid model with and without SA - Yelp Restaurants

LDim (AE)	#Layers (AE)	ES (MLP)	#Layers (MLP)	AE-MLP-SA	AE-MLP
64	1	16	5	**0.27875**	**0.28090**
		32	4	0.27921	0.28195
		64	2	0.27926	0.28118
	2	16	5	0.27984	0.28083
		32	4	0.28077	0.28095
		64	2	0.28214	0.28106
32	1	16	5	0.27921	0.28156
		32	4	0.28077	0.28209
		64	2	0.28075	0.28137
	2	16	5	0.28044	0.28153
		32	4	0.28107	0.28152
		64	2	0.28143	0.28188

Contribution of the Social Information In order to evaluate the contribution of social information on recommendation performance, we have considered in this evaluation the following weights: $\alpha = 0.6$ (favoring similarity rate rather than degree of credibility); δ

= −0.5, λ = 0.4 and γ = 0.6 (giving more priority to participation rate instead of degree of competence). We evaluated the contribution of social information on the different AE, MLP and Hybrid models, with and without SA. We present the evaluations carried out with the Hybrid model. We considered the different variants, namely: the Hybrid model combining the AE and MLP architectures without SA and social information (AE-MLP); the Hybrid model combining the AE and MLP architectures with SA and without social information (AE-MLP-SA); the Hybrid model combining the AE and MLP architectures with social information but without SA (AE-MLP-Social); and the Hybrid model combining the AE and MLP architectures with the integration of SA and SA social information (AE-MLP-Social-SA).

We can see that SA improved the performance of the hybrid AE-MLP model and that the integration of social information yielded the best performance on the AE-MLP model. Nevertheless, we noted a slight degradation in performance when we included the sentiment model and social information simultaneously. This degradation is due to the presence of fake reviews. Table 4 shows in detail the values obtained with the different parameters and architectures.

Table 4. Evaluation of the hybrid model with and without SA and social information

LDim (AE)	#Layers (AE)	ES (MLP)	#Layers (MLP)	AE-MLP	AE-MLP-SA	AE-MLP-Social	AE-MLP-Social-SA
64	1	16	5	0.28090	0.27875	0.27648	0.27875
		32	4	0.28195	0.27921	0.27625	0.27921
		64	2	0.28118	0.27926	**0.27511**	0.27926
	2	16	5	0.28083	0.27984	0.27816	0.27984
		32	4	0.28095	0.28077	0.27626	0.28077
		64	2	0.28106	0.28214	0.27666	0.28214
32	1	16	5	0.28156	0.27921	0.27825	0.27921
		32	4	0.28209	0.28077	0.27767	0.28077
		64	2	0.28137	0.28075	0.27758	0.28075
	2	16	5	0.28153	0.28044	0.27692	0.28044
		32	4	0.28152	0.28107	0.27846	0.28107
		64	2	0.28188	0.28143	0.27751	0.28143

Contribution of Fake Reviews Detection In order to evaluate the contribution of fake reviews detection, we compared our model with and without fake reviews. The results show that removing fake reviews has slightly improved the performance (see Table 5). The slight difference is due to the few number of fake reviews in these datasets.

Comparison with Related Work. We compared our model with state-of-the-art work. We tried to select a variety of models, choosing two matrix factorization models, a DL and sentiment analysis-based model and a social-based model. We can see from

Table 5. Evaluation of fake reviews detection

Models	Yelp-Restaurant		Yelp-Shopping	
	MAE	RMSE	MAE	RMSE
CASA-SR with Fake reviews	0.2872	0.3734	0.2941	0.3783
CASA-SR without Fake reviews	**0.2767**	**0.3727**	**0.2831**	**0.3778**

this table that our CASA-SR hybrid model outperformed the other models in terms of MAE and RMSE metrics. Table 6 illustrates the results obtained using both datasets Yelp-Restaurant and Yelp-Shopping.

Table 6. Performance comparison with related work

Models	Yelp-Restaurant		Yelp-Shopping	
	MAE	RMSE	MAE	RMSE
PMF	1.2238	1.4585	0.9529	1.0895
SVD++	0.7181	0.9295	0.7914	1.0235
DeepCoNN	0.3900	0.4800	0.6000	0.7500
SocialMF	0.6000	0.7800	0.7200	0.8900
CASA-SR	**0.2767**	**0.3727**	**0.2831**	**0.3778**

The comparison results show that the DL and SA-based models outperformed the standard matrix factorization models. Similarly, the models based on social information outperformed the PMF and SVD++ models. We note that the different variants of our hybrid model outperformed the state-of-the-art models, with better performance obtained with the CASA-SR model, which combines the AE and MLP architectures with social information and the confidence-aware SA module. Indeed, the results showed an improvement in performance when we eliminated the fake reviews.

5 Conclusion

In this article, we have proposed a novel social recommendation model that combines both AE and MLP models with the integration of social information. An improvement of the evaluation matrix has been performed using a hybrid confidence-aware SA model (that detects and removes fake reviews). Evaluation results on two datasets from the Yelp database demonstrated the contribution of SA and social information on the different models (AE, MLP and the hybrid AE-MLP model). The detection of fake reviews has further enhanced these performances. Furthermore, the different variants of our model outperformed the state-of-the-art models.

Future perspectives include further experiments with higher density datasets. It would be interesting to explore other DL techniques such as knowledge graph DL architectures and use the attention mechanism to improve the performance of the sentiment model. On the other hand, as the complexity of our recommendation algorithm can be significant, it would be interesting to reduce the response times.

References

1. Sun, Z., et al.: Recommender systems based on social networks. J. Syst. Softw. **99**, 109–119 (2015)
2. Guo, G., Zhang, J., Yorke-Smith, N.: TrustSVD: collaborative filtering with both the explicit and implicit influence of user trust and of item ratings. In: Proceedings of the Twenty-Ninth AAAI Conference on Artificial Intelligence, pp. 123–129 (2015)
3. Yang, B., Lei, Y., Liu, J., Li, W.: Social collaborative filtering by trust. IEEE Trans. Pattern Anal. Mach. Intell. **39**(8), 1633–1647 (2017)
4. Berkani, L., Laga, D., Aissat, A.: Social neural hybrid recommendation with deep representation learning. In: Attiogbé, C., Yahia, S.B. (eds.) Model and Data Engineering, 10th International Conference MEDI 2021, pp. 127–140. Tallinn (2021)
5. Berkani, L., Boudjenah, N.: S-SNHF: sentiment based social neural hybrid filtering. Adv. Data-Driven Eng. **52**(3), 297–325 (2023)
6. Nisha, C.C., Mohan, A.: A social recommender system using deep architecture and network embedding. Appl. Intell. **49**, 1937–1953 (2019)
7. Bathla, G., Aggarwal, H., Rani, R.: AutoTrustRec: recommender system with social trust and deep learning using AutoEncoder. Multimedia Tools Appl. **79**, 20845–20860 (2020). https://doi.org/10.1007/s11042-020-08932-4
8. Wankhade, M., Sekhara Rao, A.-C., Kulkarni, C.: A survey on sentiment analysis methods, applications, and challenges. Artif. Intell. Rev. (2022). https://doi.org/10.1007/s10462-022-10144-1
9. Dang, C.N., Moreno-García, M.N., De la Prieta, F.: An approach to integrating sentiment analysis into recommender systems. Sensors **21**, 5666 (2021). https://doi.org/10.3390/s21165666
10. Li, D., et al.: CARM: confidence-aware recommender model via review representation learning and historical rating behavior in the online platforms. Neurocomputing **455**, 283–296 (2021)
11. Birim, S.O., Kazancoglu, I., Mangla, S.K., Kahraman, A., Kumar, S., Kazancoglu, Y.: Detecting fake reviews through topic modeling. J. Bus. Res. **149**, 884–900 (2022)
12. He, X., Liao, L., Zhang, H., Nie, L., Hu, X., Chua, T.S.: Neural collaborative filtering. In: Proceedings of the 26th International Conference on World Wide Web, pp. 173–182 (2017)
13. Berkani, L., Kerboua, I., Zeghoud, S.: Recommandation Hybride basée sur l'Apprentissage Profond. Actes de la conférence EDA 2020, Revue des Nouvelles Technologies de l'Information, RNTI B.16, pp. 69–76 (2020). ISBN: 979-10-96289-13-4
14. Berkani, L., Zeghoud, S. Kerboua, I.: Chapter 19 - Neural hybrid recommendation based on GMF and hybrid MLP. In: Pandey, R., Khatri, S.K., Singh, N.K., Verma, P. (eds.) Artificial Intelligence and Machine Learning for EDGE Computing, Academic Press, pp.287–303 (2022). ISBN 9780128240540. https://doi.org/10.1016/B978-0-12-824054-0.00030-7
15. Rama, K., Kumar, P., Bhasker, B.: Deep autoencoders for feature learning with embeddings for recommendations: a novel recommender system solution. Neural Comput. Appl. **33**, 14167–14177 (2021). https://doi.org/10.1007/s00521-021-06065-9

16. Diao, Q., Qiu, M., Wu, C.Y., Smola, A.J., Jiang, J., Wang, C.: Jointly modeling aspects, ratings and sentiments for movie recommendation. In: Proceedings of the 20th ACM SIGKDD International Conference on Knowledge Discovery and Data Mining, pp.193–202 (2014)
17. Lu, Y., Dong, R., Smyth, B.: Co-evolutionary recommendation model: mutual learning between ratings and reviews. In: Proceedings of the World Wide Web conference, pp. 773–782 (2018)
18. Osman N.A., Mohd Noah, S.A., Darwich, M., Mohd, M.: Integrating contextual sentiment analysis in collaborative recommender systems. PLoS ONE **16**(3), e0248695 (2021). https://doi.org/10.1371/journal.pone.0248695
19. Avesani, P., Massa, P., Tiella, R.: Moleskiing it: a trust-aware recommender system for ski mountaineering. Int. J. Infonomics **20**, 1–10 (2005)
20. Devlin, J., Chang, M.-W., Lee, K., Toutanova, K.: BERT: pre-training of deep bidirectional transformers for language understanding. arXiv (2018). arXiv:preprint/04805
21. Jiang, L., Liu, L., Yao, J., Shi, L.: A hybrid recommendation model in social media based on deep emotion analysis and multi-source view fusion. J. Cloud Comput. Adv. Syst. Appl. **9**, 57 (2020)
22. Berkani, L., Belkacem, S., Ouafi, M., Guessoum, A.: Recommendation of users in social networks: a semantic and social based classification approach. Expert. Syst. **38**(2), e12634 (2021). https://doi.org/10.1111/exsy.12634
23. Jamali, M., Ester, M.: A matrix factorization technique with trust propagation for recommendation in social networks. In: RecSys'10 - Proceedings of the 4th ACM Conference on Recommender Systems, pp.135–142 (2010). https://doi.org/10.1145/1864708.1864736
24. Zheng, L., Noroozi, V., Yu, P.S.: Joint deep modeling of users and items using reviews for rec. arXiv:1701.04783 (2017). https://doi.org/10.48550/arXiv.1701.04783

Deep Learning Based on TensorFlow and Keras for Predictive Monitoring of Business Process Execution Delays

Walid Ben Fradj(✉), Mohamed Turki, and Faiez Gargouri

MIRACL Laboratory, ISIMS, University of Sfax, P.O. Box 242, 3021 Sfax, Tunisia
walid.benfradj@isimsf.u-sfax.tn, mohamed.turki@isims.usf.tn,
faiez.gargouri@usf.tn

Abstract. In order to enhance their performance and responsiveness, organizations must identify, manage, and monitor all business processes that involve crucial knowledge. This can be achieved through a multidisciplinary approach that combines Knowledge Management, Business Process Management, and Process Mining. To achieve these objectives, the implementation of an automated computer system for managing business processes has become paramount. In this context, we adopt the CRISP-DM approach to propose a new method called BPEDPM (Business Process Event Data Predictive Monitoring) for predictive process monitoring. This method leverages process mining techniques to exploit the execution data from a Business Process Management System (BPMS) workflow engine. Particularly, in the modeling phase, we apply deep learning based on the TensorFlow and Keras tools. To demonstrate the applicability of the BPEDPM method, we have developed an intelligent system named iBPMS4PED to predict the execution times of business processes. The research focuses on the incoming mail management process within a group health insurance context.

Keywords: Business Process Management · Process Mining · Knowledge Management · Machine Learning · Deep Learning · TensorFlow · Keras · CRISP-DM

1 Introduction

Nowadays, businesses are increasingly realizing the vital importance of optimizing their knowledge exploitation while implementing a process-oriented quality management approach. This can be achieved by adopting an interdisciplinary approach that combines Knowledge Management (KM) with Business Process Management (BPM). To improve their performance and enhance adaptability, companies must identify and master all essential processes that may involve implicit or explicit knowledge, which represents a valuable source to be leveraged. To achieve these objectives, the implementation of an automated computer system for business process automation is essential. According to [1], various information systems are linked to processes, including Enterprise Resource Planning (ERP), Workflow Management (WFM), Customer Relationship Management

M. Mosbah et al. (Eds.): MEDI 2023, LNCS 14396, pp. 156–169, 2024.
https://doi.org/10.1007/978-3-031-49333-1_12

(CRM), Supply Chain Management (SCM), Product Data Management (PDM), and Business Process Management (BPM). Most of these systems utilize execution engines, also known as workflow engines. These engines serve a dual role: firstly, they facilitate the deployment of processes, and secondly, they record business data in the form of databases, along with technical information related to process execution in the form of event logs. These data can be exploited through process mining techniques, thereby extracting new knowledge that proves valuable for process optimization and decision-making within the organization.

According to Van der Aalst [2], a pioneer in the field of process mining, several types of process mining can be distinguished, with the most commonly used ones being process discovery, conformance checking, performance analysis, comparative process mining, predictive process mining, and action-oriented process mining. Process discovery is particularly crucial in all process mining endeavors. As stated in [3], process discovery algorithms undergo a step of event log filtering, eliminating infrequent activities to retain only dominant behaviors. The resulting output is a graphical model, such as Directly Follows Graphs (DFG), Petri Nets (PN), Process Trees (PT), or Business Process Model and Notation (BPMN). This raises questions: Is it possible to extract various transitions, regardless of their frequency, and enrich a database with all paths followed by process instances? If so, can we predict the execution times of instances based on these transitions? Furthermore, despite a literature review, there is no standard scientific approach to successfully conduct a process mining project. Hence, the idea of following the well-established CRISP-DM (Cross Industry Standard Process for Data Mining) procedure used in data science is being considered.

This article focuses on predictive process monitoring and introduces an innovative method that combines process discovery and predictive process mining. Unlike extracting a graphical model, which requires removing less frequent activities, this method allows extracting all paths followed by process instances in a database. By using these paths as a basis, it predicts whether a process instance will complete its execution on time or with a delay. The adopted approach follows the CRISP-DM model and leverages event logs that record the execution data of a workflow engine, combining process mining techniques to create an intelligent system based on machine learning.

The second and third sections of this article respectively delve into the key concepts related to Knowledge Management (KM) and Business Process Management (BPM), highlighting the current trends that integrate artificial intelligence into these domains. In the fourth section, an overview of predictive process monitoring techniques for business processes is presented. The fifth section provides a detailed account of the method used to predict the execution times of business processes, following the CRISP-DM approach. The sixth section describes the implementation and evaluation of the developed IT solution, applied to a specific business process: the management of incoming mail in the health insurance management activity of the company I-WAY in Tunisia. Finally, the article concludes with a summary and outlines prospective research directions for the future.

2 KM: Knowledge Management

According to Grundstein M. (2000) in [4], capitalizing on enterprise knowledge involves identifying crucial knowledge, preserving it, and making it sustainable, while ensuring widespread sharing and usage to enhance the company's wealth. Explicit knowledge is quantifiable, understandable, directly captured, and expressed by individuals within the organization. On the other hand, tacit knowledge, also known as know-how, is specific to each individual and encompasses their informal technical expertise as well as personal beliefs and aspirations. The knowledge management process, as per the same reference, is based on five interacting facets around crucial knowledge: identification, preservation, valorization, updating, and management. Each of these facets includes sub-processes aimed at addressing associated problems. In [5], the authors address the issue of identifying sensitive business processes, focusing on the facet of crucial knowledge identification. Sensitive business processes require enhanced security management to reduce the risks of compromise and ensure business continuity. To address this, they propose a new methodology called SOPIM (Sensitive Organization Process Identification Methodology), based on a multi-criteria decision-making approach to construct a coherent family of evaluation criteria for identifying these processes. This approach mainly consists of two phases: (1) constructing a decision-maker preference model and (2) using the preference model (decision rules) to rank "potentially sensitive organizational processes." Once identified, these sensitive organizational processes will be modeled, executed, and explored. The research article discussed here specifically focuses on the exploration of these processes.

With the advancements in information technology, a new approach to knowledge management has emerged, known as Intelligent Knowledge Management (IKM) [6]. This innovative approach stands out for leveraging artificial intelligence (AI) tools to harness expert knowledge. According to [7], AI plays a potentially significant role in supporting knowledge management activities, offering various benefits such as:

- Improving predictive analysis through machine learning capabilities to anticipate future events.
- Identifying previously unknown patterns in data.
- Exploring organizational data to discover hidden relationships and correlations.
- Developing new declarative knowledge to enrich the organization's understanding.
- Efficiently collecting, classifying, organizing, storing, and searching explicit knowledge.
- Analyzing and filtering numerous content and communication channels to access relevant information.
- Facilitating knowledge reuse by teams and individuals to optimize process efficiency.
- Connecting people working on similar problems to foster weak ties and expertise sharing.
- Promoting collaborative intelligence and organizational memory sharing.
- Creating a holistic perspective on knowledge sources and potential bottlenecks.
- Establishing more connected coordination systems between different parts of the organization to encourage collaboration.
- Improving the application of localized knowledge by identifying and preparing specific knowledge sources tailored to needs.

- Ensuring equitable access to knowledge without fearing prohibitive social costs.
- And many more advantages.

3 BPM: Business Process Management

Every company, regardless of its size, operates through processes that encompass all its activities and services (e.g., Human Resources, Sales, Quality, Purchasing, etc.). According to [8], Business Process Management (BPM) adopts a process-centric approach, enabling effective monitoring of activities within the organization to improve overall performance and consequently, results. Using a BPM tool provides real-time traceability of exchanges between stakeholders involved in a process, allowing for greater responsiveness through the generation of indicators in the form of alerts or automatic notifications. This facilitates decision-making, accelerates the identification of bottlenecks, ensures adherence to deadlines, and controls production costs of products and services [9]. The BPM lifecycle consists of five phases: Design, Modeling, Execution, Monitoring, and Optimization. Firstly, the Design phase involves identifying existing processes and designing future processes. Next, the Modeling phase graphically represents the model in a manner faithful to reality. Once these preliminary steps are completed, the Execution phase takes place, where Business Process Management is implemented. This step entails interpreting business procedures by an execution engine, which coordinates all interactions between users, system tasks, and IT resources. The Monitoring phase, on the other hand, focuses on regulating individual processes, providing easy access to information about their status and delivering statistics on the performance of one or more processes. Finally, the Optimization phase allows for adjusting processes to minimize costs and optimize efficiency. Within the scope of this research article, particular attention is given to the Monitoring phase, where we position ourselves to closely examine this crucial step in the Business Process Management lifecycle.

The field of Business Process Management has witnessed significant technological advancements with the integration of artificial intelligence. In this context, the American consulting and research company Granter introduced the concept of Intelligent Business Process Management Suite (iBPMS) in 2012. iBPMS represents an evolution of BPM by combining predictive analytics, process intelligence, and emerging technologies with traditional BPM practices. This innovative approach aims to make business process management smarter, more efficient, and better suited to the current challenges faced by businesses.

4 Predictive Monitoring of Business Processes

According to reference [10], the objective of predictive monitoring of business processes is to identify and anticipate potential issues in advance. It enables the implementation of preventive measures to avoid problems and facilitates proactive decision-making. Real-time monitoring techniques allow data to be analyzed in real-time, contributing to decision-making and optimization of ongoing processes. According to [11], predictive process monitoring occurs in real-time during the execution of instances. It is important to note that during the learning phase, the input data for the predictive monitoring method includes event logs along with supplementary information. These data undergo a coding step to be interpreted by the prediction algorithm. This algorithm creates a prediction model that is applied to the ongoing process instances to determine the predicted output value for each process instance. Most predictive monitoring techniques include an offline component (which involves expensive computations) for generating the prediction model, as well as an online component (faster) that performs predictions based on the generated model.

Event logs play a crucial role as the primary data source for Process Mining techniques, particularly in the domain of predictive process monitoring. Each line in these files contains execution information related to an activity, primarily consisting of the process instance identifier and timestamp. Referring to sources [12] and [13], additional information can also be found in the log, such as the activity's cost or the name of the resource responsible for its execution. Technical characteristics describing event logs include the number of instances, the number of activities, the number of events, and the number of process variants [14].

Predictive process monitoring with a temporal dimension was first explored by [15]. Their prediction approach extracts a transition system from the event log, including additional temporal information. This approach served as a fundamental reference for the work of [12, 16-18]. According to [19], the transition system uses decision trees to predict the execution time and the next activity of a process instance. The research of [16] and [17] extended the approach of [15] by clustering event log traces based on contextual features. According to [18], an approach is used by integrating naive Bayesian regression models and support vector regressions to annotate the transition system. The addition of additional attributes has had a positive impact on the quality of predictions. However, these methods have a major drawback: they assume that the event log used for learning contains all possible process behaviors, which is generally not the case in reality. The approaches proposed by [20] and [19] share similarities. They provide a general framework to enrich the event log with derived information and discover correlations between process characteristics using decision trees. The distinction between these two approaches lies in the fact that the one proposed by [19] excludes infrequent behaviors. This latter approach is similar to the one proposed by [18], except that the process model is a reduced version of the transition system. The issue related to correlation is that numerical values must be discretized, which reduces precision. Two probabilistic methods based on Hidden Markov Model (HMM) are presented in the works of [21] and [22]. These probabilistic approaches allow predicting the probability of future activities, providing insights into the process evolution. In parallel, the approach proposed by [23] uses a generic model based on decision trees, thereby providing decision criteria tailored to the actual objectives of the process. Other approaches from various domains have been proposed for predicting delays. According to [24], the Process Mining approach relies on queuing systems. It involves building an annotated transition system and using non-linear regression algorithms to predict delays. Similarly, according to [25], the proposed approach predicts the remaining time using expressive probabilistic models and is based solely on information concerning the workflow. [26] presents a predictive model based on decision trees. This model assesses the probability of satisfying a user-defined constraint for ongoing instances. A similar approach is also explored in the work of [27].

In this approach, traces are treated as complex symbolic sequences, encoded using two methods: indexed encoding and HMM encoding. [28] employs the same encoding for their analyses. To perform clustering, the dataset is partitioned, and a predictor based on random forests is trained for each group. According to reference [29], natural language processing (NLP) is combined with various classifiers to obtain representative features for each document. Among the predictors, random forests have proven to be the most effective. Subsequently, deep neural network-based approaches emerged. Both works [30] and [22] utilize a recurrent neural network (RNN) with two hidden RNN layers, employing basic LSTM cells to predict the next event. The approach of [22] also integrates an LSTM neural network to predict both the next activity and its execution time. These two approaches do not consider additional attributes, are sensitive to hyperparameter selection, and require lengthy training periods. Furthermore, the approach of [10] utilizes artificial neural networks (ANN) to predict if a process instance exceeds the expected time. Lastly, [31] presents a comparison between two machine learning models (random forest and SVM) and two deep learning models (LSTM and DNN). The results indicate that the LSTM model performs the best.

After reviewing the existing literature, we did not find a standard scientific approach to follow for conducting a predictive process monitoring project. In our work, we opted for the well-established CRISP-DM procedure in data science. Additionally, we observed that most of the event logs used contain business data. For this study, we chose to process execution data from the workflow engine of a BPMS. Furthermore, we noticed that models are typically represented as Petri nets, with a tendency to eliminate less frequent transitions. However, in our approach, we decided to include all paths taken by process instances in the database as additional attributes. Finally, it is essential to highlight that most approaches rely on machine learning rather than deep learning. Long Short-Term Memory (LSTM) neural networks are commonly used in deep learning cases. In our study, we chose to create a deep learning model using the state-of-the-art TensorFlowntechnology to predict the execution times of business processes.

5 PMBPED (Prediction Method of Business Process Execution Delays)

During its execution, a business process can go through several stages, representing its tasks or activities. The sequence of these tasks forms a case or an instance, and all the information related to this instance is recorded in the form of an event log. For the same process, there can be multiple instances with distinct paths. The execution time of a process depends on the order of tasks and their respective durations. When a process or task exceeds a given time threshold, it may indicate the presence of bottlenecks, enabling decision-makers to optimize and improve the performance of different business processes.

The CRISP-DM (CRoss Industry Standard Process for Data Mining) is a well-established, field-proven procedure that guides data exploration activities, playing a key role in the success of data science. According to [32], CRISP-DM is a widely adopted process both in practice and research. It is an organizational process model that is not limited to a specific technology. It includes descriptions of typical project phases and tasks within each of these phases, as well as an explanation of the relationships between these tasks. It provides an overview of the data exploration lifecycle. According to [33], CRISP-DM breaks down the exploration process into six main steps: business understanding, data understanding, data preparation, modeling, evaluation, and result released (see Fig.1). In this context, using the CRISP-DM procedure as a foundation, the PBMPED method (see Fig.2) is employed to extract various paths followed by process instances from the event log. Subsequently, the execution time of each path is calculated.

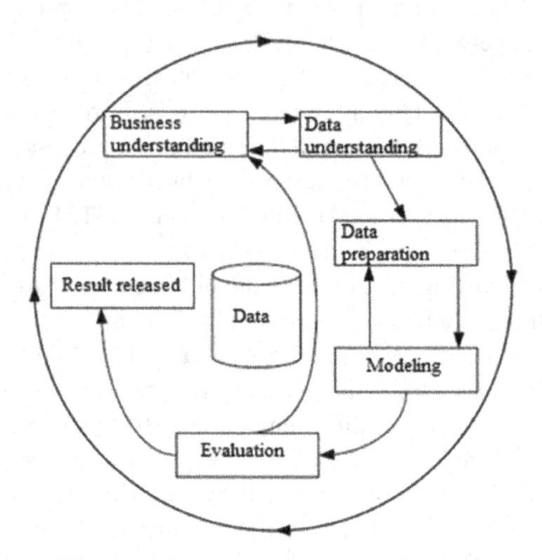

Fig. 1. CRISP-DM procedure [33] p. 216

The objective of PBMPED is to predict the execution times of business processes. This prediction can be valuable for enabling decision-makers, in the case of a semi-automatic process, or the system, in the case of an automated process, to choose the most optimal path.

In the first phase (Business understanding), our method aims to optimize overall performance and results by focusing on the business processes of the company and the flow of activities. Every company relies on a set of processes that encompass its activities. With technological advancements, companies increasingly depend on intelligent software solutions to address issues swiftly and enhance their competitiveness. Business Process Management Systems (BPMS) represent a potential solution for managing and controlling these processes. By implementing workflow engines, which are an integral part of BPMS, it becomes possible to separate business data from process execution data.

In the second phase (Data understanding), data related to process execution is recorded by the workflow engine in event logs. Retrieving this data can be done in different ways from a BPMS workflow engine: either from an integrated database within the BPMS, where tables are typically temporary and created in memory but lost upon server restart; or by configuring a REST API to connect to the integrated database and retrieve execution traces, then saving them to disk; or by configuring a separate database (e.g.,

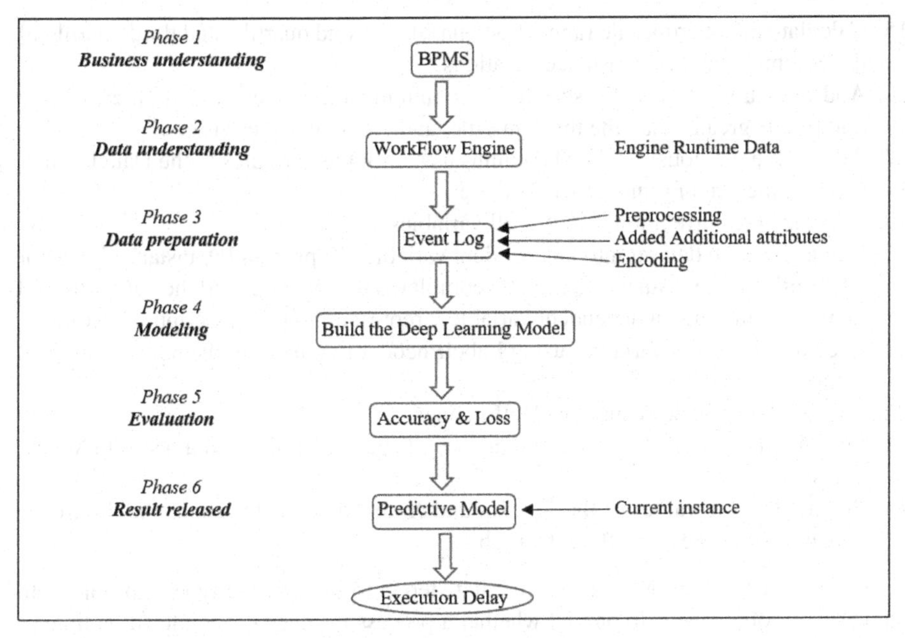

Fig. 2. PBMPED

MySQL, MongoDB, Oracle, etc.) connected to the BPMS through connectors. In our work, we use two event logs in CSV format. The content of these two files is extracted from a MySQL database configured to store execution data of a BPMS workflow engine.

In the third phase (Data preparation), the mere possession of an event log containing execution traces from a workflow engine does not guarantee that the data is ready for modeling. Indeed, the file may have redundancies, missing values, unnecessary data, and categorical variables that require encoding, among other issues. Moreover, the data needs enrichment by adding additional attributes. In the context of our PBMPED method, we propose the following approach for data preparation:

1. Import the event log (csv file) provided by the BPMS.
2. Transform the file into a Data Frame.
3. Remove empty columns.
4. Ignore indexing columns.
5. Eliminate redundant columns.
6. Replace null values with appropriate values.
7. Sort the data based on the "date" column.
8. Convert the timestamp to seconds.
9. Determine the duration of each activity.
10. Determine the execution duration of each instance.
11. For each activity, calculate the time remaining at the end of the instance.
12. For each instance, determine the path followed by the task (human or automatic).
13. Add a column "path per instance."
14. Separate each "path per instance" and distribute it into columns corresponding to the number of tasks per instance.

15. Calculate the interquartile range, first quartile, second quartile, and third quartile of the column containing instance durations.
16. Add a column named "description" containing the value "late" if the instance duration is greater than the third quartile, and "in time" otherwise.
17. Add all the previously defined columns as additional attributes to the Data Frame.
18. Convert the Data Frame into a CSV file.
19. Visualize the correlation table for all variables.
20. Decompose the dataset into independent variables X: process ID, instance ID, actor who initiated the instance, actor who completed the instance, and the path followed by the instance and a dependent variable y representing the "description" variable.
21. Encode categorical variables using LabelEncoder to transform them into numerical values.
22. Encode the variables using OneHotEncoder.
23. Split the dataset (X; y) into a training set (X_train; y_train) and a test set (X_test; y_test).
24. Standardize the values of the X_train and X_test datasets to reduce the difference between the values of different variables.

In the fourth phase (Modeling), the first step is to identify the type of problem. In our case, our objective is to predict whether a process instance is completed on time or delayed. Therefore, the prediction is a qualitative variable. Thus, we can conclude that our problem falls under the classification category in the framework of supervised learning. In our paper [34], we addressed the same topic by applying the following machine learning algorithms: KNN, Decision Tree, Random Forest, SVM, and Logistic Regression. SVM with an RBF kernel outperformed the other algorithms in terms of accuracy (84%). In this work, we create a deep learning model based on the new TensorFlow tool. We start by importing the Keras library from TensorFlow. Then, we use the sequential model to create a neural network composed of four layers: The first layer consists of 13 input neurons representing the different columns of X, the second and third layers are two hidden layers, each composed of 7 neurons, and using the relu activation function, and an output layer with a single neuron, using the sigmoid activation function for our binary classification problem. For model compilation, we use cross-entropy for error calculation, batch gradient descent with a batch size of 128 to adjust neuron weights, and accuracy for evaluation. Finally, we configure the compiler to run 300 epochs (1 epoch = propagation + Backpropagation).

In the fifth phase (Evaluation), we employ accuracy, which denotes the rate of correct classification. It quantifies how effectively instances are classified compared to the total instances.

At the end of our work, in the sixth phase (Result released), we deploy the PBMPED method for predicting the execution delays of business processes to end-users.

6 iBPMS4PED (intelligent Business Process Management System for Prediction Execution Delays)

In this section, we present iBPMS4PED (Intelligent Business Process Management System For Prediction Execution Delays), an intelligent business process management system developed using our PBMPED method. The objective of this system is to predict the execution delays of a specific business process. For our study, we chose to apply iBPMS4PED to the intelligenceWay group (https://iway-tn.com/), which uses an IT solution called I-Santé (https://i-sante.tn/). This solution offers comprehensive and tailored management for all healthcare professions, including health funds, mutuals, and insurance companies. I-Santé is built on a 100% digital platform and uses highly secure health cards. Leveraging advanced technologies such as the BPM workflow engine, ECM (Electronic Content Management), and rule engine, I-Santé meets the needs of policyholders, insurers, and healthcare professionals. In this study, we specifically focus on the "Incoming Mail" process, which is modeled on the Bonitasoft BPMS platform (see Fig. 3). Our approach aims to enhance the management of this process by predicting execution delays, which can have a significant impact on the overall efficiency of the business process management system.

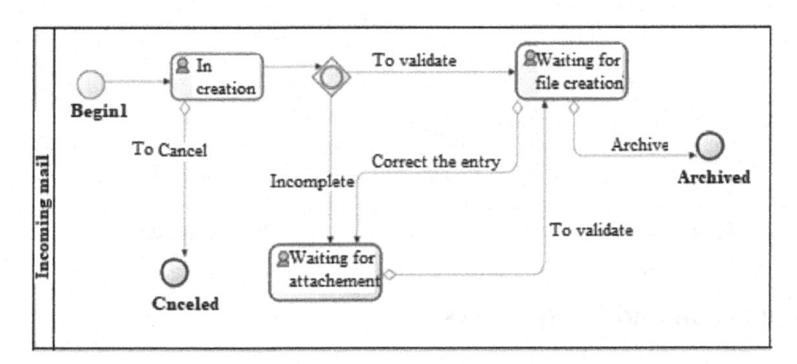

Fig. 3. Incoming mail process

In this study, we used two CSV format event log files extracted from a MySQL database configured on the I-WAY platform. These files record the execution data of a workflow engine from the BonitaSoft BPMS. The first file, "BN_PROC_INST.csv," records the execution traces for each process instance, while the second file, "BN_ACT_INST.csv," records the execution traces for each task. Using the Python programming language, we processed these two event logs following the steps described in the PBMPED method. For data preparation, we utilized the pandas, numpy, pylab, scipy-optimize, matplotlib, and seaborn libraries. The result of this processing is a single file named "Result.csv," containing 4817 rows and 24 columns. The columns ready for modeling include: [PROCESS_UUID_, INST_UUID_, START_BY, END_BY, user_act0, user_act1, user_act2, ..., useractnb, duration_INST, description]. Here's a brief explanation of these columns:

- PROCESS_UUID_: Represents the process identifier.

- INST_UUID_: Represents the instance identifier of the process.
- START_BY: Indicates the user who started the process instance.
- END_BY: Indicates the user who completed the process instance.
- user_acti: Designates the executor (person or machine) of activity number i.
- duration_INST: Gives the execution time of the process instance.
- Description: Contains a description of the observed delay in the process instance.

These data are essential for our approach to better understand and predict the execution delays of the analyzed business processes. In terms of modeling, we start by importing the Keras library from TensorFlow and selecting the sequential model. Then, we build a neural network with 13 input neurons, two hidden layers with 7 neurons each, and one output neuron. During the evaluation step, we utilize the metrics: Loss to calculate the error (see Fig. 4) and accuracy to compute the rate of correct classification (see Fig. 5). The achieved results are: accuracy = 85% and Loss = 34%.

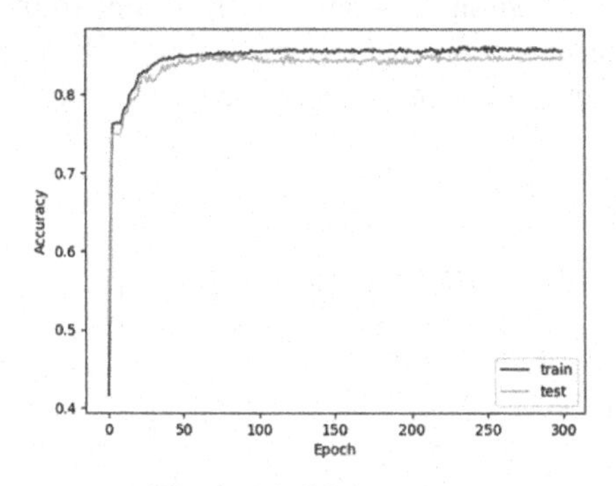

Fig. 4. Model Accuracy

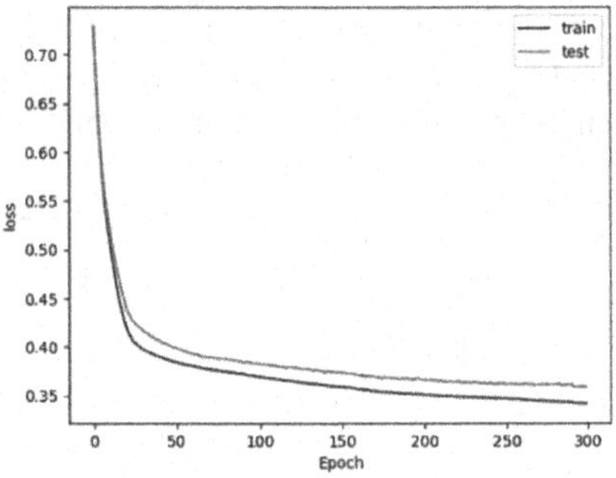

Fig. 5. Model Loss

7 Conclusion and Perspectives

In this research article, we addressed the problem of predictive monitoring of business processes, which is a current topic in the Business Process Management (BPM) domain and plays a crucial role in process-oriented organizations. To do so, we developed a method called PBMPED, which allows predicting the execution delays of a business process. We followed the CRISP-DM method, a well-known procedure in data science, to develop a Process Mining approach. We applied Process Mining techniques to event logs that record the execution data of a BPMS workflow engine. During the data preparation phase, we performed data cleaning on the recorded event logs. Then, for each instance, we defined and added additional attributes, including the relative path, execution time, and a description of the execution time. If an instance's execution time was less than or equal to the third quartile of the execution time column, we considered it as completed on time (description = "in time"); otherwise, it was considered as completed late (description = "late"). For encoding categorical variables, we used the LabelEncoder method. Next, to reduce discrepancies between their values, we standardized all the independent variables. In the modeling phase, we used TensorFlow and Keras to create a neural network for predicting the execution delays. The goal of this prediction was to determine whether a new process instance would be completed on time or late. In terms of accuracy, the evaluation demonstrated that this Deep Learning model outperformed the six Machine Learning models studied in our work [34], namely decision trees, random forest, SVM with rbf kernel, SVM with linear kernel, KNN, and logistic regression. For implementation, we applied the PBMPED method to create an intelligent Business Process Management system called iBPMS4PED, which enables predicting execution delays. This system was applied to an incoming mail management process in the health mutual insurance domain.

In addition to the contributions made in this article, some points deserve further investigation. In the short term, it would be feasible to explore other types of event logs. In the medium term, we plan to leverage business data stored in relational or NoSQL databases, in conjunction with the workflow engine's execution data, to detect bottlenecks. Furthermore, we intend to develop a monitoring system that allows visualizing performance indicators related to business processes.

References

1. van der Aalst, W.M.P., Schonenberg, M.H., Song, M.: Time prediction based on process mining. Inf. Syst. **36**(2), 450–475 (2011)
2. van der Aalst, W.M.P.: Process mining: a 360 degree overview. In: van der Aalst, W.M.P., Carmona, J. (eds.) Process Mining Handbook. Lecture Notes in Business Information Processing, vol. 448, pp. 3–34. Springer: Cham (2022). https://doi.org/10.1007/978-3-031-08848-3_1
3. van der Aalst, W.M.P.: Foundations of process discovery. In: van der Aalst, W.M.P., Carmona, J. (eds.) Process Mining Handbook. Lecture Notes in Business Information Processing, vol. 448, pp. 37–75. Springer, Cham 37-75. (2022). https://doi.org/10.1007/978-3-031-08848-3_2
4. Morey, D., Maybury, M.T., Thuraisingham, B.M.: Knowledge Management: Classic and Contemporary Works. MIT Press, Cambridge (2002)
5. Turki, M., Saad, I., Gargouri, F., Kassel, G.: A business process evaluation methodology for knowledge management based on multicriteria decision-making approach. In: Information Systems for Knowledge Management, pp. 249–277. Wiley, Hoboken (2014)
6. Sanzogni, L., Guzman, G., Busch, P.: Artificial intelligence and knowledge management: questioning the tacit dimension. Prometheus **35**(1), 37–56 (2017)
7. Jarrahi, M.H., Askay, D., Eshraghi, A., Smith, P.: Artificial intelligence and knowledge management: a partnership between human and AI. Bus. Horiz. **66**(1), 87–99 (2023)
8. Dumas, M., La Rosa, M., Mendling, J., Reijers, H.A.: Fundamentals of business process management. In: Fundamentals of Business Process Management, pp. 371–412. Springer, Heidelberg (2018). https://doi.org/10.1007/978-3-662-56509-4_10
9. Ko, R.K., Lee, S.S., Wah Lee, E.: Business process management (BPM) standards: a survey. Bus. Process Manage. J. **15**(5), 744–791 (2009)
10. Teinemaa, I., Dumas, M., Rosa, M.L., Maggi, F.M.: Outcome-oriented predictive process monitoring: review and benchmark. ACM Trans. Knowl. Discov. Data **13**(2), 1–57 (2019)
11. Marquez-Chamorro, A.E., Resinas, M., Ruiz-Cortes, A.: Predictive monitoring of business processes: a survey. IEEE Trans. Serv. Comput. **11**(6), 962–977 (2018)
12. Aalst, W.M.P.: Data scientist: the engineer of the future. In: Mertins, K., Bénaben, F., Poler, R., Bourrières, J.-P. (eds.) Enterprise Interoperability VI. PIC, vol. 7, pp. 13–26. Springer, Cham (2014). https://doi.org/10.1007/978-3-319-04948-9_2

13. vom Brocke, J., Zelt, S., Schmiedel, T.: On the role of context in business process management. Int. J. Inf. Manage. **36**(3), 486–495 (2016)
14. Augusto, A., et al.: Automated discovery of process models from event logs: review and benchmark. IEEE Trans. Knowl. Data Eng. **31**(4), 686–705 (2019)
15. van der Aalst, W., et al.: Process mining manifesto. In: Daniel, F., Barkaoui, K., Dustdar, S. (eds.) BPM 2011. LNBIP, vol. 99, pp. 169–194. Springer, Heidelberg (2012). https://doi.org/10.1007/978-3-642-28108-2_19
16. Folino, F., Greco, G., Guzzo, A., Pontieri, L.: Mining usage scenarios in business processes: outlier-aware discovery and run-time prediction. Data Knowl. Eng. **70**(12), 1005–1029 (2011)
17. Folino, F., Guarascio, M., Pontieri, L.: Discovering context-aware models for predicting business process performances. In: Meersman, R., et al. (eds.) On the Move to Meaningful Internet Systems: OTM 2012, vol. 7565, pp. 287–304. Springer, Heidelberg (2012). https://doi.org/10.1007/978-3-642-33606-5_18
18. Polato, M., Sperduti, A., Burattin, A., de Leoni, M.: Data-aware remaining time prediction of business process instances. In: 2014 International Joint Conference on Neural Networks (IJCNN), Beijing, China, pp. 816–823 (2014)
19. Ceci, M., Lanotte, P.F., Fumarola, F., Cavallo, D.P., Malerba, D.: Completion time and next activity prediction of processes using sequential pattern mining. In: Džeroski, S., Panov, P., Kocev, D., Todorovski, L. (eds.) DS 2014. LNCS (LNAI), vol. 8777, pp. 49–61. Springer, Cham (2014). https://doi.org/10.1007/978-3-319-11812-3_5
20. de Leoni, M., van der Aalst, W.M.P., Dees, M.: A General framework for correlating business process characteristics. In: Sadiq, S., Soffer, P., Völzer, H. (eds.) BPM 2014. LNCS, vol. 8659, pp. 250–266. Springer, Cham (2014). https://doi.org/10.1007/978-3-319-10172-9_16
21. Lakshmanan, G.T., Shamsi, D., Doganata, Y.N., Unuvar, M., Khalaf, R.: A Markov prediction model for data-driven semi-structured business processes. Knowl. Inf. Syst. **42**(1), 97–126 (2015)
22. Tax, N., Verenich, I., La Rosa, M., Dumas, M.: Predictive business process monitoring with LSTM neural networks. In: Dubois, E., Pohl, K. (eds.) CAiSE 2017. LNCS, vol. 10253, pp. 477–492. Springer, Cham (2017). https://doi.org/10.1007/978-3-319-59536-8_30
23. Ghattas, J., Soffer, P., Peleg, M.: Improving business process decision making based on past experience. Decis. Support Syst. **59**, 93-107 (2014)
24. Senderovich, A., Weidlich, M., Gal, A., Mandelbaum, A.: Queue mining for delay prediction in multi-class service processes. Inf. Syst. **53**, 278–295 (2015)
25. Rogge-Solti, A., Weske, M.: Prediction of business process durations using non-Markovian stochastic Petri nets. Inf. Syst. **54**, 1–14 (2015)
26. Maggi, F.M., Di Francescomarino, C., Dumas, M., Ghidini, C.: Predictive monitoring of business processes (2013)
27. Leontjeva, A., Conforti, R., Di Francescomarino, C., Dumas, M., Maggi, F.M.: Complex symbolic sequence encodings for predictive monitoring of business processes. In: Motahari-Nezhad, H.R., Recker, J., Weidlich, M. (eds.) BPM 2015. LNCS, vol. 9253, pp. 297–313. Springer, Cham (2015). https://doi.org/10.1007/978-3-319-23063-4_21
28. Di Francescomarino, C., Ghidini, C., Maggi, F.M., Milani, F.: Predictive process monitoring methods: which one suits me best? In: Weske, M., Montali, M., Weber, I., vom Brocke, J. (eds.) BPM 2018. LNCS, vol. 11080, pp. 462–479. Springer, Cham (2018). https://doi.org/10.1007/978-3-319-98648-7_27
29. Teinemaa, I., Dumas, M., Maggi, F.M., Di Francescomarino, C.: Predictive business process monitoring with structured and unstructured data. In: La Rosa, M., Loos, P., Pastor, O. (eds.) BPM 2016. LNCS, vol. 9850, pp. 401–417. Springer, Cham (2016). https://doi.org/10.1007/978-3-319-45348-4_23

30. Evermann, J., Rehse, J.-R., Fettke, P.: A deep learning approach for predicting process behaviour at runtime. In: Dumas, M., Fantinato, M. (eds.) BPM 2016. LNBIP, vol. 281, pp. 327–338. Springer, Cham (2017). https://doi.org/10.1007/978-3-319-58457-7_24
31. Kratsch, W., Manderscheid, J., Röglinger, M., Seyfried, J.: Machine learning in business process monitoring: a comparison of deep learning and classical approaches used for outcome prediction. Bus. Inf. Syst. Eng. **63**(3), 261–276 (2021)
32. Schröer, C., Kruse, F., Gómez, J.M.: A systematic literature review on applying CRISP-DM process model. Procedia Comput. Sci. **181**, 526–534 (2021)
33. Zhang, Y.: Sales forecasting of promotion activities based on the cross-industry standard process for data mining of E-commerce promotional information and support vector regression. **32**(1), 212–225 (2021)
34. Ben Fradj, W., Turki, M.: Prediction of business process execution time. In: Abraham, A., Pllana, S., Casalino, G., Ma, K., Bajaj, A. (eds.) ISDA 2022. LNCS, vol. 715, pp. 105–114. Springer, Cham (2023). https://doi.org/10.1007/978-3-031-35507-3_11

Ontology and Database Systems

OntoFD: A Generic Social Media Fake News Ontology

Fériel Ben Fraj[1(✉)] and Nourhène Nouri[2]

[1] RIADI, National School of Computer Sciences, Manouba University, Manouba, Tunisia
feriel.benfraj@riadi.rnu.tn

[2] SMART, Higher Institute of Management, University of Tunis, Tunis, Tunisia

Abstract. Fake news has increased dramatically in recent years as a result of social media's rapid and explosive expansion. To lessen their negative consequences, detecting them has become essential and a very busy and dynamic area of research. Despite the fact that much progress has been made in this field, it remains a difficult task due to its complexity. Indeed, there are various types of fake news. The news itself is made up of several data components of various types (textual, graphical, multimedia, network, social, psychological, and so on). It also involves multiple actors: a creator, a victim, and a target community. To handle all of this knowledge and improve the ability to recognize fake news on social media, this study proposes an ontological structure for its depiction. The result is an ontology with the name OntoFD. It shows a hierarchy of concepts and relations that can be used to assess whether information propagating on social media is credible or not.

Keywords: Fake News · Ontology · Detection · Social Media · Protégé

1 Introduction

In recent years, social media has grown incredibly quickly and explosively. This has resulted in an astounding rise in the number of false news stories that are currently a problem, especially given the volume of information that is being circulated online. They are simultaneously obtrusive, overbearing, distracting, and aggravating. Indeed, social networks like Facebook and Twitter have ingrained themselves into our daily lives. We are interested in posting our ideas and opinions on these networks, as well as sharing our thoughts on products and services through posts and comments. However, the use of these social media platforms has resulted in various dishonest acts, such as the dissemination of false information [1].

Because of this, finding fake news on social media has recently drawn a lot of interest and is now considered to be a growing area of study [2]. Fake news comes in a wide variety of forms, including hoaxes, propaganda, satire, parody, and click-bait. They can also be audiovisual, textual, or pictorial, among

M. Mosbah et al. (Eds.): MEDI 2023, LNCS 14396, pp. 173–185, 2024.
https://doi.org/10.1007/978-3-031-49333-1_13

other formats. Various communities, as well as governments, politicians, and celebrities, may be impacted by this type of content.

Several fields of study, including NLP (Natural Language Processing), artificial intelligence (AI), among others, have been applied to this problem. Despite the fact that much progress has been made in this field, the complexity of the task still makes it difficult. As a result, in order to carry out this detection properly, it is required to fully explore and incorporate other approaches.

Given the interference of numerous actors in its life cycle, such as its creator, victim(s), and target, as well as its essential components, such as its content, diffusion, public reactions, and so on, it is imperative to have an efficient detection system to identify false information. Such a system must ensure the modeling and application of its semantic data. We choose ontology formalism to model false news knowledge.

Indeed, researchers use ontologies to represent the various sorts of knowledge. Since the beginning of the 1990s, the term "ontology" has been used in the field of Artificial Intelligence, particularly in relation to knowledge engineering and knowledge representation. Ontology is a rapidly expanding discipline that is one of the current academic topics [3].

An ontology is a formal representation of knowledge as a set of concepts within a domain and the relations between them. To enable such a description, we must formally describe components such as individuals (object instances), classes, properties, and relations as well as constraints, rules, and axioms. As a result, ontologies offer the possibility of incorporating extra domain-specific knowledge as well as a reusable and transferable knowledge representation.

In this context, the work discussed in this paper, which entails developing an ontology to organize the various aspects, is carried out. The ontology's name is OntoFD. It aims to provide the knowledge base needed to determine whether or not the news spreading on social media is fake and to investigate it.

The rest of the paper is structured as follows. Section 2 addresses the reasons for conducting this research. We pay special attention to data-type classification when studying prior research to identify false news. The suggested ontology is described in Sect. 3. We emphasize the knowledge organization in particular. Section 4 outlines significant comparable research and suggests a comparison of their respective aspects to our ontoFD. Section 5 addresses the conclusion and possible subsequent works.

2 Motivations

Fake news is one of the most difficult problems with a negative effect on societies. Social networks are widely used for interactive information sharing, communication, and dissemination, making them a very attractive environment for those who wish to spread false information. In fact, spreading false information on social media is simple. Therefore, it is essential to locate them right away in order to lessen the damage they cause. However, due to their complexity, they are difficult to find. As shown in Fig. 1 which is an example of fake news that pretends the death of a well-known Egyptian figure. False information in social networks, unlike traditional fake news, has multiple components and is typically multi-modal rather than being solely text-based.

Fig. 1. A caption of an example of Fake News published on Facebook. It consists of several components: (a) the title, (b) the creator and creation date, (c) the body(video), (d) the context (emotional reactions, comments and shares) and (e) the victim.

To address this issue, many approaches have been put out in recent years. Approaches are categorised according to the types of data involved in detection [4]. Thus, there are:

- Textual data approaches that primarily rely on style-based [5], linguistic [6], syntactic [7], emotional and sentimental [8,9], and semantic [10] aspects, and they solely use text content.
- Image [11,12].
- Video [13,14].

- Social-based [15–17], that especially rely on the fact that harmful people share a common behavioural pattern to make it easy to spot and mark their messages as fake.
- Multi-modal approaches that combine several types of data to increase detection effectiveness [18–20].

As a result, the fake news detection problem is difficult because it involves NLP, opinion analysis, and social network analysis. Therefore, we must use a variety of heterogeneous knowledge to discern between them, including textual, pictorial, multimedia, emotional, graphical, social, and so on. All of this knowledge must be represented using an expressive formalism, such as ontology, in order for their links to be expressed.

3 Proposed Ontology

To express all of the knowledge that can be helpful when detecting fake news, we opt for the ontology formalism. Our choice wasn't made at random. In fact, regardless of how heterogeneous or homogeneous knowledge may be, an ontology can represent it. It provides efficient support for managing, processing, and storing knowledge. As a result, it is simple and quick to find solutions to technical and operator queries. Ontologies are thus particularly helpful tools for modelling a portion of reality due to their generic nature and the range of expression they offer.

3.1 Concepts' Identification

Creating an ontology requires identifying concepts and their relationships. These concepts must be applied to the initial purpose of the ontology, which is to detect incorrect information. We made an attempt to read a number of research articles [4,21,22] for this work, and they were all in agreement that detection is dependent on a variety of parameters, but primarily on various characteristics. This resulted in the discovery of a substantial number of concepts for the OntoFD, which could be divided into four major categories: Information source, Content, Context, and Target victim. The hierarchy that falls under each category is depicted in Fig. 2.

Information Source. A news item that has been posted must have a creator and a source from which it was first supplied. The source in our case is the social network where the news is first shared, such as Facebook, Twitter, TikTok, and so on.

Humans and robots are the two basic types of creators. Organizations and governments can also produce news. In a social media context, the creator provides several aspects, especially his profile information. Indeed, looking into the creator's background can assist assess whether he or she has a track record of

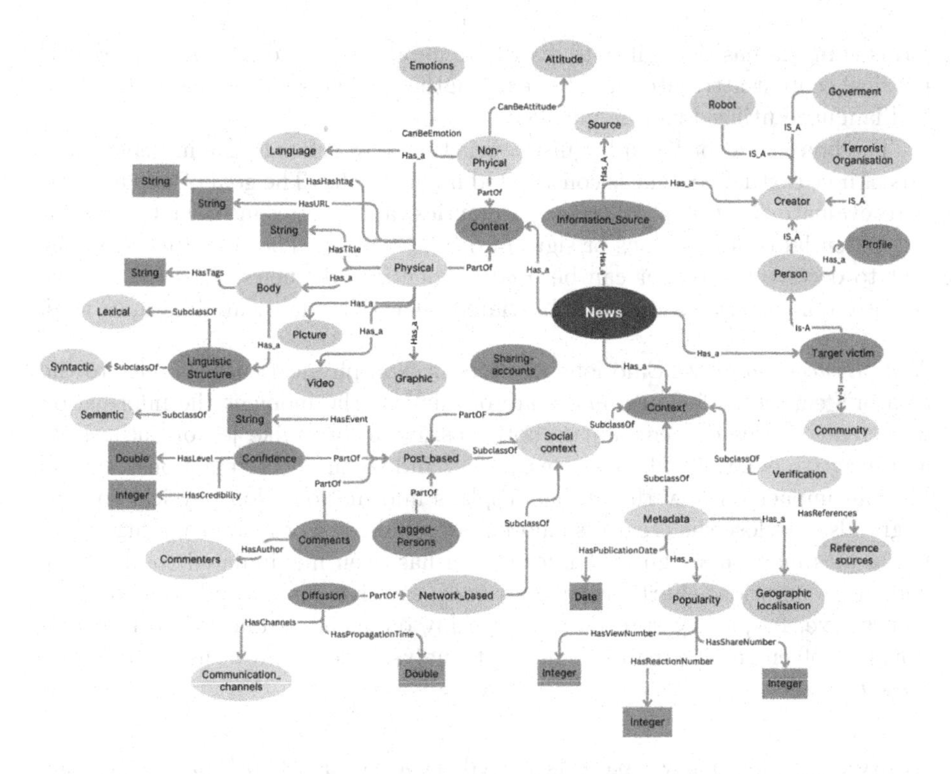

Fig. 2. Conceptual specification of the different concepts of OntoFD.

creating accurate and reputable material. If the creator has a history of spreading inaccurate information, this may be an indication of suspicion. It is also vital to uncover any potential bias in the creator's profile. For example, if the author has recognized political, economic, or ideological links, their presentation of information may be influenced. Furthermore, an examination of the creator's language in his or her writings can reveal hints as to his or her neutrality and possibility of propagating erroneous or misleading information.

Content. This concept is fundamental to comprehending and assessing rumors, information, and disinformation. To illustrate fake news, multiple bits of information are used. The content will be divided into two groups (See Fig. 3).

Physical Content. This concept covers all information, media, and pieces that are disseminated, or published across several platforms. It consists primarily of the title and body text. The title is frequently an important component of the content since it summarizes the subject or theme of the information. It can be used to draw attention and impact how people perceive information. In written content, the body reflects the core content. Images, videos, hyperlinks, audio files, graphics, and other forms of media can all be included. Each of these forms of media has the ability to distribute information and rumors. As a result, physical content attributes such as text length, image resolution, video duration, and language utilized can be included.

The body concept has a linguistic structure that refers to the in-depth analysis of how textual content is constructed linguistically. The goal of this analysis is to evaluate the text's coherence, semantics, and syntax in order to look for any potential contradictions or signs of disinformation. The construction of the text to deceive the reader can be used to identify fake news. This can involve the use of incorrect reasoning, abbreviated quotations, fabricated data, or tags.

Non-physical Content. Emotional aspects in Non-physical content refer to the creator's emotions, feelings, and state of mind at the moment the information was created. This can encompass both positive feelings like joy or enthusiasm and negative emotions like rage, fear, grief, and so on. These emotions can also have an impact on how the author displays information. Non-physical content might also disclose the creator's emotional prejudices. For instance, a high emotional bias may be a sign that information has been manipulated to affect the audience's emotions. Furthermore, Strongly emotive content, whether positive or negative, frequently spreads more quickly on social media due to the emotional involvement it creates. Sentiment analysis tools can be used to extract these elements.

Context. When a news piece is posted on a social network, a context may exist. The context concept gives critical information about the context in which information is conveyed, allowing for a more comprehensive and exact analysis. Investigating the context means looking into the circumstances that influenced the creation, diffusion, and reception of this news. The social context, metadata, and verification are all addressed by the context.

Social Context. It has two levels: the Post level and Network level, which are sub-concepts [15–17]. For the Post-level concept, we considered the following sub-concepts:

- Tagged Persons helps to search for specific people who were mentioned or tagged in the post. This can be useful for evaluating who is actively disseminating information, who is associated with the event or subject under debate, and whether or not credible or reputable sources are involved.
- Comments (Commenters): Comments are the reactions or dialogues triggered by the post. The idea of "Comments," which gives rise to the concept of "Commenters," enables tracking of interactions related to the publication. Comment analysis can reveal public emotions, multiple points of view, questions, concerns, and content evaluations.
- Post-event: The post is associated with a specific event. It is beneficial for contextualizing the content of the post by linking it to a real-world event or news story. This can help determine whether the post is relevant to the event to which it refers.

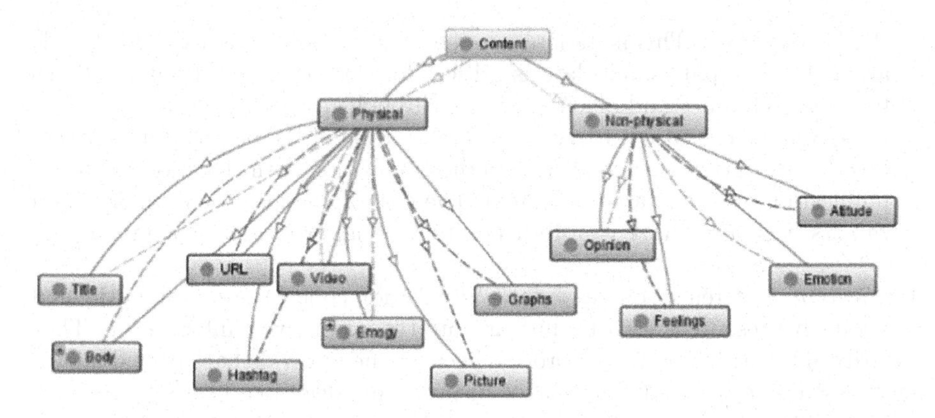

Fig. 3. Sub-concepts of the Content class extracted from OntoFD.

- Sharing-accounts: Users or accounts who have shared the post on their own platforms or social networks are referred to. By identifying these accounts, we may learn how information moves across different networks and which individuals or organizations contribute to its spread.
- Confidence: The concept of "Confidence" allows to assess the dependability and reliability of the post or the author of the post. This can include things like the source of the post, the author's reputation, the quality of the content, and so on. Confidence in our ontology results in two measures: confidence level and credibility.

The Diffusion concept at the network level makes it possible to study how information flows across social networks and online media. It is advantageous to comprehend the dynamics of dissemination, including the propagation time, the range of the dissemination, and the primary communication channels. These elements work together to provide valuable information on how information spreads and has an impact across diverse networks and online communities. This makes determining the reliability and influence of the information within the social network simpler.

Communication Channels is a sub-concept of the diffusion concept. It refers to websites, social media platforms, forums, blogs, and other methods for disseminating information. Understanding these channels can help to better understand how and where information is most influential. A piece of information, for instance, might be widely disseminated on Twitter yet barely noticeable on Facebook.

Metadata. The "Metadata" concept in OntoFD encompasses essential information associated with a publication or piece of information, including:

- Popularity: This is a measure of a publication's renown or visibility, which is typically based on the number of views, likes, shares, or interactions. The popularity of information can suggest its impact and prospective influence on the population.

- Publication date: This metadata specifies when a publication was first made available. The publishing date is critical for determining the information's temporal relevance and placing it in its chronological context.
- Geographic localisation: It relates to the location of the publication, whether it is the site of the event referenced in the publication or the location where the publication was generated or shared. The geographic location of information can assist contextualize it in terms of its regional or local importance.

Verification. Reference sources for the Verification concept are trustworthy and authoritative resources used to authenticate the accuracy of information. They are critical to the fact-checking process because they provide background knowledge or strong evidence to support or disprove an allegation.

Target Victims. It is possible to identify the people, groups, or organizations that are most likely to be harmed by a specific rumor or piece of information by using this concept. This covers attendees, those who have been the focus of disinformation campaigns, and misinformation victims. Victims come in two forms: those who are the subject of the information and those who allow it to shape their opinions. Misinformation typically leaves out explicit identification of its victims.

3.2 Relations' Description

Our ontology's "News" instances are structured and enhanced by relations, which makes it simpler to examine, confirm, and identify erroneous information. Figure 2 shows the hierarchical organization of concepts using conceptual relations:

- Has-a or PartOf present the composition, as in the instance of the two concepts Physical and Non-Physical, which are parts of the concept Content.
- Is-A represents inheritance, just as the concepts Robot, Person, Government, and Terrorist-organization do.
- SubclassOf, which introduces specialization in the same way that the Context concept's subclasses Social-context, Matadata, and Verification do.

Furthermore, there is a diverse collection of qualities. Table 1 lists them in detail.

3.3 Rules of OntoFD

The rules in an ontology are logical statements that allow to infer information or draw conclusions based on existing facts and assumptions. Here are some examples of rules that we have developed for our rumor detection ontology:

- Verification infraction rule: If a new item has been verified by a reliable source, it is considered true.

Table 1. Main OntoFD properties and their descriptions.

Property	Description
HasTitle	A data property that associates a title with a "Physical" instance to reflect the information's title.
HasURL	A data property to associate URLs with an instance of "Physical".
HasHashtag	A data property for linking hashtags with an instance of "Physical".
HasTags	It's a data property that relates values (in this case, tags or keywords) with a Body concept instance, which is its domain. Its field type is string.
CanBeEmotion	This characteristic is important for describing abstract components of content such as emotions and feelings. It enables the capture of the emotional dimension of non-physical material in information content.
CanBeAttitude	This property is useful for modeling attitudes in the context of Non-physical material, which can be useful for analyzing ideas or perspectives stated in content.
HasEvent	The "HasEvent" property is used to associate an event with a Post-Based concept instance. This property's field is string.
HasLevel	"HasLevel" quantifies the level of credibility, resulting in a more realistic representation of information credibility rating.
HasCredibility	This property measures the credibility associated with a certain level of confidence. For example, a value of 1 for "HasCredibility" may suggest high credibility, whilst a value of 0 may indicate low credibility.
HasAuthor	This relation enables comment attribution, analysis, and management in the social environment by tracking and attributing comments to the individuals or entities who posted them.
HasChannels	This property establishes a connection between the information dissemination process and the communication channels that are used. Understanding how a rumor or piece of information circulates through different media or communication channels is crucial.
HasPropagationTime	It is possible to analyze how quickly disinformation or news spread by keeping track of the time it takes for information to spread using this object property, which is helpful for monitoring and tracking the process.
HasPublicationDate	It is a crucial property for keeping track of and classifying information according to its chronological order.
HasViewNumber	It creates a link between the information's polarity and the number of views or visualizations it has had. The visibility of information can be measured using this property as a function of its polarity.
HasReactionNumber	It can be used to measure how polarized the public's response to a piece of information is. It might, for instance, provide the proportion of favorable or unfavorable reactions to a given piece of information.
HasShareNumber	This property can be used to quantify information spread or sharing based on its polarity.
HasReferences	It is required for recording and documenting the trustworthy sources utilized in the verification process. It contributes to the ontology's traceability and transparency of information verification, allowing the veracity of information to be evaluated.

- Social propagation rule: If a new post is shared a specific number of times, it is considered potentially false.
- Linguistic coherence rule: If the news has linguistic incoherence detected in its syntax, it is considered dubious.
- Rule of Creator Reputation: If the creator of a news has a bad reputation, the new is suspicious.

3.4 OntoFD Characteristics

To construct and modify our ontology, we used the academic platform Protégé, which provides a simple graphical interface that allows users to create and modify ontologies using conceptual diagrams. Protégé is RDF (Resource Description Framework) and OWL (Web Ontology Language) compliant, making it simple to integrate into semantic web-based applications. As a result, the resulting OntoFD ontology formalizes knowledge as a set of concepts and their relationships.

OntoFD contains general knowledge that is less abstract than high-level ontologies while yet being broad enough to be relevant in a range of fields. It can be factual knowledge; for example, we can learn who is responsible for fake news in our circumstance, regardless of domain or language.

Furthermore, our contribution provides a vast vocabulary, ensuring that the issue domain is thoroughly described. It should be noted that as the granularity of the model increases, the modelled concepts will correlate to increasingly particular notions. In our case, OntoFD encompasses concepts, properties, and even a complex hierarchy of subclasses. Natural language-operational ontologies, such as OntoFD, are informal ontologies. OntoFD stores all forms of data (textual, numerical, video, image, and so on). Otherwise, the OntoFD knowledge is heterogeneous.

Let us not forget that the goal of developing OntoFD is to detect misleading information on social networks. To complete this objective, we must first collect a big dataset of real and false news examples. This data can be improved by referencing to the metadata in our ontology before going through the appropriate linguistic pre-processing, such as removing punctuation marks and stopwords. Following this cleaning, an OntoFD-based structured representation of the news is constructed. The structured data is then used to train a deep learning model. Finally, this model will go on to identify the veracity of the news.

4 Related Works and Discussion

Ontologies are now commonly utilized as models to conceptualize applications in KR and/or NLP. We found two ontologies for describing fake news in the literature: namely Pheme [23] and Fane [24].

In order to automatically verify Internet rumors as they spread around the world through social media and online networks, the Pheme project aims to create new techniques [23]. Pheme is a domain ontology. This kind of ontology controls a collection of terms and ideas that characterize a target domain. The two application domains of Pheme are healthcare and journalism.

FANE (FAke NEws ontology) uses the web annotation standard in conjunction with suitable existing ontologies, such as the Prov-O ontology, to represent provenance information [24]. FakeNewsTagType, FakeNewsTag (the Fake News Tag linked with the annotation), and FakeNewsTag (a content item labeled as fake news by a human or machine) are only a few of the classes and aspects of the FANE ontology that are presented by FANE (the type of Fake News Tag). The application ontology Fane has concepts that frequently correlate to the roles that domain entities perform when a particular activity is being carried out. For instance, it can only pick up on satire.

When compared to these two previous works, OntoFD is more generic because it is domain independent. This is not true for the other two works because Pheme is a domain ontology that deals with health and journalism domains. FANE, on the other hand, is an application ontology whose concepts frequently correspond to the roles that domain entities play during the execution of a specific activity. The FANE, for example, can only detect satire. It remains silent about other types of fake information.

Furthermore, OntoFD is fine-granulated. A high granularity ontology, on the other hand, has a less detailed vocabulary and provides a more general description of the problem domain. Pheme and FANE stand in for this. FANE, for example, only considers FakeNewsFeedback for satire identification. Pheme is limited to the sources of postings and replies to postings for the detection of fake news.

In addition, heavyweight ontologies such as OntoFD and Pheme define the advanced features of their concepts, allowing for inferences and deductions. For example, we can deduct from our ontology that the creator can create several pieces of content. Whereas, FANE belongs to the lightweight ontologies because it only describes the hierarchy of concepts and their interactions (In Fane, there are only 5 concepts and 3 relationships). To conclude this discussion, Table 2 summarizes this comparison based on the previously established criteria.

Table 2. OntoFD vs FANE and Pheme.

Ontology	Conceptualization Objective	Granularity Level	Completeness Level	Level of formalism of representation	Ontology weight
OntoFD	Generic	Fine	Operational	Informal	Heavy
FANE	Application	High	Semantic	Formal	Light
Pheme	Domain	High	Semantic	Informal	Heavy

5 Conclusion and Future Work

Because not all news has the same writing style, research has shown that many existing machine learning algorithms cannot detect fake news. The complexity of the fake news detection problem stems from the fact that it involves a variety of heterogeneous knowledges, including textual, photographic, multimedia, emotive, graphic, network, etc. The relationships between them must be expressed using an expressive formalism. OntoFD is our proposed to deal with such an issue. It contains generic knowledge that is less abstract than high-level ontologies but general enough to be reused in multiple fields. It is an ontology with a very detailed vocabulary, which ensures a very detailed description of the problem domain.

Every contribution has flaws. Given this, as well as the fact that this is a new study avenue, we recommend the following as further research perspectives that complement our contribution. We specifically note the addition of more information to the ontology concepts and properties, as well as the creation of technologies to bring real-time information from social networks to OntoFD and store it in a database for subsequent use.

References

1. Kumar, S., West, R., Leskovec, J.: Disinformation on the web: impact, characteristics, and detection of Wikipedia hoaxes. In: 25th International Proceedings on World Wide Web, pp. 591–602. International World Wide Web Conferences Steering CommitteeRepublic and Canton of GenevaSwitzerland, Montréal, Québec, Canada (2016)
2. Wu, L., Morstatter, F., Carley, K., Liu, H.: Misinformation in social media: definition, manipulation, and detection. ACM SIGKDD Explor. Newslett. **21**(2), 80–90 (2019)
3. Staab, S., Studer, R. (eds.): Handbook on Ontologies. IHIS, Springer, Heidelberg (2009). https://doi.org/10.1007/978-3-540-92673-3
4. Rastogi, S., Bansal, D.: A review on fake news detection 3T's: typology, time of detection and taxonomies. Int. J. Inf. Secur. **22**(1), 177–212 (2023)
5. Ahmed, H., Traore, I., Saad, S.: Detection of online fake news using n-gram analysis and machine learning techniques. In: Traore, I., Woungang, I., Awad, A. (eds.) ISDDC 2017. LNCS, vol. 10618, pp. 127–138. Springer, Cham (2017). https://doi.org/10.1007/978-3-319-69155-8_9
6. Pérez-Rosas, V., Kleinberg, B., Lefevre, A., Mihalcea, R.: Automatic detection of fake news. In: International Conference in Computer Linguistics, pp. 1–2. arXiv preprint arXiv:1708.07104 (2017)
7. Varma, R., Verma, Y., Vijayvargiya, P., Churi, P.P.: A systematic survey on deep learning and machine learning approaches of fake news detection in the pre- and post-COVID-19 pandemic. Int. J. Intell. Comput. Cybernet. **14**(4), 617–646 (2021)
8. Horne, B., Adali, S.: This just. In: fake news packs a lot in title, uses simpler, repetitive content in text body, more similar to satire than real news. In: 11th Proceedings of the International AAAI Conference on Web and Social Media, pp. 759–766. Publisher, Québec, Canada (2017)
9. Ahuja, R., Bansal, S., Prakash, S., Venkataraman, K., Banga, A.: Comparative study of different sarcasm detection algorithms based on behavioural approach. Procedia Comput. Sci. **143**, 411–418 (2018)

10. Klyuev, V.: Fake news filtering: semantic approaches. In: 7th International Proceedings on Reliability, Infocom Technologies and Optimization (Trends and Future Directions) (ICRITO), pp. 9–15. IEEE, Noida, India (2018). https://doi.org/10.1109/ICRITO.2018.8748506
11. Marra, F., Gragnaniello, D., Cozzolino, D., Verdoliva, L.: Fake news filtering: semantic approaches. In: 2018 IEEE Conference on Multimedia Information Processing and Retrieval (MIPR), pp. 384–389. IEEE, Miami, FL, USA (2018)
12. Hsu, C.C., Zhuang, Y.X., Lee, C.Y.: Deep fake image detection based on pairwise learning. Appl. Sci. **10**(1), 370 (2020)
13. Papadopoulou, O., Zampoglou, M., Papadopoulos, S., Kompatsiaris, I.: A corpus of debunked and verified user-generated videos. Online Inf. Rev. **43**(1), 72–88 (2019)
14. Varshney, D., Vishwakarma, D.K.: A unified approach for detection of Clickbait videos on YouTube using cognitive evidences. Appl. Intell. **51**(7), 4214–4235 (2021)
15. Zhou, X., Zafarani, R.: Network-based fake news detection. ACM SIGKDD Explor. Newsl. **21**(2), 48–60 (2019)
16. Ren, Y; Jiawei, Z.: HGAT: hierarchical graph attention network for fake news detection. arXiv preprint arXiv:2002.04397 (2020)
17. Raza, S., Ding, C.: Fake news detection based on news content and social contexts: a transformer-based approach. Int. J. Data Sci. Analytics **13**(4), 335–362 (2022)
18. Yang, Y., Zheng, L., Zhang, J., Cui, Q., Li, Z., Yu, P.S.: TI-CNN: convolutional neural networks for fake news detection. arXiv preprint arXiv:1806.00749 (2018)
19. Li, D., Guo, H., Wang, Z., Zheng, Z.: Unsupervised fake news detection based on autoencoder. IEEE Access **9**, 29356–29365 (2021)
20. Raza, S., Ding, C.: Multimodal fake news detection. Information **13**(6), 284 (2022)
21. Ali, I., Bin Ayub, M.N., Shivakumara, P., Noor, N.F.B.M.: Fake news detection techniques on social media: a survey. Wirel. Commun. Mob. Comput. (2022). https://doi.org/10.1155/2022/6072084
22. Zhang, X., Ghorbani, A.A.: An overview of online fake news: characterization, detection, and discussion. Information Processing and Management **57**(2), 102025 (2020). https://doi.org/10.1016/j.ipm.2019.03.004
23. Declerck, T., Osenova, P., Georgiev, G., Lendvai, P.: Ontological modelling of rumors. In: Trandabăţ, D., Gîfu, D. (eds.) RUMOUR 2015. CCIS, vol. 588, pp. 3–17. Springer, Cham (2016). https://doi.org/10.1007/978-3-319-32942-0_1
24. Rehm, G., Moreno-Schneider, J., Bourgonje, P.: Automatic and manual web annotations in an infrastructure to handle fake news and other online media phenomena. In: 11th International Proceedings on Language Resources and Evaluation (LREC 2018), pp. 1–2. European Language Resources Association (ELRA), Miyazaki, Japan (2018)

Finding a Second Wind: Speeding Up Graph Traversal Queries in RDBMSs Using Column-Oriented Processing

Mikhail Firsov, Michael Polyntsov, Kirill Smirnov, and George Chernishev(✉)

Saint-Petersburg University, Saint Petersburg, Russia
chernishev@gmail.com

Abstract. Recursive queries and recursive derived tables constitute an important part of the SQL standard. Their efficient processing is important for many real-life applications that rely on graph or hierarchy traversal. Position-enabled column-stores offer a novel opportunity to improve run times for this type of queries. Such systems allow the engine to explicitly use data positions (row ids) inside its core and thus, enable novel efficient implementations of query plan operators.

In this paper, we present an approach that significantly speeds up recursive query processing inside RDBMSes. Its core idea is to employ a particular aspect of column-store technology (late materialization) which enables the query engine to manipulate data positions during query execution. Based on it, we propose two sets of Volcano-style operators intended to process different query cases.

In order to validate our ideas, we have implemented the proposed approach in PosDB, an RDBMS column-store with SQL support. We experimentally demonstrate the viability of our approach by providing a comparison with PostgreSQL. Experiments show that for breadth-first search: 1) our position-based approach yields up to 6x better results than PostgreSQL, 2) our tuple-based one results in only 3× improvement when using a special rewriting technique, but it can work in a larger number of cases, and 3) both approaches can't be emulated in row-stores efficiently.

Keywords: Query Processing · Column-stores · Recursive Queries · Late Materialization · Breadth-First Search · PosDB

1 Introduction

The ANSI'99 SQL standard introduced the concept of recursion into SQL with syntactic constructs that define recursive views and recursive derived tables. This allowed users to store graph data in a tabular form and to express some graph queries using CTEs and recursive syntax. The admissible subset is rather limited compared to specialized graph systems, but it is sufficient for solving a number of common tasks. Such tasks originate from many real-life applications and usually

M. Mosbah et al. (Eds.): MEDI 2023, LNCS 14396, pp. 186–199, 2024.
https://doi.org/10.1007/978-3-031-49333-1_14

concern some hierarchy traversal which can be considered as a breadth-first search computation.

In this paper, we present another outlook on RDBMS architecture that significantly improves system performance at least for some types of graph queries expressed by recursive SQL. More specifically, we present a column-oriented approach that improves run times for queries that perform breadth-first search.

Having emerged about fifteen years ago, column-stores quickly became ubiquitous in analytic processing of relational data. Their idea is simple: store data in a columnar form in order to read only the columns necessary for evaluating the query. Such an approach also provides better data compression rates [1], improves CPU cache utilization, facilitates SIMD-enabled data processing, and offers other benefits. However, some column-stores additionally allow the query engine to explicitly use data positions during query execution. This made way for a number of optimizations and techniques that offered various benefits for query processing. Thus, we differentiate the "position-enabled" column-stores from the rest as the column-stores that are able to reap benefits from explicit position manipulation inside their engine. We believe that such an approach can give RDBMs a second wind in handling graph queries.

Positions (also called row ids or offsets) are integers that refer to some record or individual attribute value inside a table. Operating on data positions allows query engine to achieve savings by deferring switching to data values. This group of techniques is called late materialization and it was successfully employed for various query plan operators [2,12,18,20].

We employ this technique to design two sets of Volcano-style [7] operators intended to handle different query cases that involve recursive processing. We have implemented them inside a position-enabled column-store PosDB [3,4]. Next, in order to evaluate them we run experiments with queries that perform breadth first search. We have also performed the comparison of our approach with PostgreSQL.

The overall contribution of this paper is the following:

1. A survey of existing query processing techniques in RDBMSs that concern recursive queries.
2. A design of two query operators for position-enabled column store that speed up recursive query evaluation.
3. An experimental evaluation of proposed techniques and a comparison with state-of-the-art row-store RDBMS.

This paper is organized as follows. In Sect. 2, we survey various aspects of implementation and usage of recursive queries inside relational DBMSs. Then, in Sect. 3 we present the main features of PosDB and discuss its query processing internals. After this, in Sect. 4 we describe implementation details of the proposed recursive operators and their use in the existing query plan model of PosDB. Section 5 contains an evaluation that compares PosDB with PostgreSQL using a series of experiments on trees of different additional payload. Finally, in Sect. 6 we conclude this paper and discuss future work.

2 Related Work and Motivation

2.1 Related Work

In this section, we review existing papers that address graph query processing in SQL-supporting systems, paying special attention to recursive evaluation.

One of the earliest papers that addressed the problem of recursive query evaluation was [11]. In it, the author introduces several query optimizations for recursive queries with graphs, namely early evaluation of row selection conditions, elimination of duplicate rows in intermediate tables, and using an index to accelerate join computation.

The authors of the papers [5,6] describe several issues with query optimization in relational databases in the implementation of recursive queries. The first approach they mention is the full feedback approach (FFB), which provides the optimizer with the demographics of each recursion iteration so that it can generate a new plan for the subsequent iteration. However, FFB interrupts potential pipelining and cannot take advantage of global query optimizations, making it unsuitable for parallel DBMS. The next approach, look ahead with feedback (LAWF), generates plans for the subsequent k iterations in advance, with k depending on the query planning cost and propagation of join estimation errors. The authors present a dynamic feedback mechanism based on passive monitoring to collect feedback and to determine when re-planning is necessary. The LAWF method supports both pipelining and global query optimization.

In the paper [14] the authors consider two graph problems: transitive closure computation and adjacency matrix multiplication. In order to solve them, they study the optimization of queries that involve recursive joins and recursive aggregations in column- and row-oriented DBMS. They evaluate the impact of several query optimization techniques, such as pushing aggregation through recursion and using ORDER BY with merge joins in column-store instead of hash joins. The authors assess the effects of indexing columns that contain vertices and effects of sorting rows in a row-store to evaluate the iteration of k joins.

In the paper [13] the author evaluates various recursive query optimizations for the plan generator. The paper considers five techniques: storage and indexing for efficient join computation, early selection, early evaluation of non-recursive joins, pushing duplicate row elimination, and pushing aggregation. The author uses four types of graphs: tree, list, cyclic, and complete graphs. However, similarly to previous work [14], the author uses a sequence of SQL commands (including INSERTs) to implement the proposed optimizations. Such an approach may suffer from various overheads, as opposed to implementing an operator node in the engine source code. This, in turn, may lead to inaccurate results.

The Recursive-aggregate-SQL (RaSQL) [8] system extends the current SQL standard to support aggregates in recursion. It can express powerful queries and declarative algorithms, such as graph and data mining algorithms. The RaSQL compiler allows mapping declarative queries into one basic fixpoint operator supporting aggregates in recursive queries. The aggregate-in-recursion optimization brought by the PreM property and other improvements make the RaSQL system more performant than other similar systems.

In the paper [17] the authors address the problem of storing large property data graphs inside a relational DBMS. They adapt the SQLGraph [19] approach to reduce the disk volume and increase processing speeds. They evaluate their schema using the PostgreSQL on LDBC-SNB and show that their schema not only performs better on read-only queries, but also performs better on workloads that include update operations.

Graph databases are good for storing and querying provenance data. One of the earliest papers that evaluated this possibility was the study [23]. The authors compare relational and graph databases on different types of queries. This study demonstrated that for traversal queries, graph databases were clearly faster, sometimes by a factor of 10. This result was expected since relational databases are not designed to perform traversals such as standard breadth-first-search.

Another paper that concerned graph databases in data provenance domain was the study [16]. The authors propose an improved version of the DPHQ framework for capturing and querying provenance data. They conclude that graph databases offer significant performance gain over relational databases for executing multi-depth queries on provenance. The performance gain becomes much more pronounced with the increase in traversal depth and data volume.

2.2 Motivation

The related works discussed above have highlighted the popularity and relevance of graph queries and graph database systems. However, they have also showed that there is only a handful of studies that address processing of graph queries (BFS, transitive closure) using recursion in SQL-supporting systems.

Moreover, despite the existence of studies which touch upon the processing of recursive SQL queries in column-stores (e.g. [14]), there are no studies that propose to leverage data positions. However, we propose an in-depth operator redesign based on this idea.

3 Background

PosDB is a disk-based distributed column-store which features explicit position manipulation, i.e., it is a “position-enabled” system. In this regard, it is close to the ideas of early systems such as the C-Store [18] and the MonetDB [10].

PosDB uses the pull-based Volcano model [7] with block-oriented processing. Its core idea is to employ two types of intermediate representations: tuple- and position-based. In the tuple-based representation, operators exchange blocks of value tuples. This type of representation is similar to most existing DBMSs. Conversely, position-based representation is a characteristic feature of PosDB. In the positional form, intermediates are represented by a generalized join index [22] which is presented in Fig. 1a. The join index stores an array of record indices, i.e., positions, for each table it covers (top of Fig. 1a). Tuples are encoded using

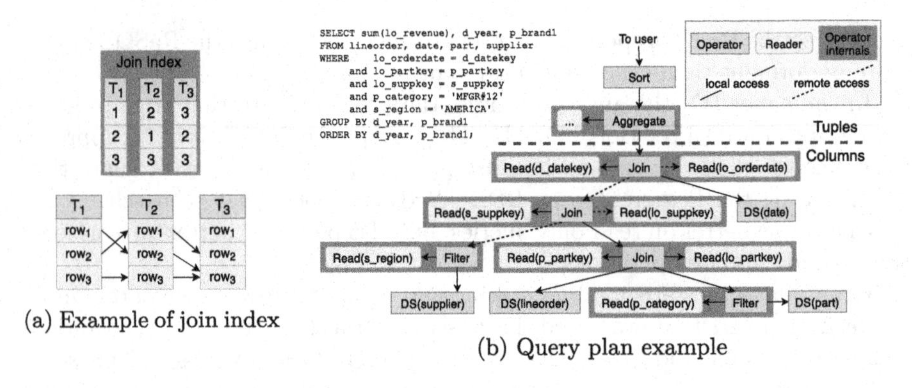

(a) Example of join index

(b) Query plan example

Fig. 1. PosDB internals

rows in the join index. Most operators in PosDB are either positional or tuple-based, with positional ones having specialized Reader entities for reading values of individual table attributes. The query plan in PosDB is divided into positional and tuple parts, and the moment of converting positions into tuples is called materialization. Materialization is to be performed at some moment of query plan, since the user needs tuples and not positions. It can be performed by either a special `Materialize` operator or by some operators, such as an aggregation operator.

In the query plan presented in Fig. 1b, the materialization point is indicated by a brown dotted line. Below the line uses positional representation and above the line uses tuple representation. In the latest version of PosDB, a query plan may contain several materialization points, in such a manner that every leaf-root path will have one.

This architecture leads to several different classes of query plans which are discussed in [4]. Operating on positions instead of tuple values allows to achieve significant cost savings for some queries. For example, in case of a filtering join, it is possible to reduce the total amount of data read from disk if the join will be performed on positions first, and then the rest of necessary columns will be read. This is a general idea of late materialization and it was extensively used for implementing many [1,9,12,15,21] operators and their combinations. At the same time, in PosDB, it is possible to build plans equivalent to naive [1] column-stores, i.e. which will read only necessary columns, construct tuples and continue as it was row-store. In this paper, we are going to discuss an application of this technique for processing recursive queries.

PosDB is a large project and it has many features and implementation details. However, they are out of the scope of this work and are not necessary for its understanding. A detailed description of baseline architecture can be found in paper [3], and the recent additions are described in [4]. Finally, an interactive demo of PosDB can be seen at the following link[1].

[1] https://pos-db.com/.

4 Proposed Approach

In order to implement recursive queries in the PosDB, we have introduced two new operator groups into its operator set. These groups share the same use pattern and differ only in the used data representation (rows and positions). Their generalized operation flow is presented in Fig. 2, which is as follows:

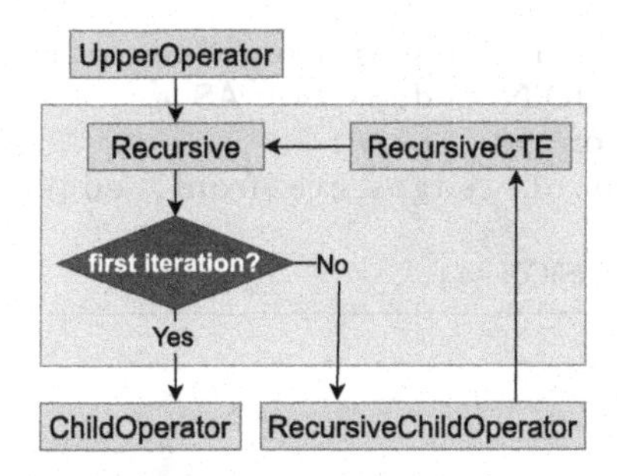

Fig. 2. Sample query plan representation using Recursive and RecursiveCTE

- The `Recursive` operator stores pointers to `RecursiveCTE`, `ChildOperator` and `RecursiveChildOperator`. `ChildOperator` is used for the non-recursive part of the query, with its help PosDB gets initial rows or initial positions. The `RecursiveChildOperator` is a regular operator, but internally it either explicitly or implicitly (via several intermediate operators) receives data from `RecursiveCTE`.
- `RecursiveCTE` stores a pointer to the Recursive, from which it asks for new records to be passed by the Next method in `RecursiveChildOperator`.

Recall that there are two types of intermediate data representation in PosDB: tuple-based and positional. This results in two sets of operators:

- `TRecursive` and `TRecursiveCTE` that only work with blocks of tuples.
- `PRecursive` and `PRecursiveCTE` that only work with position blocks.

We have designed only these two sets, each focusing on one particular data representation, either tuple-based or positional. However, the first thing which comes to mind is to use a combination of tuple-based and positional operators. For example, consider a case when `ChildOperator` and `RecursiveCTE` return a position block and `RecursiveChildOperator` returns a tuple block. In this case, the query engine will have to translate the tuples received from `RecursiveChildOperator` back to positions in order to use them for the second and subsequent steps of the recursion. However, this may be impossible in certain circumstances, if, for example, a generated attribute (e.g. $value * 2 + 1$) is present in the tuple block. In this case, there will be no original column which may be pointed to by a position.

Let us work through an example for clarity. Consider the following recursive query to find all neighbors of a vertex with id = 0 up to depth 4:

Listing 1.1. Recursive query example

```
1  WITH RECURSIVE edges_cte (id, from, to) AS
2       (SELECT edges.id, edges.from, edges.to
3        FROM edges WHERE edges.from = 0
4        UNION ALL
5        SELECT edges.id, edges.from, edges.to
6        FROM edges JOIN  edges_cte AS e
7        ON edges.from = e.to_v)
8  SELECT edges_cte.id, edges_cte.from, edges_cte.to
9  FROM edges_cte
10 OPTION (MAXRECURSION 4);
```

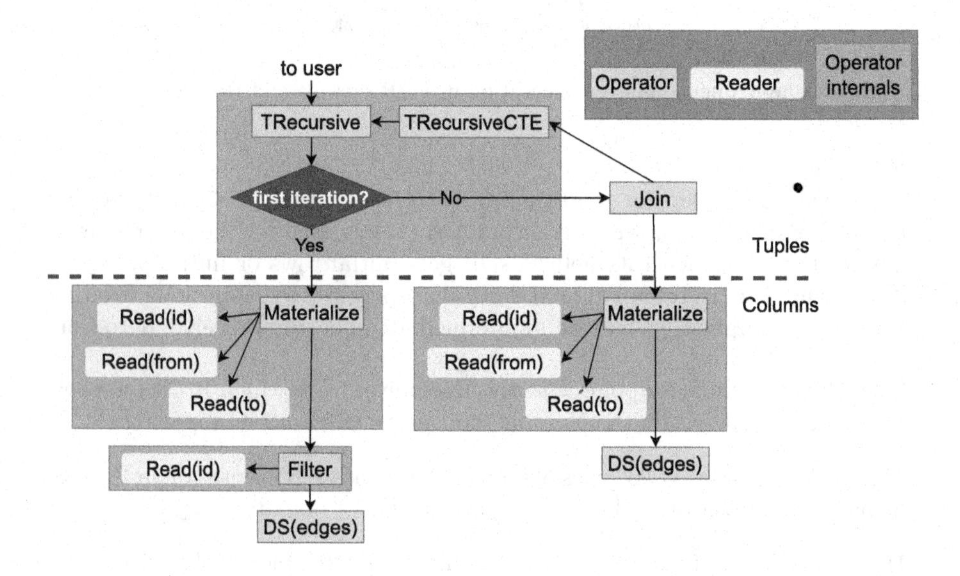

Fig. 3. Query tree using TRecursive and TRecursiveCTE

The plan of this query with the introduced structures is presented in the diagram in Fig. 3. In this figure:

- The left `Materialize` operator is a `ChildOperator`: it will be executed once in order to initialize the initial set of tuples.
- The Join is a `RecursiveChildOperator`.
- The set of tuple blocks of the current recursion step is stored inside `TRecursive`: we will call it `curLevel`. In addition, `TRecursive` will store the position of the block in `curLevel`, which should be passed to `TRecursiveCTE` next time.

The evaluation itself is as follows:

1. `TRecursive` requests blocks from the left `Materialize` as long as they are not empty and stores them in `curLevel`.
2. `TRecursive` passes all blocks from `curLevel` up until it reaches the end of `curLevel`.
3. To get the block of the new recursion step, `TRecursive` requests the block from `Join`.
4. `Join` requests blocks from `TRecursiveCTE` and from the right `Materialize`.
5. `TRecursiveCTE` asks `TRecursive` for blocks.
6. `TRecursive` increments the internal counter and passes `TRecursiveCTE` in response to its requests for blocks from `curLevel`.
7. After typing a new block of a certain size, `Join` gives it to `TRecursive`. If it is a non-empty block, then `TRecursive` will store it in `nextLevel` temporary storage and proceed to Step 3. If it is an empty block, then `curLevel` is replaced with `nextLevel`. If now `curLevel` is an empty set of blocks, then we say that `TRecursive` has finished its work, otherwise `TRecursive` proceeds to Step 2.

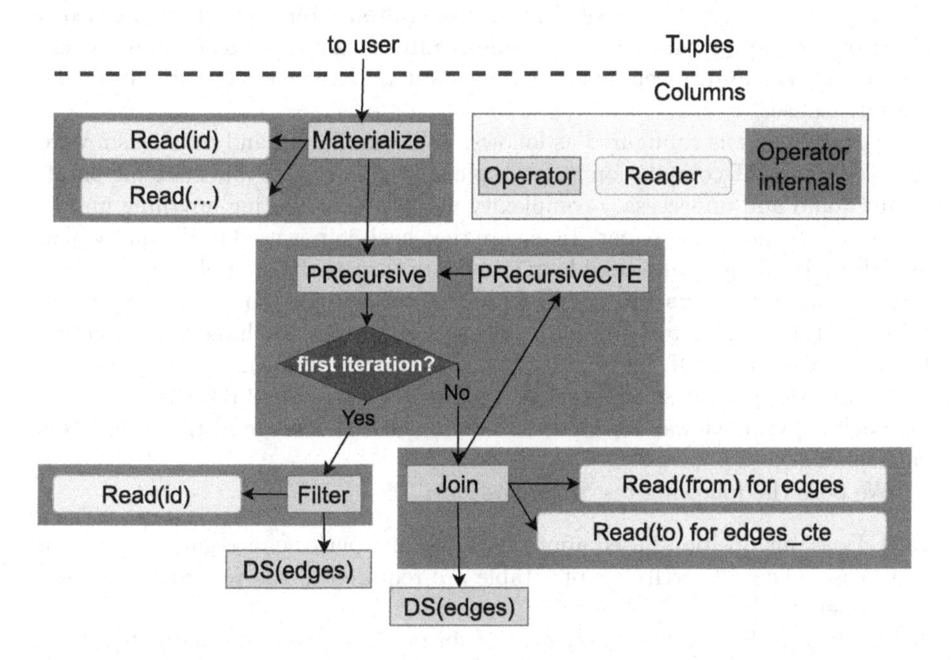

Fig. 4. Query tree using PRecursive and PRecursiveCTE

The plan of the same query using `PRecursive` operator can be represented via the diagram in Fig. 4. Evaluation of this query tree will proceed in a similar way. An important limitation here is that we can only work with positions of the same table, which means that the Join operator must return the positions of the same table as the left Filter operator. However, curLevel stores not tuple blocks, but positions blocks. In all other respects, the logic of the PRecursive and PRecursiveCTE operators is completely identical to the logic of their tuple-based counterparts.

5 Evaluation

5.1 Methodology

We evaluate our implementations using hierarchical recursive queries on a tree graph. We generated the datasets for the experiments with a simple script[2]. All evaluated queries solve the task of finding all nodes that lie at a distance of n hops from the root using the BFS algorithm. A test graph was stored in PosDB and PostgreSQL as an edge list. Columns are of the following types: id, from, to are int (4 bytes); name is varchar(15) (32 bytes); each additional column in the second and third test sets is varchar(20) (42 bytes). The number of table rows is indicated above the figures with test results.

In order to evaluate our solution, we have selected a baseline of PostgreSQL. Our choice is based on the following considerations. First of all, we needed a classic row-store system in order to demonstrate the advantages of our approach. Second, PostgreSQL meets another important requirement: it is free from the DeWitt clause[3].

PostgreSQL was configured as follows: JIT compilation and parallelism were disabled since JIT compilation is not implemented in PosDB and enabling parallelism would add unnecessary complexity without contributing anything important in the scope of this paper. To ensure that hash join is used in the query plan as in PosDB, merge and nested loop joins were turned off using planner parameters. The temp_buffers and work_mem parameters were set to values that ensure that any table under test fits into memory. To prevent caching from affecting the results, PostgreSQL internal caches were cleared between runs.

PosDB buffer manager was set to 1 GB (32K pages of 32 KB size).

Each experiment was repeated 10 times, and the average of the results was calculated.

We pose the following research questions:

RQ1 Does our position-based approach bring any performance gain in a special case when all attributes of a table are required in the recursive part of a query?

RQ2 What performance advantage does our position-based approach offer when introducing additional payload through auxiliary attributes used in projection?

---

[2] https://github.com/Firsov62121/tree_generator.

[3] https://www.brentozar.com/archive/2018/05/the-dewitt-clause-why-you-rarely-see-database-benchmarks/.

RQ3 Is it possible to emulate our approach inside a row-store by rewriting a query in such a way that it will keep a minimum number of columns inside the recursive core and then join the rest?

To answer these questions, we have designed the following experiments for each RQ:

1. For the first experiment, we used a BFS query with the table consisting only of attributes required for the traversal, giving no benefits to PosDB (see Listing 1.1 and corresponding plans in Figs. 3 and 4).
2. For the second experiment, we used the query from the first experiment with payload attributes added to the input table itself and to all projections. SQL queries of the following type were used for PostgreSQL and PosDB:

```
1  WITH RECURSIVE edges_cte (id, from, to, column1,
2        ..., columnN, depth) AS
3        (SELECT edges.id, edges.from, edges.to,
4            edges.column1, ..., edges.columnN, 0
5         FROM edges WHERE edges.from = 0
6         UNION ALL
7         SELECT edges.id, edges.from, edges.to,
8            edges.column1, ..., edges.columnN,
9         e.depth + 1 FROM edges JOIN edges_cte AS e
10        ON edges.from = e.to AND e.depth < DEPTH_VAL)
11 SELECT edges_cte.id, edges_cte.from, edges_cte.to,
12       edges_cte.column1, ..., edges_cte.columnN
13 FROM edges_cte;
```

3. For the third experiment, we have created a special type of query:

```
1  WITH RECURSIVE edges_cte(id, to, depth) AS
2        (SELECT edges.id, edges.to, 0 FROM edges
3        WHERE edges.from = 0
4        UNION ALL
5        SELECT edges.id, edges.to, e.depth + 1
6        FROM edges JOIN edges_cte AS e
7        ON edges.from = e.to AND e.depth < DEPTH_VAL)
8  SELECT edges.id, edges.to, edges.from,
9        column1, ..., columnN FROM edges JOIN
10       edges_cte ON edges.id = edges_cte.id;
```

In all plans of all evaluated systems, the `edges_cte` was hashed in the hash join (default PostgreSQL behavior).

5.2 Experimental Setup

Experiments were performed using the following hardware and software configuration. Hardware: Intel® Corete™ i7-8550U CPU @ 1.80 GHz (8 cores), 16 GiB RAM, 500 GB KINGSTON SA2000M8500G. Software: Ubuntu 22.04.3 LTS x86_64, Kernel 6.2.0-32-generic, gcc 11.4.0, PostgreSQL 14.9.

5.3 Experiments and Discussion

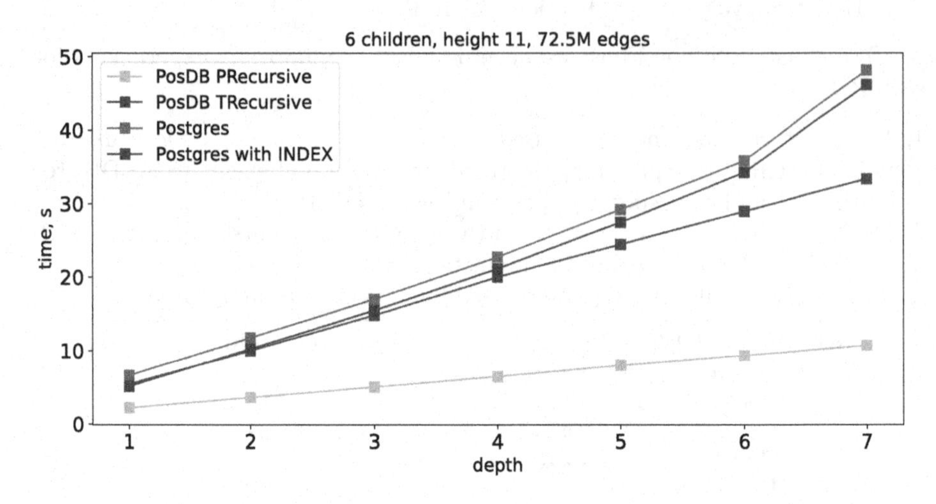

Fig. 5. Experiment 1 results

Experiment 1. The results are presented in Fig. 5. The `TRecursive` approach exhibits performance that is similar to that of PostgreSQL, as expected, since the underlying query engines perform identical operations. Meanwhile, `PRecursive` outperforms `TRecursive` significantly, because it uses only two out of four attributes (`from` and `to`) during the search, and materializes values of the third attribute (`id`) only when the required table rows are known. The number of rows scheduled to be materialized is much smaller than the total number of rows in the table (by roughly 200 times), resulting in operators passing around intermediate results of much smaller size. The index in PostgreSQL was built over `from`, since it is used to find edges in the join in the recursive part of the query. As we can see, this helps improve PostgreSQL performance, although it is a small improvement. Moreover, with the increase in depth, TRecursive demonstrates slightly better results compared to PostgreSQL with Index.

Experiment 2. The queries considered in this experiment were executed on a dataset with an additional parameter `N`, which corresponds to the number of added columns. The query plans for PosDB and PostgreSQL remain almost identical to those used in the first experiment, with the addition of ancillary columns to the `Materialize` operators in PosDB and to the projections in PostgreSQL, respectively.

Due to the space constraints, we did not include `PostgreSQL with INDEX` in this and subsequent experiments, as its behavior is equivalent to that of PostgreSQL when changing the parameter `N`. Furthermore, results of `PRecursive` were only included for the maximum number of additional columns, as the time taken by `PRecursive` was predictably found to be almost independent of `N`.

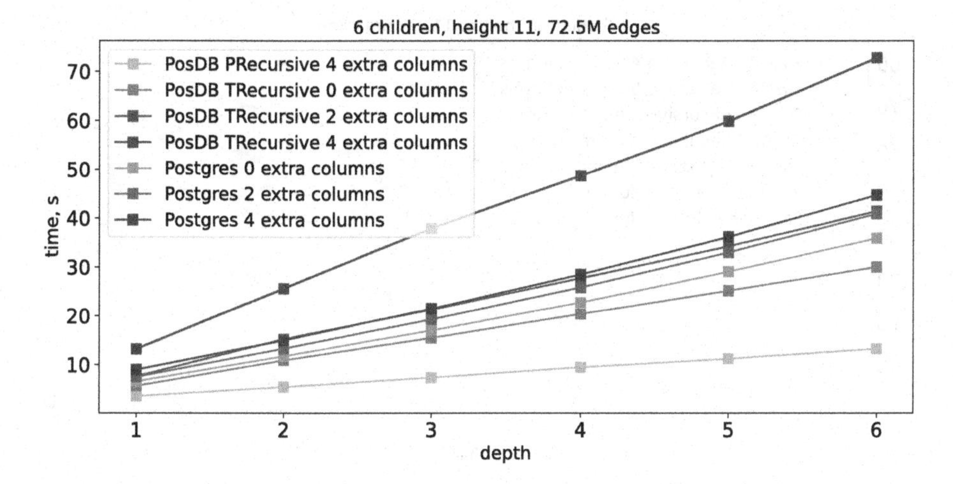

Fig. 6. Experiment 2 results

The run times of the queries depending on the depth of the traversal are presented in Fig. 6. As we can see, with increasing depth, the gap in the runtime on tables of different "widths" grows. PosDB `PRecursive` outperforms all other approaches. This is due to late materialization reducing sizes of intermediate results significantly. In this experiment, it matters even more due to the substantial overhead associated with passing "wide" intermediates (all columns) between operators, even though only two of them are required for the recursive part (`from` and `to`). It is important to mention that as the "width" of passed row grows, PosDB `TRecursive` falls behind PostgreSQL. This happens because of columnar nature of PosDB. With `TRecursive` it requires more disk accesses (one for each column) to retrieve a single row from a table. In contrast, PostgreSQL can do this with a single access since all the data for table rows is stored together.

Experiment 3. This experiment is similar to the previous one, but now we are trying to conserve space in `edges_cte` by reducing the size of the intermediates. We only store the data necessary for reconstructing the original table row and navigating through the tree.

The run times of the queries depending on the depth of the traversal are presented in Fig. 7. It can be seen that the performance of `PRecursive` is similar to its performance in the previous experiment, and it is the best in this experiment too among all compared approaches. `TRecursive`, however, beats PostgreSQL in this experiment.

As we can see by the performance improvement of `TRecursive`, this method helps reduce disk access overhead of row reconstruction inside `TRecursive`. In PosDB, unnecessary columns are now only materialized once at the very end. Whereas in `PostgreSQL`, the internal hash join involves inefficient sequential data reads that discard unnecessary columns. Finally, this experiment shows that our approach cannot be emulated inside PostgreSQL via join.

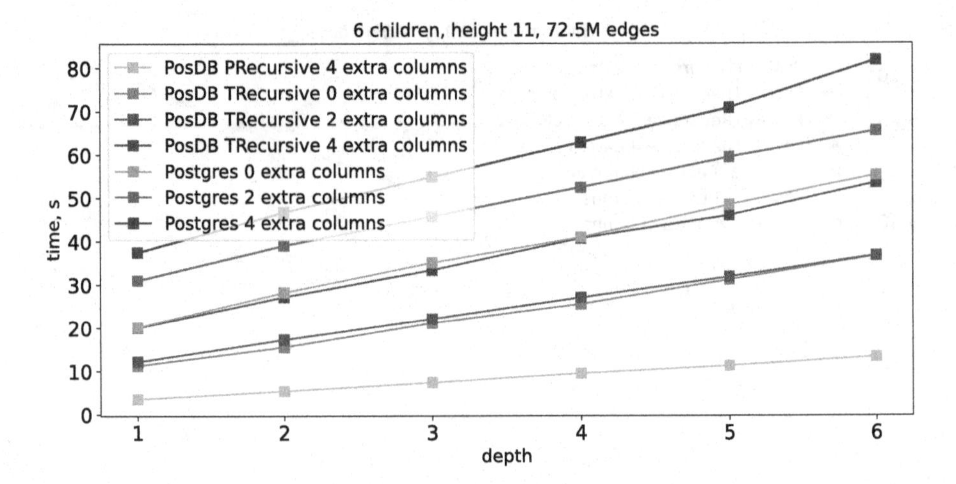

Fig. 7. Experiment 3 results

6 Conclusion

In this paper, we proposed two approaches to implementing recursive queries in a position-enabled column-oriented DBMS: `TRecursive` and `PRecursive`, with the latter utilizing positions to implement late materialization. We implemented two sets of operators: 1) a tuple-based set which is similar to the operators that can be found in classical row-stores but leveraging columnar data access, and 2) a positional-based set which is the main contribution of the paper.

We conducted experiments to evaluate the performance of the proposed approaches and used PostgreSQL as the baseline. Our experiments have demonstrated that both approaches offer improvement, with `PRecursive` offering up to 6 times performance gain over PostgreSQL and 3 times over `TRecursive`. However, `TRecursive` remains the only option if there are two (or more) distinct tables in the `RECURSIVE` part used, due to implementation-related restrictions. Also, `TRecursive` yields up to 3× improvement over PostgreSQL when an additional payload columns which are not required in the `RECURSIVE` part exist and the query can be rewritten to project them only in the top-level. Finally, we have shown that it is not possible to emulate our approach inside a row-store efficiently.

References

1. Abadi, D., Boncz, P., Harizopoulos, S.: The Design and Implementation of Modern Column-Oriented Database Systems. Now Publishers Inc. (2013)
2. Boncz, P.A., Kersten, M.L.: MIL primitives for querying a fragmented world. VLDB J. **8**(2), 101–119 (1999). https://doi.org/10.1007/s007780050076
3. Chernishev, G.A., Galaktionov, V.A., Grigorev, V.D., Klyuchikov, E.S., Smirnov, K.K.: PosDB: an architecture overview. Program. Comput. Softw. **44**(1), 62–74 (2018). https://doi.org/10.1134/S0361768818010024

4. Chernishev, G.A., Galaktionov, V., Grigorev, V.V., Klyuchikov, E., Smirnov, K.: A comprehensive study of late materialization strategies for a disk-based column-store. In: DOLAP@EDBT 2022, vol. 3130, pp. 21–30. CEUR (2022)
5. Ghazal, A., Crolotte, A., Seid, D.: Recursive SQL query optimization with k-iteration lookahead. In: Bressan, S., Küng, J., Wagner, R. (eds.) DEXA 2006. LNCS, vol. 4080, pp. 348–357. Springer, Heidelberg (2006). https://doi.org/10.1007/11827405_34
6. Ghazal, A., Seid, D., Crolotte, A., Al-Kateb, M.: Adaptive optimizations of recursive queries in teradata. In: SIGMOD 2012, pp. 851–860 (2012)
7. Graefe, G.: Query evaluation techniques for large databases. ACM Comput. Surv. **25**(2), 73–169 (1993)
8. Gu, J., et al.: RaSQL: greater power and performance for big data analytics with recursive-aggregate-SQL on spark. In: SIGMOD 2019, pp. 467–484 (2019)
9. Harizopoulos, S., Abadi, D., Boncz, P.: Column-oriented database systems, VLDB 2009 tutorial (2009). http://nms.csail.mit.edu/~stavros/pubs/tutorial2009-column_stores.pdf
10. Idreos, S., et al.: MonetDB: two decades of research in column-oriented database architectures. IEEE Data Eng. Bull. **35**(1), 40–45 (2012)
11. Jachiet, L., Genevès, P., Gesbert, N., Layaida, N.: On the optimization of recursive relational queries: application to graph queries. In: SIGMOD 2020, pp. 681–697 (2020)
12. Mukhaleva, N., Grigorev, V., Chernishev, G.: Implementing window functions in a column-store with late materialization. In: Schewe, K.-D., Singh, N.K. (eds.) MEDI 2019. LNCS, vol. 11815, pp. 303–313. Springer, Cham (2019). https://doi.org/10.1007/978-3-030-32065-2_21
13. Ordonez, C.: Optimization of linear recursive queries in SQL. IEEE Trans. Knowl. Data Eng. **22**(2), 264–277 (2010)
14. Ordonez, C., Gurram, A., Rai, N.: Recursive query evaluation in a column DBMS to analyze large graphs. In: DOLAP 2014, pp. 71–80 (2014)
15. Polyntsov, M., Grigorev, V., Smirnov, K., Chernishev, G.: Implementing the comparison-based external sort. In: Chiusano, S., et al. (eds.) ADBIS 2022. CCIS, vol. 1652, pp. 500–511. Springer, Cham (2022). https://doi.org/10.1007/978-3-031-15743-1_46
16. Rani, A., Goyal, N., Gadia, S.K.: Efficient multi-depth querying on provenance of relational queries using graph database. In: Proceedings of the 9th Annual ACM India Conference, pp. 11–20 (2016)
17. Schmid, M.: On efficiently storing huge property graphs in relational database management systems. In: iiWAS 2019, pp. 344–352 (2019)
18. Stonebraker, M., et al.: C-store: a column-oriented DBMS. In: VLDB 2005, VLDB Endowment, pp. 553–564 (2005)
19. Sun, W., et al.: SQLGraph: an efficient relational-based property graph store. In: SIGMOD 2015, pp. 1887–1901 (2015)
20. Tsirogiannis, D., Harizopoulos, S., Shah, M.A., Wiener, J.L., Graefe, G.: Query processing techniques for solid state drives. In: SIGMOD 2009, pp. 59–72 (2009)
21. Tuchina, A., Grigorev, V., Chernishev, G.: On-the-fly filtering of aggregation results in column-stores. In: SEIM 2018, No. 2135 in CEUR, pp. 53–60 (2018)
22. Valduriez, P.: Join indices. ACM Trans. Database Syst. **12**(2), 218–246 (1987)
23. Vicknair, C., et al.: A comparison of a graph database and a relational database: a data provenance perspective. In: Proceedings of the 48th Annual Southeast Regional Conference (2010)

Ontology Matching Using Multi-head Attention Graph Isomorphism Network

Samira Oulefki[1,3](✉), Lamia Berkani[2,3], Nassim Boudjenah[3], Imad Eddine Kenai[3], and Aicha Mokhtari[1,3]

[1] Laboratory for Research in Intelligent Informatics, Mathematics and Applications (RIIMA), USTHB University, 16111 Bab Ezzouar, Algiers, Algeria

[2] Laboratory for Research in Artificial Intelligence (LRIA), USTHB University, 6111 Bab Ezzouar, Algiers, Algeria

[3] Department of Artificial Intelligence and Data Sciences, Faculty of Informatics, USTHB University, 16111 Bab Ezzouar, Algiers, Algeria

{soulefki,lberkani,amokhtari}@usthb.dz

Abstract. Ontology matching is a widely used solution to the semantic heterogeneity problem in data integration or sharing. It consists of establishing mappings between entities that belong to different ontologies. However, as the number of ontologies is increasing for a given domain and the overlap between ontologies grows proportionally, it becomes crucial to develop more reliable and accurate techniques for the automation of this task. While traditional ontology mapping approaches are based on string metrics and structure analysis, some recent methods are using deep neural networks. In this article, we propose a novel approach for ontology matching based on Graph Neural Networks (GNN) as graph representations are helpful for entity and graph comparisons. Our approach is more precisely based on Multi-Head Attention Graph Isomorphism Network (MHA-GIN). The results of experiments demonstrate the effectiveness of our approach compared with existing methods.

Keywords: Ontology matching · deep learning · graph neural network · attention mechanism · multi-head attention · graph isomorphism network

1 Introduction

Over the last decade, ontologies are providing a shared understanding of common domains to meet the need for knowledge sharing between people and/or systems [1]. They provided a mechanism for representing concepts and their relationships within a domain or between different domains. However, given the large number and the variety of ontologies developed (by different developers) for a given domain, how to manage the heterogeneity of ontologies has attracted a considerable attention in recent years. An effective solution to the semantic heterogeneity problem is known as ontology matching [2]. This process refers to the task of establishing correspondences between semantically related entities (i.e. classes/properties) from different ontologies.

M. Mosbah et al. (Eds.): MEDI 2023, LNCS 14396, pp. 200–213, 2024.
https://doi.org/10.1007/978-3-031-49333-1_15

Most ontology matching works, such as LogMap [3] and AML [4], relies on logical reasoning and rule-based methods to extract various sophisticated features from ontologies. These terminological and structural features are then used to calculate the similarities between ontological entities that determine ontology mappings. However, features from one ontology are often not transferred to other ontologies. Recently, deep learning (DL) has been widely used in various fields, including ontology matching. Zhang et al. [5] were the first to apply deep learning techniques to ontology matching. The proposed strategy is based on concept similarity. This method based on pre-trained semantic embeddings provided the basic idea for other researches such as OntoEmma [6] and VeeAlign [7]. Some recent approaches are using graph neural networks (GNN) such as BioHAN [8], a GNN-based ontology matching framework with an attention mechanism. However, we noticed a lack of models using graph isomorphism network (GIN) [9], which was proposed as a powerful GNN for graph classification. Accordingly, we propose in this article a novel DL-based ontology matching approach using a multi-head attention graph isomorphism network. Experiments were conducted using two different datasets.

The remainder of this article is organized as follows. Section 2 presents the related work about ontology matching using DL techniques in general and GNN in particular. Section 3 provides the preliminary attached to our context. Section 4 proposes our method for ontology matching. In Sect. 5 experiments are presented and discussed. Finally, the conclusion highlights the most important results with some perspectives.

2 Related Work

The most popular rules-based matching systems are LogMap [3] and AML [4], which have been highly ranked on the Ontology Alignment Evaluation Initiative (OAEI[1]) tracks. By considering two input ontologies, LogMap built their lexical indexing and efficiently calculated initial anchor mappings, i.e. exact mappings. The anchor mappings were then used as a starting point for discovering additional mappings. AML is a scalable ontology matching system, which directly uses an external ontology as a mediator between input ontologies.

Recently, DL has been widely used in various fields, including ontology matching. LogMap or AML output mappings are often used as training data to train supervised methods. Zhang et al. [5] applied DL techniques to ontology matching based on the concept similarity and Natural Language Processing (NLP) techniques that have been integrated to enhance the semantic information of concept embeddings. VeeAlign [7], uses a supervised DL approach to discover alignments. In particular, it uses a two-step model of attention combined with multi-faceted context representation to produce contextualized representations of concepts, which aids matching based on semantic and structural properties of an ontology. Bento et al. [10] used convolutional neural networks to perform string matching between class labels using character embeddings. To further improve the alignment the authors rely on the set of super-classes. He et al. [11] proposed the BERTMap approach that supports both semi-supervised and unsupervised

[1] http://oaei.ontologymatching.org.

configurations. BERT model is used as it can learn robust contextual embeddings and requires few resources for fine-tuning. The corpus used is composed of pairs of synonymous labels and pairs of non-synonymous labels. The classifier consists of a linear layer that takes as input the token embeddings produced by BERT's output layer and transforms them into two-dimensional vectors.

Recently, graph representation learning has emerged as an effective method for learning vector representations of graph-structured data. Moreover, as an ontology can be seen as a graph structure with semantics, thus some initiatives are applying GNN for ontology matching. For instance, Wang et al. [8] proposed the BioOntGCN approach that directly learns embeddings of ontology-pairs for biomedical ontology matching through two steps: (1) a convolutional neural network to extract the similarity feature vectors of nodes; and 2) a graph convolutional network to propagate the similarity features and obtain the final embeddings of concept-pairs. Wang and Hu [8] proposed BioHAN through a hybrid graph attention network. First, an ontology-enriching method is proposed to refine and enrich the ontologies through axioms and external resources. Then, a hyperbolic graph attention layers is used to encode hierarchical concepts in a unified hyperbolic space. Finally, the authors aggregated the features of both the direct and distant neighbors with a graph attention network. However, even if these methods were competitive with the state-of-the-art ontology matching methods, we noticed the lack of GIN-based matching models. This motivates us to explore the use of this type of GNN for improving mappings between ontologies.

3 Preliminaries

Graph Neural Network (GNN). GNN is a type of neural network architecture specifically designed to operate on graph-structured data. It uses the inherent structure of graphs and node features to learn representation vectors of nodes. The process performs through iterative updates where in each iteration, the representation of a node is refined by combining the representations of its neighboring nodes. After k-iterations of this aggregation and updating process, a node's representation encapsulates the structural information present in its k-hop neighborhood [9]. We can formalize the update representation on a node, $\boldsymbol{v}$, in the $\boldsymbol{l}$- th layer as follows:

$$h_v^{(l)} = \text{COMBINE}(AGG^{(l)}\left(\left\{m_u^{(l)}, u\epsilon N(v)\right\}\right), h_v^{(l-1)}) \quad (1)$$

where $h_v^{(l)}$ is the feature vector of the node v in the l- th layer, COMBINE and AGG are the combination and the aggregation functions respectively and $N(v)$ is the set of nodes that are directly attached to node v, i.e., its neighborhood. Formally, $N(v) = \{u : (u, v)\epsilon E$ or $(v, u)\epsilon E\}$ and E is the set of edges in the graph.

GNNs have seen several versions and advancements in these recent years. They differ mainly in their implementation of the aggregation and the node representation update functions. In Xu et al. [9], the authors analyzed the ability of the variants of GNNs to capture different graph structures. Their results show that the existing GNNs cannot learn to distinguish graph structures. Subsequently, they proposed a new architecture called Graph Isomorphism Networks (GIN) that is as powerful as the Weisfeiler-Lehman graph isomorphism test [9].

Weisfeiler-Lehman Test. It is used to determine if two graphs are isomorphic, meaning they have the same structure, but their nodes may have different labels. The Weisfeiler-Lehman (WL) algorithm is an iterative procedure that refines the node labels of a graph by aggregating the neighborhoods of each node and assigning a new label based on the aggregated information. If the algorithm converges to the same node labeling for both graphs, they are considered as isomorphic. Otherwise, if there is a divergence in the node labeling, the graphs are deemed non-isomorphic [9].

Graph Isomorphism Network (GIN). GIN [9] is designed to be provably the most expressive among the class of GNNs. It is also considered as powerful as the Weisfeiler-Lehman graph isomorphism test. It applies a multi-layer perceptron (MLP) on node features and uses summation as aggregation. The node representation updating formula of GIN is as follow:

$$h_v^{(l)} = \mathrm{MLP}^{(l)}((1+\epsilon^{(l)}).h_v^{(l-1)} + \sum_{u \in N(v)} h_u^{(l-1)}) \tag{2}$$

where ϵ is a learnable parameter. GIN has traditionally been used for graph classification, but it provably achieves excellent performance for node classification tasks [12].

Ontology as Graph. An ontology $\boldsymbol{O}$ can be represented with a directed graph such as: $\boldsymbol{O} = (\boldsymbol{C}, \boldsymbol{E}, \boldsymbol{R})$ where $\boldsymbol{C}$, which is the set of concepts, is represented by the nodes' set of the graph; $\boldsymbol{E}$, the set of relations holding between the concepts, is represented by the edges' set of the graph and $\boldsymbol{R}$ is the set of types of relations.

Ontology Matching. In this era of ever-expanding data and interconnected systems, the need to effectively integrate and harmonize information from multiple ontologies has become necessary. To enable using simultaneously several ontologies, the most common approach consists of creating mappings between their entities. The process allowing to find this mapping is called ontology matching. Formally, we can define ontology matching as a function $\boldsymbol{f}$ taking as input two ontologies and outputs mappings that represent a set of the correspondences holding between their concepts. It is worth noting that in this paper, we consider only the "is-a" relation type between concepts.

4 Our Approach

As shown in Fig. 1, our proposed approach for ontology matching, named MHAGINOM, is based on Multi-Head Attention Graph Isomorphism Network. This approach takes as input two OWL ontologies and performs four phases to output a set of correspondences between their concepts:

- *Preprocessing module*: This module involves reading the OWL files, creating an RDF (Resource Description Framework) graph and extracting relevant information.
- *Semantic embedding generation module.* The objective in this module is to generate semantic embeddings using the Bidirectional Encoder Representations from Transformers (BERT) model.
- *MHAGIN module*: We propose a new GNN variant that combines the expressive power of GIN and the benefits of the attention mechanism.

- *Matching module*: This module uses an MLP network to find mappings between the concepts of the input ontologies.

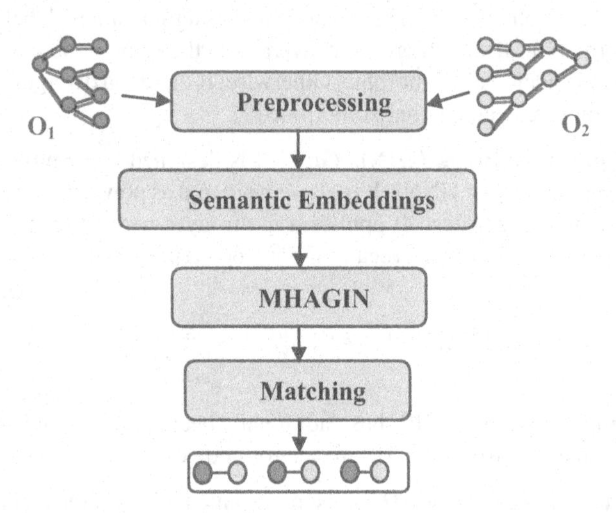

Fig. 1. MHAGINOM system architecture

4.1 Preprocessing Module

The main objective of this module is to prepare raw data to be compatible as input to our MHAGIN proposed model. The process can be described in two steps:

1. *Reading OWL files and creating an RDF graph*: The RDFLib[2] library is used to read the OWL ontology files. This library also transforms the data into an RDF graph, where concepts are represented by nodes and relations by edges.
2. *Extraction of Labels and Synonyms from Ontologies*: For each concept in the input ontologies, we extract the main label (name) and its relevant synonyms. Subsequently, the extracted labels and associated synonyms are concatenated to form a comprehensive list of terms representing each concept. This step ensures that different expressions related to a concept are considered, enhancing the robustness of the generated embeddings.

4.2 Semantic Embedding Generation Module

To generate semantic representations of ontology entities, we use the Sentence-BERT model [21]. Every expression resulting from the concatenation of labels and synonyms in the previous module is passed through the Sentence-BERT model which generates semantic embeddings.

[2] https://github.com/RDFLib/rdflib/blob/6.2.0/CHANGELOG.md.

4.3 MHAGIN Module

We build a new GNN variant that brings together the expressive power of GIN and the benefits of the multi-head attention mechanism. The MHAGIN model takes as input a set of initial node features vectors $h = \left\{\overrightarrow{h}_1, \overrightarrow{h}_2, \ldots. \overrightarrow{h}_n\right\}$ such as $\overrightarrow{h}_i \epsilon R^F$ and F is the dimension of each node features vector. It performs several transformations on the set h to produce a new set $h' = \left\{\vec{h}'_1, \vec{h}'_2, \ldots.\vec{h}'_n\right\}$ where $\vec{h}'_i \epsilon \mathrm{R}^{\mathrm{F}'}$. As depicted in Fig. 2, in each layer, the process of these transformations is phased in four steps:

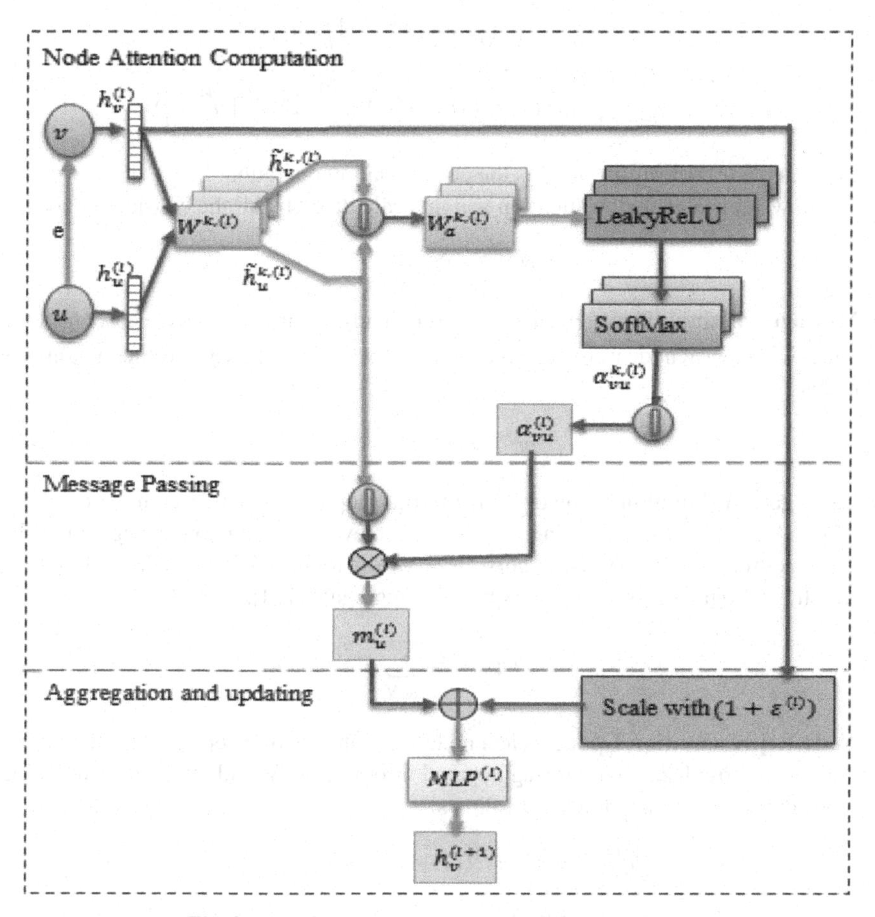

Fig. 2. Illustration of our proposed MHAGIN Model

a) **Node attention computation:** We utilize a multi-head attentional setup following closely the work of Veličković et al. [13]. We use K independent attention mechanisms that will be concatenated. Firstly, each head k applies, on each node v, a shared linear transformation parameterized by a weight matrix $W^k \epsilon R^{F \times F'}$, such that:

$$\tilde{h}_v^{k,(l)} = W^{k,(l)} \overrightarrow{h}_v^{(l)} \tag{3}$$

Then, a shared attentional mechanism $a^k : R^{F'} \times R^{F'} \rightarrow R$ is performed:

$$e_{vu}^{k,(l)} = a^{k,(l)}(\tilde{h}_v^{k,(l)}, \tilde{h}_u^{k,(l)}) \tag{4}$$

where e_{vu}^k indicates the importance of the node u to the node v by using the head k.

Then, we utilized a single-layer feed-forward neural network parameterized by a weight vector $\overrightarrow{W}_a^k$ to represent the attention mechanism a^k and we applied the LeakyReLU nonlinearity as follows [12]:

$$\alpha_{vu}^{k,(l)} = \frac{\exp(LeakyReLU(\overrightarrow{W}_a^{k,(l)^T}\left[\tilde{h}_v^{k,(l)} \| \tilde{h}_u^{k,(l)}\right])}{\sum_{j \in N(v)} \exp(LeakyReLU(\overrightarrow{W}_a^{k,(l)^T}\left[\tilde{h}_v^{k,(l)} \| \tilde{h}_j^{k,(l)}\right])} \tag{5}$$

where $.^T$ is the transposition and $\|$ is the concatenation operation.

Then, we concatenate the attention weights calculated by all the considered head:

$$\alpha_{vu}^{(l)} = \|_{k=1}^{K} \alpha_{vu}^{k,(l)} \tag{6}$$

b) **Message Passing:** This step involves exchanging information between connected nodes where each node v creates a message $m_v^{(l)}$ which will be sent to other neighbor's nodes:

$$m_v^{(l)} = \alpha_{vu}^{(l)} \|_{k=1}^{K} \tilde{h}_v^{k,(l)} \tag{7}$$

c) **Aggregation Functions:** The objective of this step is to enable each node v to integrate information from all its neighbors. To achieve this, each node v aggregates the messages received from its neighbors. Here, we opt for utilizing the SUM aggregation, which is recognized as the most expressive aggregator [14].

$$h_{agg_v}^{(l)} = \sum_{u \in N_v} m_u^{(l)} \tag{8}$$

d) **Node Representation Update:** each node v updates its own representation by combining its current state with the aggregated information. We adopt the GIN updating representation for each node v as follows:

$$h_v^{(l+1)} = MLP^{(l)}((1 + \varepsilon^{(l)}).h_v^{(l)} + h_{agg_v}^{(l)}) \tag{9}$$

where ε is a learnable parameter.

4.4 Matching Module

Based on the concept representations $h'^{(l)}$ learned from our proposed MHAGIN, the matching module takes as input pairs of concept embeddings $h_c^{O_1}, h_{c'}^{O_2}$ from O_1 and O_2 and try to predict the correspondence through a calculated score [15]:

$$M\left(h_c^{O_1}, h_{c'}^{O_2}\right) = \sigma(W_2.\gamma(W_1(h_c^{O_1} \| h_{c'}^{O_2}) + b_1) + b_2) \tag{10}$$

As the reference alignment provided in the datasets is composed by equivalence, our matching module can only predict this type of relations between concepts.

To train the matching module, we use the following loss formula:

$$\mathcal{L}^M = \sum_{(i,j)\in M^+} M\left(h_i, h_j\right) + \sum_{(i',j')\in M^-} \omega\left[\lambda - M\left(h_{i'}, h_{j'}\right)\right]_+ \tag{11}$$

where: M^+ denotes the positive matches between O_1 and O_2, M^- denotes a set of negative samples, λ is the margin value, ω is a balance hyper-parameter, and $[\cdot]_+ = \max(0, \cdot)$.

5 Experiments

5.1 Datasets

In He et al. [16], the authors point out limitations in the Ontology Alignment Evaluation Initiative (OAEI) tracks, particularly for ML-based systems. To address this issue, they proposed a novel machine learning-friendly track[3] based on Mondo[4] and UMLS[5] resources. To train and test our approach, we used Mondo datasets involving the following ontologies: OMIM [22], ORDO [23], NCIT (National Cancer Institute Thesaurus) and DOID (Human Disease Ontology). For every task, a reference alignment is manually constructed containing pairs of positive examples. To obtain a balanced dataset, we generated the same number of negative examples, by replacing randomly one of the concepts in the positive sample pairs.

Dataset Splitting. Each dataset is divided into three sets as follow [16]: the first one corresponding to the train set represent 20%, while the second partition witch correspond to the validation set is fixed to 10%; the remaining 70% are for the test set.

5.2 Evaluation Metrics

To evaluate the performance of our proposed ontology matching approach, we adopt the same evaluation metrics given in [16].

Let m be a correspondence such that $m = (c, c')$ where $c \epsilon O_1$ and $c' \epsilon O_2$. Let $\mathcal{M}_m$ be the set of negative correspondences. Because of our approach is ML-based, we use Hits@K and MRR metrics defined as follow:

$$Hits@K = \frac{\left\{m \in \mathcal{M}_{ref} | Rank(m) \leq K\right\}}{\left|\mathcal{M}_{ref}\right|} \tag{12}$$

$$MRR = \frac{\sum_{m \epsilon \mathcal{M}_{ref}} Rank(m)^{-1}}{\left|\mathcal{M}_{ref}\right|} \tag{13}$$

[3] https://www.cs.ox.ac.uk/isg/projects/ConCur/oaei/.

[4] https://mondo.monarchinitiative.org/.

[5] https://www.nlm.nih.gov/research/umls/index.html.

where $\mathcal{M}_{ref}$ denote the reference alignment and *Rank(m)* the position of m in the set $\mathcal{M}_m \cup \{m\}$ ordered by the score of its elements.

We also use general metrics that are Precision (P), Recall (R) and F-score (F_β).

$$P = \frac{|\mathcal{M}_{out} \cap \mathcal{M}_{ref}|}{|\mathcal{M}_{out}|} \tag{14}$$

$$R = \frac{|\mathcal{M}_{out} \cap \mathcal{M}_{ref}|}{|\mathcal{M}_{ref}|} \tag{15}$$

$$F_\beta = \left(1 + \beta^2\right) . \frac{P.R}{\beta^2.P + R} \tag{16}$$

where $\mathcal{M}_{out}$ is the alignment provided by the system under evaluation and β is a weighting for Precision and Recall.

5.3 Baselines

To assess the effectiveness of our proposed approach, we conducted a comparison with the following systems:

- **LogMap.** Is a state-of-the-art rule-based ontology matching system. It employs a hybrid approach that combines both linguistic and structural techniques to discover semantic correspondences between entities in different ontologies [3].
- **AgreementMakerLight (AML).** It is a leading rule-based ontology matching system. It is an efficient tool that focuses on attribute-based matching for RDF data [4].
- **BERTMap.** a ML-based ontology matching system which adapt the BERT model on a corpus of concept labels extracted from the ontologies to be aligned [11].
- **EditSim.** According to He et al. [16], it is reasonable to consider the simple edit distance between concept labels as baseline. It represents the minimum number of single-character edits (insertions, deletions, or substitutions) required to transform one string into another [17].

5.4 Experimental Configurations

To set the hyper-parameters of our system, we used the ADAM optimizer [18] with initial learning rate 0.001. We started by varying the number of layers and epochs of the MHAGIN model according to all the metrics. Then we varied the number of MLP layers and batch size, in terms of precision, recall and F1 metrics. The following tables show the results obtained using the OMIM-ORDO dataset (Tables 1, 2, 3, 4).

According to the results obtained, we will consider 2 layers for MHAGIN and 6 layers for MLP. While the best batch size obtained is equal to 64, considering the best recall and F1 score, even if the best precision was obtained with a batch size equal to 32. For the number of epochs, we'll consider the value 1000, which gave the best performance according to all the metrics.

Table 1. Varying the number of epochs of MHAGIN

# Epoch	P	R	F1	MMR	HIT@1
100	0.747	0.489	0.591	0.805	0.781
500	0.757	0.499	0.601	0.822	0.792
1000	**0.765**	**0.505**	**0.608**	**0.840**	**0.799**

Table 2. Varying the number of layers of MHAGIN

# Layers	P	R	F1	MMR	HIT@1
1	**0.765**	0.505	0.608	0.840	0.799
2	0.757	**0.525**	**0.620**	**0.847**	**0.802**
3	0.762	0.496	0.601	0.836	0.794

Table 3. Varying the number of layers of the MLP

Number of layers	P	R	F1
1	0.426	0.526	0.471
2	0.511	0.398	0.505
4	0.572	0.451	0.505
6	**0.590**	**0.570**	**0.580**

Table 4. Varying the batch size of the MLP

# Batch size	P	R	F1
32	**0.590**	0.570	0.580
64	0.579	**0.598**	**0.588**
128	0.586	0.516	0.549
256	0.579	0.508	0.541

5.5 Evaluation Results

Evaluation of the Different Variants of the Multi-head Attention. We have considered the different variants (with/without the integration of synonyms), namely: the MHAGINOM and the MHAGINOM without Synonyms (MHAGINOM-S). Table 5 shows that the MHAGINOM model outperformed the MHAGINOM-S model for both datasets, demonstrating the contribution of integrating synonyms. Accordingly, this model will be used in the rest of our evaluations.

Table 5. Contribution of synonyms

Dataset	Model	P	R	F1	MMR	HIT@1
OMIM-ORDO	MHAGINOM-S	0.757	0.545	0.634	0.871	0.814
	MHAGINOM	**0.773**	**0.560**	**0.650**	**0.887**	**0.827**
DOID-NCIT	MHAGINOM-S	0.884	0.814	0.847	0.971	0.966
	MHAGINOM	**0.889**	**0.836**	**0.862**	**0.975**	**0.968**

Contribution of Multi-head Attention with GIN for Ontology Matching. To evaluate the performance of our model, we compared it with the GIN model (without attention) and the GIN model with self-attention (SAGIN-OM). In addition, to assess the contribution of GIN, we added a comparison with the GAT graph model. According to [19], higher MMR and HIT@k scores reflect improved performance when evaluating entity matching methods. Table 6 shows the results obtained using both datasets, in terms of MMR and HIT metrics:

Table 6. Evaluation of MHAGINOM in terms of MMR and HIT

Model	OMIM-ORDO		DOID-NCIT	
	MMR	HIT@1	MMR	HIT@1
GAT-OM	0.827	0.791	0.893	0.873
GIN-OM	0.847	0.802	0,954	0,941
SAGIN-OM	0.869	0.811	0,969	0,959
MHAGINOM	**0.887**	**0.827**	**0.975**	**0.968**

We can notice that our approach outperformed the other methods for both datasets. Furthermore, we evaluated the ontology matching performance in terms of P, R and F1 metrics, as illustrated in Table 7.

Table 7. Evaluation of MHAGINOM in terms of P, R and F1

Model	OMIM-ORDO			DOID-NCIT		
	P	R	F1	P	R	F1
GAT-OM	0.728	0.503	0.595	0.867	0.765	0.813
GIN-OM	0.757	0.525	0.620	0,885	0,803	0,842
SAGIN-OM	0.764	0.537	0.631	**0,896**	0,823	0,858
MHAGINOM	**0.773**	**0.560**	**0.650**	0.889	**0.836**	**0.862**

The results obtained show that our approach outperforms the other models, achieving a better precision, recall and F1 score with OMIM-ORDO dataset. However, with DOID-NCIT dataset, our model performed better in terms of recall and F1 score but the best precision was obtained with SAGIN-OM model. We note that we have only used HIT@1, as the values obtained with a variation of K (K = 5 and 10) are almost equal to 1 with all the models.

Comparison with Related Work. We have compared the performance of our model with the state-of-the-art using both OMIM-ORDO and DOID-NCIT datasets. The following tables illustrate the results obtained, given that we have reused the results of related work from He et al. [16] since we are using the same datasets he shared (Tables 8 and 9):

Table 8. Comparison with related work using the OMIM-ORDO dataset

Model	P	R	F1	MMR	HIT@1
LogMap	**0.788**	0.501	0.612	0.805	0.744
AML	0.702	0.517	0.596	NA	NA
BERTMap	0.762	0.548	0.637	0.877	0.823
EditSim	0.781	0.507	0.615	0.777	0.727
MHAGINOM	0.773	**0.560**	**0.650**	**0.887**	**0.827**

Table 9. Comparison with related work using the DOID-NCIT dataset

Model	P	R	F1	MMR	HIT@1
LogMap	**0.896**	0.661	0.761	0.559	0.363
AML	0.841	0.770	0.804	NA	NA
BERTMap	0.823	**0.887**	0.854	0.968	0.955
EditSim	0.889	0.771	0.826	0.903	0.883
MHAGINOM	0.889	0.836	**0.862**	**0.975**	**0.968**

The comparison results show that with the OMIM-ORDO dataset, our model outperformed the state-of-the-art models in terms of recall, F1 score, MMR and HIT@1 metrics. For precision, the best value obtained was 0.788 with LogMap against a value of 0.773 with our model. However, with the DOID-NCIT dataset our model outperformed the other models in terms of F1 score, MMR and HIT@1 metrics. For precision, the best value obtained was 0.896 with LogMap against a value of 0.889 with our model. Similarly, for recall, the best value obtained was 0.887 with BERTMap against a value of 0.836 with our model.

5.6 Discussion and Analysis

Experiments carried out on two different datasets demonstrate that the ontology matching based on multi-head with attention mechanism and using GIN outperformed the model based on self-attention GIN as well as the GIN or other types of knowledge graph models such as the GAT model.

Moreover, the results obtained prove that our model outperformed the state-of-the-art models in terms of F1 score, MMR and HIT@1 metrics. We can notice that the most precise model is LogMap, but our model yielded a better recall and F1 score than this model for both datasets. On the other hand, although BERTMap provided a better recall with the DOID-NCIT dataset, our model was more precise with both datasets and also yielded a better recall and F1 score than this model. Therefore, these evaluations prove that our model is the most effective for ontology matching.

However, the major shortcomings of our contribution lies in the fact that we have considered only one type of mapping, namely the equivalence, and one type of relation, the "is-a" relation. It would be interesting to extend our approach by considering other types of mapping, such as the subsumption [20], and other types of semantic relations, such as "part-of", "has-a" and so on.

6 Conclusion

We proposed in this article a multi-head attention graph isomorphism network for ontology matching. Our method consists of three steps: ontology pre-processing, generating embeddings with BERT; the MHAGIN module, proposing a novel GNN variant that combines the expressive power of GIN and the benefits of the attention mechanism; and the matching module, which uses an MLP network to find mappings between the concepts of the input ontologies. The experiment conducted on two different datasets showed that our approach outperformed existing state-of-the-art methods, proving the contribution of multi-head GIN and attention mechanism along with the integration of synonyms.

As future work, we will consider different types of relations not only the "is-a" relation and other types of mapping such as the subsumption. Furthermore, we plan to deploy our solution in a real-world environment, developing applications in a variety of contexts including e-learning, e-tourism and bioinformatics.

References

1. Gruber, T.R.: A translation approach to portable ontology specifications. Knowl. Acquis.. Acquis. **5**(2), 199–220 (1993)
2. Kalfoglou, Y., Schorlemmer, M.: Ontology mapping: the state of the art. Knowl. Eng. Rev.. Eng. Rev. **18**(1), 1–31 (2003). https://doi.org/10.1017/S0269888903000651
3. Jiménez-Ruiz, E., Grau, B.C., Zhou, Y., Horrocks, I.: Large-scale interactive ontology matching: algorithms and implementation. ECAI **242**, 444–449 (2012)
4. Faria, D., Pesquita, C., Santos, E., Palmonari, M., Cruz, I.F., Couto, F.M.: The AgreementMakerLight ontology matching system. In: Meersman, R., Panetto, H., Dillon, T., Eder, J., Bellahsene, Z., Ritter, N., De Leenheer, P., Dou, D. (eds.) OTM 2013. LNCS, vol. 8185, pp. 527–541. Springer, Heidelberg (2013). https://doi.org/10.1007/978-3-642-41030-7_38

5. Zhang, Y., et al.: Ontology matching with word embeddings. In: Sun, M., Liu, Y., Zhao, J. (eds.) CCL/NLP-NABD -2014. LNCS (LNAI), vol. 8801, pp. 34–45. Springer, Cham (2014). https://doi.org/10.1007/978-3-319-12277-9_4
6. Wang, L., Bhagavatula, C., Neumann, M., Lo, K., Wilhelm, C., Ammar, W.: Ontology alignment in the biomedical domain using entity definitions and context. In: Proceedings of the BioNLP 2018 Workshop, pp. 47–55 (2018). https://doi.org/10.18653/v1/w18-2306
7. Iyer, V., Agarwal, A., Kumar, H.: Veealign: a supervised deep learning approach to ontology alignment. In: Proceedings of the 15th International Workshop on Ontology Matching co-located with the 19th International Semantic Web Conference (ISWC 2020), Virtual conference (originally planned to be in Athens, Greece), Vol. 2788 of CEUR Workshop Proceedings, CEUR-WS.org, pp. 216–224 (2020)
8. Wang, P., Hu, Y.: Matching biomedical ontologies via a hybrid graph attention network. Front. Genet. **13**, 893409 (2022). https://doi.org/10.3389/fgene.2022.893409
9. Xu, K., Hu, W., Leskovec, J., Jegelka, S.: How powerful are graph neural networks?. In: International Conf. on Learning Representations (ICLR) (2018). arXiv 2018:00826.1810
10. Bento, A., Zouaq, A., Gagnon, M.: Ontology matching using convolutional neural networks. In: Proceedings of the Twelfth Language Resources and Evaluation Conference, pp. 5648–5653, Marseille, France. European Language Resources Association (2020)
11. He, Y., Chen, J., Antonyrajah, D., Horrocks, I.: BERTMap: a BERT-based ontology alignment system. In: Proceedings of the AAAI Conference on Artificial Intelligence (2022)
12. Hu, Y., Tang, Y., Huang, H., He, L.: A graph isomorphism network with weighted multiple aggregators for speech emotion recognition. Proc. Interspeech, 4705–4709 (2022)
13. Veličković, P., Cucurull, G., Casanova, A., Romero, A., Lio, P., Bengio, Y.: Graph attention networks (2017). arXiv preprint arXiv:1710.10903
14. You, J., Ying, Z., Leskovec, J.: Design space for graph neural networks. In: Advances in Neural Information Processing Systems, vol. 33, pp. 17009–17021 (2020)
15. Hao, J., et al.: MEDTO: medical data to ontology matching using hybrid graph neural networks. In: Proceedings of the 27th ACM SIGKDD Conference on Knowledge Discovery and Data Mining (KDD 2021) (2021)
16. He, Y., Chen, J., Dong, H., Jiménez-Ruiz, E., Hadian, A., Horrocks, I.: Machine learning-friendly biomedical datasets for equivalence and subsumption ontology matching. arXiv preprint arXiv:2205.03447 (2022)
17. Lê Bach T.: Building a Multi-Perspective Semantic Web. PhD thesis in Computer Science. École des Mines de Nice at Sophia Antipolis (2006)
18. Kingma, D., Adam, B.J.: A method for stochastic optimization. In: Proceedings of the 3rd International Conference on Learning Representations (ICLR 2015) (2015)
19. Cai, W., Ma, W., Zhan, J., Jiang, Y.: Entity alignment with reliable path reasoning and relation-aware heterogeneous graph transformer. In: International Joint Conference on Artificial Intelligence (2022)
20. Chen, J., He, Y., Geng, Y., Jimenez-Ruiz, E., Dong, H., Horrocks, I.: Contextual semantic embeddings for ontology subsumption prediction. World Wide Web, pp. 1–23 (2023)
21. Reimers, N., Gurevych, I.: Sentence-BERT: Sentence embeddings using siamese BERT-networks. In: Proceedings of Conference on Empirical Methods in Natural Language Processing. Association for Computational Linguistics (2019). arXiv preprint arXiv:1908.10084
22. Amberger, J.S., Bocchini, C.A., Schiettecatte, F., Scott, A.F., Hamosh, A.: OMIM.org: Online Mendelian Inheritance in Man (OMIM®), an online catalog of human genes and genetic disorders. Nucleic Acids Res. **43**(D1), D789–D798 (2015)
23. Vasant, D., et al.: ORDO: an ontology connecting rare disease, epidemiology and genetic data. In: Proceedings of ISMB, vol. 30 (2014)

Investigating the Perceived Usability of Entity-Relationship Quality Frameworks for NoSQL Databases

Chaimae Asaad[1,2(✉)], Karim Baïna[1], and Mounir Ghogho[2,3]

[1] Alqualsadi, Rabat IT Center, ENSIAS, Mohammed V University in Rabat, Rabat, Morocco
chaimae.asaad@uir.ac.ma
[2] TicLab, Faculty of Engineering and Architecture, International University of Rabat, Rabat, Morocco
[3] University of Leeds, Leeds, UK

Abstract. Quality assessment of data models can be a challenging task due to its subjective nature. For the schemaless, heterogeneous and diverse group of databases falling under the NoSQL umbrella, quality is generally operation and performance oriented, and no quality assessment framework exists. As a first step in shaping our understanding of NoSQL database model quality, this paper investigates the perceived usability of quality evaluation frameworks adopted from Entity Relationship (ER) modeling to the context of NoSQL databases. A first evaluation is performed on the three most widely used ER quality frameworks, where they are assessed for their usefulness, ease of use and suitability in the context of NoSQL databases. Based on the results of this assessment, a second evaluation is performed on the best scoring framework. This evaluation is comprised of a real use case adoption of the framework to assess the quality of NoSQL database models. This paper merges targeted crowdsourcing, Stack overflow data mining and white-box classification to gain insights into the concept of NoSQL database model quality, its characterizing features and the trade-offs it involves. This work illustrates the first investigation of ER-defined quality framework to NoSQL on a sample of diverse NoSQL schemas and using both industrial and academic participants. A decision tree is utilized to describe the heuristics of data model assessment, and an analysis is performed to identify inter-annotator disagreement, quality criterion importance, and quality trade-offs. In the absence of works approaching NoSQL data model quality assessment, this paper aims to lay groundwork and present preliminary insights on quality characterization in the context of NoSQL, as well as highlight current gaps, limitations and potential improvements.

Keywords: NoSQL Databases · Data Model Quality Assessment · Crowdsourcing · Decision Trees

M. Mosbah et al. (Eds.): MEDI 2023, LNCS 14396, pp. 214–227, 2024.
https://doi.org/10.1007/978-3-031-49333-1_16

1 Introduction

Quality assessment represents a salient phase for information systems [1–4], software design and engineering [5–7], business Process models [8–10], conceptual [6,10–13] and logical design [14,15]. In the process of database design, frameworks have been proposed to assess quality of conceptual [11,12], logical [14,16] and physical schemas [17]. NoSQL database quality evaluation mainly focuses on aggregate-oriented NoSQL databases (Key-Value, Document, Column Family), and targets physical-level attributes such as transaction performance, partition-tolerance and data model mapping [18,19], availability and security [17], and consistency, performance, scalability [20]. In addition to availability and security, popularity, maturity, query possibilities, concurrency control and conflict resolution were considered as quality attributes in a proposed framework aiming to assist IT departments align perceived risks of NoSQL database adoption [17]. Works approaching NoSQL data model quality from a logical design perspective are scarce, and often focus on one category of NoSQL database. For instance, a set of metrics including types, collections, nesting depth, width of documents, referencing rate and redundancy were proposed to characterize aspects of the complexity of document-oriented schemas with the aim of facilitating schema analysis and comparison [21]. The scarcity of literature relating to the task of evaluating quality for NoSQL database models is due to the flexible characteristics of NoSQL databases [22]. An a priori defined schema is often not required for ingesting data into a NoSQL database. Additionally, NoSQL databases are highly heterogeneous and differ in type, features and underlying data model [22]. The lack of standardization further complicates quality assessment [23]. In the absence of a framework addressing database-agnostic quality for NoSQL databases, the gap in literature remains. In an effort to approach the quality assessment of data models for NoSQL databases, this paper investigates the potential applicability of Entity-Relationship (ER) quality frameworks to evaluate NoSQL database models. Several works have proposed frameworks and quality criteria aiming to evaluate Entity-Relationship models. For instance, Genero et al. [24] proposed and validated different ERD structural complexity measures such as number of entities, number of derived attributes, number of composite attributes, etc. In this work, we evaluate three ER quality frameworks identified in a study presented by Krogstie et al. [25] as the most cited frameworks: the Moody and Shanks framework [26], and the Batini and Scannapieco framework [16]. We add to the evaluation the Kesh Someswar framework [14] to allow for further comparison. In the remainder of this paper, we refer to these three frameworks as the MS (Moody Shanks), BS (Batini Scannapieco) and KS (Kesh Someswar) frameworks. In this paper, we first perform an experiment aiming to evaluate the perceived usability of the three aforementioned ER quality framework for NoSQL database models[1] using experts from both industrial and

[1] In this work, database model, data model and schema are used interchangeably given the context of physical design in NoSQL databases. These concepts are, however, not equivalent in other contexts of database design.

academic communities. We defined perceived usability based on three variables: perceived ease of use (PEU), perceived usefulness (PU) and perceived suitability (PS). Based on the results of the first experiment, we identified the most well scored framework and performed a real use-case adoption of the framework and an analysis of its quality criteria using data mining on Stack Overflow and decision trees. Obtaining ratings from participants has been previously employed in the literature. For instance, quality sub-characteristics of Entity-Relationship diagrams were rated using a scale of 7 linguistic labels by a group of subjects [24]. Stack Overflow questions have previously been mined in the literature to investigate the main challenges and issues faced by developers of NoSQL databases [27]. In machine learning, crowdsourcing is a popular method to acquire ground truth [28]. However, in many of its applications areas, extreme difficulties can be met in obtaining such ground truth, in most part due to high costs and task subjectivity [29]. Questionnaires were previously used in the literature in similar contexts. For instance, in [24], a questionnaire was used for the evaluation of participants' level of understanding of the entity relationship diagrams to be rated. The contributions of this paper are manifold and include a first investigation of potential adoption and adaptability of ER quality frameworks to the NoSQL context, an analysis of an ER framework's applicability in a real case scenario, the use of a hybrid method comprised of data mining, targeted crowdsourcing and decision trees for the identification of quality criterion importance, quality trade-offs and quality characterization heuristics. A presentation of threats to validity and potential improvements is included to guide future scaled up experiments. The remainder of this paper is organized as follows: Sect. 2 presents our the first experiment proposed, along with its results and analyses performed. In Sect. 3, the testing of the best scoring quality framework is conducted, and results are highlighted. In Sect. 4, discussions into potential applicability, improvements as well as threats to validity are detailed. Conclusions and future work are presented in Sect. 5.

2 Perceived Usability of the MS, BS and KS Frameworks

In this section, we propose and perform an experiment using targeted crowdsourcing to gauge the 'perceived usability' of three Entity-Relationship (ER) frameworks, Moody Shanks (MS) [26], Batini Scannapieco (BS) [16], and Kesh Someswar (KS) [14], in a NoSQL database context.

2.1 Background

The Moody-Shanks data model quality framework [26,30] was conceptualized for the quality evaluation of Entity Relationship (ER) data models. This framework includes seven quality criteria: Correctness, Completeness, Simplicity, Flexibility, Integration, Understandability, and Implementability. In the context of Entity-Relationship data models, the Moody-Shanks framework defined *correctness* as the "conformity to the modeling rules and techniques", *completeness* as "the representation of all information included in user requirements", *simplicity* as the "minimality of representative structures and lack of unnecessary components", *flexibility* as the "model's adaptability to future changes", *understandability* as "the ease of understanding of the model", *integration* as the "consistency of the model within the context it is defined", and *implementability* as "the ease, cost, and time consumption facets of model implementation" [26,30]. The Batini Scannapieco framework [16] approached the issue of improving the quality of a database schema, in addition to schema documentation, implementability and maintenance, and provided a set of quality criteria: Completeness, correctness, minimality, expressiveness, readability, self explanation, extensibility and normality. In this framework, a schema is complete when "it represents all relevant features of the application domain", correct when "it properly uses the concepts of the ER model, syntactically and semantically", minimal when "every aspect of the requirements appears only once in the schema and no concept can be deleted without loss of information", expressive when "it represents requirements in a natural way and can be easily understood", readable when "it respects certain aesthetic criteria that make the diagram graceful" (e.g., drawn in a grid, minimal number of crossings, etc.), self explanatory when "a large number of properties can be represented using the conceptual model itself, without other formalisms", extensible when "it is easily adapted to changing requirements", and normalized based on the theory of normalization associated with the relational model. The Kesh Someswar framework [14] differentiated between ontological and behavioral attributes in quality assessment of an E-R model. In this context, ontological quality is defined based on two facets, quality of structure and quality of content, where the former includes as criteria suitability, soundness, consistency and conciseness, and the latter includes completeness, cohesiveness and validity. Behavioral quality, on the other hand, includes user usability, designer usability, maintainability, accuracy and performance. In the context of this framework, suitability of structure refers to "the fact that form should follow structure", soundness represents "adherence to technical design principles", consistency and conciseness represent respectively the lack of contradictions in the model" and "redundant relationships". Completeness refers to the inclusion of "all attributes of the entities", cohesiveness to the "closeness of attributes", validity to the "correct representation of descriptions and properties of the attributes". Usability is defined for users and designers respectively as the extent to which "users will feel confident from their diagram that requirements were taken care" and for "designers to proceed to the next stage". Maintainability represents the "ease with which the model can be modified, corrected and extended", while accuracy represents "the reliability of the model" and performance reflects its "efficiency".

2.2 Experiment

In order to gauge the perceived usability of the three aforementioned quality frameworks for the NoSQL context, we propose the following experiment.

We first define perceived usability based on three variables: perceived ease of use (PEU), perceived usefulness (PU) and perceived suitability (PS). We adopt the formalization of perceived usefulness and perceived ease of use from the application of the Technology Acceptance Model [31,32] which describes how the actual system's usage depends on the attitude of users and defines measurement scales based on which these variables can be evaluated, and which is widely applied for different information systems [33]. Additionally, we define the perceived suitability as an extra measure to take into account the framework's potential suitability for the NoSQL context. Together, these three variables reflect perceived usability. For PU, we define 6 characterizing items, based on application of the Technology Acceptance Model [31,32]: work more quickly, job performance, increase productivity, effectiveness, makes job easier, useful. For PEU, we define 6 characterizing items, based on application of the Technology Acceptance Model [31,32]: easy to learn, controllable, clear and understandable, flexible, easy to become skillful, easy to use. For PS, we define 3 characterizing items: relevance to NoSQL, representativeness of NoSQL, willingness to apply in real case scenarios. The next step includes using targeted crowdsourcing, we use volunteering annotators to score the three frameworks for each of the characterizing items of PEU, PU and PS. Unlike its conventional counterpart, targeted crowdsourcing requires a specific type of workers for tasks that are either subjective or knowledge intensive [34]. In software engineering, expert opinion is the most frequently used validation method [28]. In the case of quality criteria annotation, the task is subjective both by design and by nature, and thus the use of multiple sources of annotations presents a perfect opportunity for a "natural shift from the traditional reliance on a single domain expert [29]". Given the subjectivity and high expertise required in data model quality and quality criteria annotation, multi-annotator targeted crowdsourcing was selected as the most appropriate method. Given the different stakeholder perspectives involved in the design process of any application or data model, quality evaluation is highly dependant on which perspective is taken into consideration and which stakeholder's satisfaction is prioritized. In this perceived usability study, we exclusively focus on the architect's and builder's perspectives. The architect's perspective represents the view of the data modeler or analyst responsible for developing the data model and ensuring its conformance to data modeling practices [26]. The builder's perspective, on the other hand, represents the view of the 'application' developer responsible for the implementation of the data model in a particular technology [26]. These two perspectives constitute what we denote "expert perspective", and represent the view of the expert responsible for the data modeling and implementation of the data model in a particular NoSQL database. For NoSQL databases, the modeling phase is integrated into the "implementation" phase given the schema-less nature of NoSQL [22], and so, the perspectives of the architect and builder are often enmeshed. In order to gain expert PEU, expert PU and expert PS, we use a skills filtering process given that a high level of expertise is required. To diversify the pool of annotators participating in this experiment, we include both the industrial and academic communities. In order to objectively capture level of expertise, we use classifications provided by both the 'Multilingual Classification of

European Skills, Competences, Qualifications and Occupations (ESCO)[2]' [35] and SkillsDB[3]. Data obtained from ESCO and SkillsDB respectively allows for a mapping between academic and industrial fulfillment of required NoSQL expertise. We evaluated the experts involved in the experiment based on these skill classifications. Experts were required to have background and experience corresponding to at least 80% of these skills to qualify for participation. As a result, we include 37 volunteer annotators, 17 of which are from the academic community and 17 from the industrial community. A 5 Likert scale is used in this experiment to score each characterizing item of perceived ease of use (PEU), perceived usefulness (PU) and perceived suitability (PS), reflecting the following scale: extremely unlikely, slightly unlikely, neither likely or unlikely, slightly likely, extremely likely. To annotate, the participants are instructed to formulate the question as "Would using the quality framework [X_i] allow for [characterizing_item]?", where X_i refers respectively to the MS, BS and KS framework. To avoid bias propagation and experiment contamination, the expert annotators were instructed not to discuss the experiment or the results of the annotation with one another.

2.3 Results

Upon the performance of the annotation experiment, we found that the average annotations for the MS framework are consistently higher in most characterizing features comprising the definition of perceived ease of use, perceived usefulness and perceived usability. In order to assess the statistical significance of these results, we perform the Mann-Whitney U test comparing the distributions of the annotations for Group "MS" and the combined groups "BS" and "KS" over the groups PEU, PU, and PS. In this case, we find that the p-values for all three feature groups (PEU, PU, and PS) are below 0.05, indicating that there is a significant difference between the distributions of the annotations for the frameworks MS and the combination of BS and KS. This result is consistent for the two sample T test comparing the means of the annotations of MS and the combined BS and KS groups over feature groups PEU, PU and PS. The same result is found when the U test is performed on the characterizing items. For most of the PEU features and some of the U and S features, the p-values are below 0.05, indicating that there is a significant difference between the distributions of the annotations, however, for some features (PU1, PU2, PU3, PU6, PS1, PS2), the p-values are above 0.05, suggesting that there is no significant difference between the distributions of the annotations for these features.

3 MS as a Quality Framework for NoSQL Databases

Based on the results of the previous section, we identify that the Moody Shanks MS framework was annotated at a higher score across most characterizing items

[2] https://esco.ec.europa.eu/.
[3] https://www.skillsdb.net/.

with a statistical significance. This means that the expert annotators found the MS framework to be easy to use, useful and suitable for the NoSQL context. In order to further investigate its applicability in a real use case, we conduct the following experiment.

3.1 Experiment

In this experiment, our aim is to put the MS framework to the test and explore its potential applicability by real practitioners in a quality assessment scenario. To that end, publicly available NoSQL schemas were collected from different sources such as database websites, modeling websites, etc. The NoSQL schema dataset is constituted of 35 schemas of different NoSQL databases. The database is comprised of 51.4% document oriented database models, 28.6% graph oriented database models, 14.3% key value database models and 5.7% column-family database models. These models are implemented in different databases: Cassandra, Couchbase, DynamoDB, MongoDB and Neo4j.

These models will be scored with respect to the 7 quality criteria outlined in the MS framework: Correctness, Completeness, Simplicity, Flexibility, Integration, Understandability and Implementability. A Likert scale ranging from 1 to 10 is used to score each quality criterion based on the annotator's level of agreement with the schema's fulfillment of the given criterion. We use 30 of the same annotators to score these measures, while the remaining 7 score the schemas for a measure of *overall quality*, defined to the participants as "as-is use", i.e., "would you use this data model as it is without making any changes, to operate and apply queries, for this particular universe of discourse". Two iterations of this experiment are conducted. The first where the annotators are given the schemas as JSON files and visualizations using Hackolade[4], along with documentation. And the second iteration where concise definitions of the quality criteria along with application examples are provided to the participants. In order to concisely define the quality criteria of the Moody-Shanks MS quality framework, we conduct a data collection process on two fronts. First, we collect modeling guidelines and good (or bad) practices from various NoSQL database providers' websites and technical reports. These guidelines are subsequently organized in documents and are presented to the participants in the form of 'cheat sheets' or 'quick recaps' for data modeling in multiple NoSQL database categories and databases (e.g., MongoDB, Cassandra). The second data collection consists of web scraping Stack Overflow[5] for Questions and Answers (Q&A) related to the NoSQL tag/topic. The collection of good/bad practices, coupled with the collected Stack Overflow Q&A, are used to illustrate the mapping from the Moody-Shanks quality criteria to real use case application examples. The collected Stack Overflow Q&A data was collected by an initial crawling of the #nosql tag, resulting in a dataset of 1020 rows comprised of the following cells: question, answer, author of answer,

[4] https://hackolade.com/.

[5] Stack Overflow is a public platform that provides a community-based space to find and contribute answers to technical challenges. Link: https://StackOverflow.com/.

Table 1. Quality Criteria Definitions and Examples Used in Second Iteration of the Experiment

Quality Criterion	Adapted Redefinition	Application Examples
Correctness	Relates to the appropriate use of structures within data model	- *Using embedding to model an object that will be accessed on its own (MongoDB)* - *Uniqueness of column key and row key to avoid accidental overwriting (Cassandra)*
Completeness	Inclusion of minimum requirements to fulfill functionality	- *Absence of relationships in graph database (Neo4j)* - *Unrepresented access patterns and keys (Cassandra)*
Simplicity	Conciseness and minimality of data model	- *Excessive number of duplicated properties that represent complex data instead of one shared node representing the property (Neo4j)* - *Structural redundancy in embedding documents leading to issues in coherency (MongoDB)*
Flexibility	Existence of structural qualities in the data model enabling (or hindering) ease of evolution	Some design choices potentially hindering flexibility: - *An embedding of sub-documents implying a pre-join (MongoDB).* - *Nested columns or supercolumns and their nesting depth (Cassandra).* - *Storing data in two different buckets or in a single one affects access since sub-objects are not supported (Riak).* - *Ease in schema evolution in terms of node additions, deletions or merging (Neo4j).*
Integration	Relates to the absence of contradictions within data model structures	- *An unconstrained relationship between two nodes contradicting type (Neo4j).* - *Representing a document embedding explicitly contradicting another (MongoDB).*
Understandability	Clarity of the data model for relevant stakeholders	- *Trade-off between modeling complex data and impact on explainability of model*
Implementability	Estimated time constraints and effectiveness of data model in realizing access patterns	- *Modeling data to reduce depth of downstream traversal access path (Neo4j)* - *Modeling downstream data as a relationship instead of node label or property (Neo4j)*

badge(s) of author of answer, votes of answer. This data contained varying percentages of mentions to different NoSQL databases, namely, 27.1% of questions pertaining to mongoDB, 10.29 % questions related to cassandra, around 10% pertaining to dynamoDB (5%) and Redis (5.3%). This dataset was then further extended using the Neo4j Stack Overflow dump database[6] comprised of 10.000.488 programming questions, 16.548.187 solutions and 138.390.250 comments and edits. Because of the diverse perspectives of stakeholders, quality criteria often have varying descriptions and definitions. Additionally, the chosen design level impacts how quality criteria are defined. If the evaluation is done on a requirements level, then the quality criteria definitions will relate to the requirement. Similarly, at the data level, quality criteria will reflect aspects of data quality. In this paper, the scope of the quality assessment is strictly defined

[6] https://archive.org/download/stackexchange.

with respect to the design and implementation level (which we consider as one, given the nature of NoSQL). The requirements analysis and deployment levels, which incorporate requirement quality, meta-model quality, modeling quality and data quality [3] are beyond the scope of this paper. Additionally, all stakeholder perspectives besides the expert perspective (combining the architect's and builder's perspectives) are beyond the scope of this paper. And thus, these are the parameters of evaluation of the adapted quality criteria. The result of this process is illustrated in the table below where each MS quality criterion is concisely presented and highlighted by application examples, thus providing a learning-by-example process for the participants in their annotation task. Statistical analysis is then conducted to compare the results of the two iterations, and a decision tree is used to infer feature importance and quality trade-offs between the criteria.

3.2 Results

To identify the features that are most agreed upon by the annotators, we generated a heatmap and identified the cells with the lowest standard deviation, indicating that the annotations for those features have less variation among the annotators. When taking the example of a schema corresponding to "Buzzfeed", the features "Integration" and "Implementability" have the lowest disagreement among annotators. For the schema "DynamoDB examples", the features "Integration" and "Completeness" have the lowest disagreement among annotators. For the schema "Kansas City Fountains" the features "Understandability" and "Simplicity" have the lowest disagreement among annotators. This pattern continues for the other schemas as well. It's important to note that the standard deviation is a measure of dispersion, so a lower value indicates that the annotations are closer to the mean, and hence, there is more agreement among the annotators. To get an overall view of the features most agreed upon across the entire dataset, we can look at the average standard deviation for each feature. A lower average standard deviation indicates that the annotations for that feature have less variation among the annotators, and thus, there is more agreement. Results show the features "Completeness" and "Flexibility" have the lowest average standard deviation, indicating that these features have the most agreement among annotators across the entire dataset. On the other hand, the feature "Correctness" has the highest average standard deviation, indicating that it has the least agreement among annotators.

Disagreement Analysis. After introducing the concise definitions of the MS quality criteria and the application examples mapped using Stack overflow data (Table 1), the features "Flexibility" and "Implementability" have the lowest average standard deviation in the perturbed data, indicating that these features have the most agreement among annotators. On the other hand, the feature "Simplicity" has now the highest average standard deviation, indicating that it has the least agreement among annotators. The second iteration's average standard deviations are now for flexibility, implementability, integration, completeness, correctness, understandability and simplicity, 0.917, 1.407, 1.418, 1.427, 1.441, 1.454 and 1.457, respectively. In analyzing the box plot, we found that the variability in annotations has significantly decreased, which can further highlight the benefits of a concise definition of quality criteria and the high impact of having real use case application examples to guide the quality evaluation process.

Decisions Trees for White-box modeling of Expert Opinion. Traditionally, supervised learning employs a domain expert fulfilling the 'teacher' role and thus providing necessary supervision [29]. The most common case being one where expert annotations serve as data point labels in classification problems [29]. In this experiment, labels were obtained through targeted crowdsourcing and subsequently fed to a decision tree model. The objective behind the use of decision trees is not the classification task, but rather the white box (WB) modeling of the heuristics of expert quality assessment. The visual, interpretable, explainable and transparent characteristics of WB models motivated the use of decision trees in this experiment. In the case of NoSQL data model quality assessment experiment, a decision tree is fed the annotations of all 30 participants as data points, while a majority-vote of the "overall quality" annotations of the 7 participants is used to generate labels. The seven quality criteria included in the Moody-Shanks framework and used as features for the decision tree have varying levels of importance and characterize quality with a 75% accuracy. Integration, simplicity, completeness and correctness were found to be the most important features, with importance percentages of 17.39%, 16.26%, 16.16% and 16.08% respectively, while Understandability, Implementability and Flexibility have importance percentages of 12.23%, 10.99% and 10.89% respectively. Using the results of the decision tree, different scenarios can be constructed to illustrate how the degree of fulfillment of these quality criteria affects the overall quality of the data model. In one trade-off example, Implementability seems to outweigh completeness, i.e., a data model can still be of good overall quality even when completeness is not fulfilled, as long as implementability is. In contrast, a data model would be of bad overall quality, if it is neither flexible nor understandable, even if completeness is fulfilled. These trade-offs speak to the interactions between various combinations of quality criteria and overall quality of the data model, and give some insight on the process of quality characterization and assessment for NoSQL databases.

4 Discussion

In this paper, an experiment was proposed for the investigation of the adaptation of three Entity-Relationship quality frameworks for NoSQL database model quality assessment. Preliminary results describing the experiment were presented. A learning-by-example process enabled by Stack Overflow question/answer collection and mapping with good practices and guidelines was tested for its ability and was found to decrease inter-annotator disagreement, even in a subjective task such as quality assessment. Expert annotations were conducted on publicly available collected NoSQL schemas and a decision tree was leveraged for feature importance and trade-offs. In highly subjective annotation tasks, analysis of inter-annotator agreements and disagreements might make it possible to build classifiers that embody the inter-subjective overlap between the mental conceptions of the annotators [36]. Experiment replication is necessary in order to draw any further conclusions on this aspect of annotation. These preliminary results enable some elucidation of the heuristics used by experts in assessing quality in the context of NoSQL database models, and allow us to shed light on potential improvements to the adaptation of the Moody-Shanks framework in the context of NoSQL databases. Although a concise definition and example illustration of Moody-Shanks quality criteria was conducted in this work, a quantification of these criteria as metrics would add objectivity and allow for an empirical approach to quality assessment. Such metrics need to be NoSQL-specific. A few works in the literature mentioned NoSQL-specific characteristics such as schema size and denormalized schema state [37], embedding and nesting 'levels' [38], normalization and embedding [39] and aggregation [23]. These characteristics can potentially be metrics illustrating quality criteria such as flexibility and correctness. Another perspective on the improvement of the framework delineates the potential completion of the quality criteria of Moody-Shanks framework by additional criteria from other quality frameworks used in conceptual modeling and software design upon adapting them to the NoSQL context. Additionally, linking the quality criteria to universe of discourse requirements, query-first design and data quality aspects may potentially transform the quality framework and allow it to be functional in real use-cases. The findings of this study are to be considered in light of a number of limitations. The work conducted in this paper represents a first investigation of a perceived usability study and is an ongoing effort, and therefore further experiment replication, statistical analysis and hypothesis testing is paramount for any potential generalizability of findings. The subjectivity of the target variables as well as potential annotation bias represent substantial limitations, and although quality evaluation can often be subject to bias, we contend that this work is an effort to understand such subjectivity and not to eliminate it. Various threats to both internal and external validity can be highlighted. Aspects related to internal validity include (i) difference in level of expertise amongst subjects. In this experiment, all participants were evaluated with the condition of fulfilling a minimum of skills outlined by the skill classification knowledge bases used (ESCO and SkillDB) in order to ensure equivalence in skill. The challenge of (ii) knowledge of the universe of discourse was addressed by providing documentation to all participants and including details and additional context on each schema. Threats to external validity include materials used, dataset size and number of participants. These limitations, however, are relevant to the experiment and not the methodology proposed. In the context of NoSQL database model quality assessment, the gap in literature is significant, and consequently, this work aims to lay ground work and present a first effort to approach quality using a methodology combining learning-by-example, targeted crowdsourcing and white-box modeling.

5 Conclusions and Future Work

Characterizing NoSQL data model quality using Entity Relationship frameworks is a novel idea. In his work, we gauged the concept of perceived usability by defining it as the sum of perceived ease of use, perceived usefulness and perceived suitability. Results led to a real use case test of the applicability of the Moody-Shanks framework, with and without explicit application examples. These experiments have yielded insights potentially delineating the process of quality evaluation conducted by experts. Targeted crowdsourcing allowed testing of the usability, ease of use and validity of the framework in the context of NoSQL databases. White box model decision trees allowed for the determination of quality criterion importance and construction of feature trade-off diagrams. Using Stack Overflow question/answer data and mapping them as examples to illustrate the quality criteria allowed for a NoSQL-specific formalization of the Moody-Shanks framework and added a layer of relative objectivity and standardization of the annotation process. Threats to validity were explicitly highlighted and ongoing efforts are aiming to enrich criteria definitions with quantifiable metrics thus allowing for less subjectivity in the framework as well as potential automation.

Acknowledgements. Authors would like to extend thanks to all the participants for their time, efforts and contributions to the experiment.

Data Availibility Statement. All data, schemas, documentation, annotations and further illustrations and graphs used in experimentation are available online: https://github.com/ChaiAsaad/NoSQL-Quality-Experiment.

References

1. Dedeke, A.: A conceptual framework for developing quality measures for information systems. In: IQ, pp. 126–128 (2000)
2. Moody, D.L.: The method evaluation model: a theoretical model for validating information systems design methods. In: ECIS 2003 Proceedings (2003)
3. Thi, T.T.P., Helfert, M.: A review of quality frameworks in information systems. arXiv preprint arXiv:1706.03030 (2017)
4. Batini, C., Scannapieco, M.: Data Quality: Concepts, Methodologies and Techniques. Springer, Heidelberg (2006)
5. Lourenço, J.R., Abramova, V., Vieira, M., Cabral, B., Bernardino, J.: NoSQL databases: a software engineering perspective. In: Rocha, A., Correia, A.M., Costanzo, S., Reis, L.P. (eds.) New Contributions in Information Systems and Technologies. AISC, vol. 353, pp. 741–750. Springer, Cham (2015). https://doi.org/10.1007/978-3-319-16486-1_73
6. Moody, D.L., Sindre, G., Brasethvik, T., Solvberg, A.: Evaluating the quality of information models: empirical testing of a conceptual model quality framework. In: Proceedings of the 25th International Conference on Software Engineering, pp. 295–305. IEEE (2003)
7. Blin, M.-J., Tsoukiàs, A.: Multi-criteria methodology contribution to the software quality evaluation. Softw. Qual. J. **9**(2), 113–132 (2001)

8. Sánchez-González, L., García, F., Ruiz, F., Piattini, M.: Toward a quality framework for business process models. Int. J. Cooperative Inf. Syst. **22**(01), 1350003 (2013)
9. Moody, D.L., Sindre, G., Brasethvik, T., Sølvberg, A.: Evaluating the quality of process models: empirical testing of a quality framework. In: Spaccapietra, S., March, S.T., Kambayashi, Y. (eds.) ER 2002. LNCS, vol. 2503, pp. 380–396. Springer, Heidelberg (2002). https://doi.org/10.1007/3-540-45816-6_36
10. Moody, D.L.: Theoretical and practical issues in evaluating the quality of conceptual models: current state and future directions. Data Knowl. Eng. **55**(3), 243–276 (2005)
11. Eick, C.F.: A methodology for the design and transformation of conceptual schemas. In: VLDB, vol. 91, pp. 25–34 (1991)
12. Cherfi, S.S.-S., Akoka, J., Comyn-Wattiau, I.: Measuring UML conceptual modeling quality, method and implementation. In: Pucheral, P. (ed.) Proceedings of the BDA Conference, Collection INT, France (2002)
13. Shanks, G., et al.: Conceptual data modelling: an empirical study of expert and novice data modellers. Australas. J. Inf. Syst. **4**(2) (1997)
14. Kesh, S.: Evaluating the quality of entity relationship models. Inf. Softw. Technol. **37**(12), 681–689 (1995)
15. Moody, D.L., Shanks, G.G., Darke, P.: Improving the quality of entity relationship models—experience in research and practice. In: Ling, T.-W., Ram, S., Li Lee, M. (eds.) ER 1998. LNCS, vol. 1507, pp. 255–276. Springer, Heidelberg (1998). https://doi.org/10.1007/978-3-540-49524-6_21
16. Batini, C., Ceri, S., Navathe, S.B., et al.: Conceptual Database Design: An Entity-Relationship Approach, vol. 116. Benjamin/Cummings, Redwood City (1992)
17. Mackin, H., Perez, G., Tappert, C.C.: Adopting NoSQL Databases Using a Quality Attribute Framework and Risks Analysis. SCITEPRESS - Science and Technology Publications, Lda. (2016)
18. Klein, J., Gorton, I., Ernst, N., Donohoe, P., Pham, K., Matser, C.: Quality attribute-guided evaluation of NoSQL databases: an experience report. Technical report, Carnegie-Mellon Univ Pittsburgh PA Software Engineering Inst (2014)
19. Klein, J., Gorton, I., Ernst, N., Donohoe, P., Pham, K., Matser, C.: Performance evaluation of NoSQL databases: a case study. In: Proceedings of the 1st Workshop on Performance Analysis of Big Data Systems, pp. 5–10. ACM (2015)
20. Klein, J., Gorton, I.: Design assistant for NoSQL technology selection. In: 2015 1st International Workshop on Future of Software Architecture Design Assistants (FoSADA), pp. 1–6. IEEE (2015)
21. Gómez, P., Roncancio, C., Casallas, R.: Towards quality analysis for document oriented bases. In: Trujillo, J.C., et al. (eds.) ER 2018. LNCS, vol. 11157, pp. 200–216. Springer, Cham (2018). https://doi.org/10.1007/978-3-030-00847-5_16
22. Asaad, C., Baïna, K., Ghogho, M.: NoSQL databases: yearning for disambiguation. arXiv e-prints arXiv:2003.04074 (2020)
23. Asaad, C., Baïna, K.: NoSQL databases – seek for a design methodology. In: Abdelwahed, E.H., Bellatreche, L., Golfarelli, M., Méry, D., Ordonez, C. (eds.) MEDI 2018. LNCS, vol. 11163, pp. 25–40. Springer, Cham (2018). https://doi.org/10.1007/978-3-030-00856-7_2
24. Genero, M., Piattini, M., Calero, C.: Assurance of conceptual data model quality based on early measures. In: Proceedings Second Asia-Pacific Conference on Quality Software, pp. 97–103. IEEE (2001)
25. Krogstie, J.: Quality of conceptual data models. In: ICISO 2013 (2013)

26. Moody, D.L., Shanks, G.G.: What makes a good data model? Evaluating the quality of entity relationship models. In: Loucopoulos, P. (ed.) ER 1994. LNCS, vol. 881, pp. 94–111. Springer, Heidelberg (1994). https://doi.org/10.1007/3-540-58786-1_75
27. Islam, S., Hasan, K., Shahriyar, R.: Mining developer questions about major NoSQL databases. Int. J. Comput. Appl. **975**, 8887 (2021)
28. Yan, M., Xia, X., Zhang, X., Xu, L., Yang, D.: A systematic mapping study of quality assessment models for software products. In: 2017 International Conference on Software Analysis, Testing and Evolution (SATE), pp. 63–71. IEEE (2017)
29. Yan, Y., et al.: Modeling annotator expertise: learning when everybody knows a bit of something. In: Proceedings of the Thirteenth International Conference on Artificial Intelligence and Statistics, pp. 932–939. JMLR Workshop and Conference Proceedings (2010)
30. Moody, D.L., Shanks, G.G.: What makes a good data model? A framework for evaluating and improving the quality of entity relationship models. Aust. Comput. J. **30**(3), 97–110 (1998)
31. Davis, F.D.: A technology acceptance model for empirically testing new end-user information systems: theory and results. Ph.D. thesis, Massachusetts Institute of Technology (1985)
32. Davis, F.D.: Perceived usefulness, perceived ease of use, and user acceptance of information technology. MIS Q. 319–340 (1989)
33. Jalali, A.: Evaluating user acceptance of knowledge-intensive business process modeling languages. Softw. Syst. Model. 1–24 (2023)
34. Yang, J., Drake, T., Damianou, A., Maarek, Y.: Leveraging crowdsourcing data for deep active learning an application: learning intents in Alexa. In: Proceedings of the 2018 World Wide Web Conference, pp. 23–32. International World Wide Web Conferences Steering Committee (2018)
35. De Smedt, J., le Vrang, M., Papantoniou, A.: ESCO: towards a semantic web for the European labor market. In: Ldow@ WWW (2015)
36. Reidsma, D., op den Akker, R.: Exploiting 'subjective' annotations. In: Coling 2008: Proceedings of the workshop on Human Judgements in Computational Linguistics, pp. 8–16 (2008)
37. Scherzinger, S., Sidortschuck, S.: An empirical study on the design and evolution of NoSQL database schemas. arXiv preprint arXiv:2003.00054 (2020)
38. Vera, H., Boaventura, W., Holanda, M., Guimaraes, V., Hondo, F.: Data modeling for NoSQL document-oriented databases. In: CEUR Workshop Proceedings, vol. 1478, pp. 129–135 (2015)
39. Kanade, A., Gopal, A., Kanade, S.: A study of normalization and embedding in MongoDB. In: 2014 IEEE International Advance Computing Conference (IACC), pp. 416–421. IEEE (2014)

Fuzzy HealthIoT Ontology for Comorbidity Treatment

Ahlem Rhayem[1(✉)], Ishak Riali[2(✉)], Mohamed Ben Ahmed Mhiri[3], Messaouda Fareh[2], Raúl García-Castro[1], and Faiez Gargouri[3]

[1] Ontology Engineering Group (OEG), Universidad Politécnica de Madrid, Madrid, Spain
ahlemrhayem@gmail.com
[2] LRDSI Laboratory, Faculty of Sciences, University of Blida 1, Ouled Yaïch, Algeria
ishakriali@gmail.com
[3] Miracl Laboratory, University of Sfax, Sfax, Tunisia

Abstract. The utilization of Internet of Things (IoT) technologies in the medical field has resulted in the development of numerous intelligent applications and devices for health monitoring. These devices generate a large amount of data, which is collected in various formats and often exhibits uncertainty. As a consequence, interpreting and sharing these data among various medical systems poses a significant challenge. To address this challenge, ontologies, particularly fuzzy ontologies, have been employed to ensure semantic interoperability among these systems and enable them to comprehend, share, and effectively utilize fuzzy data. Therefore, to address these issues, the main objective of this paper is the fuzzification of the HealthIoT ontology. Fuzzification includes concepts related to the medical field and the IoT domain (connected objects). We showcased the application of the Fuzzy-HealthIoT ontology in a specific use case in healthcare, specifically focusing on patient comorbidity management.

Keywords: HealthIoT ontology · Fuzzy health data · Fuzzy ontology · Comorbidity management · Internet of Medical Things

1 Introduction

In recent years, we have witnessed a remarkable technological evolution that has affected various sectors such as industry, agriculture, and education. This advancement has led to a convergence of these fields, giving rise to what is now known as the Internet of Things (IoT) [22]. One particular domain where IoT has made significant strides is the medical sector, leading to the emergence of the Internet of Medical Things (IoMT) [16]. The IoMT refers to the interconnection of medical devices and their integration into healthcare networks to improve service quality [2]. Consequently, numerous heterogeneous systems and medical applications based on connected objects have been developed recently [14,15].

M. Mosbah et al. (Eds.): MEDI 2023, LNCS 14396, pp. 228–241, 2024.
https://doi.org/10.1007/978-3-031-49333-1_17

As a result of the proliferation of Medical Connected Objects (MCOs), a massive amount of heterogeneous data has been generated. These data exhibit semantic heterogeneity as they are acquired in various formats and originate from various MCOs. Moreover, many of these MCOs are mobile, changing their deployment contexts over time based on different criteria such as time and location. This dynamic nature of MCOs implies changes in their descriptions and the data they produce.

These characteristics have presented a complex challenge in the design of interoperable medical systems capable of effectively communicating and exchanging data. To ensure interoperability, it becomes crucial to have a semantic representation of MCOs, their data, and their deployment contexts, which can serve as a unified and shareable model. Ontology has emerged as a promising and efficient solution for explicitly representing IoMT knowledge and relationships.

Although various works have been proposed to address this need [6,18] one significant challenge that has often been overlooked is the uncertainty associated with the received data. In fact, the data captured by the device may exhibit fuzziness due to various factors such as sensor noise, measurement errors, or physiological variations.

For example, when monitoring a patient's heart rate, the device might record values that fluctuate slightly even when the patient is at rest. This variability can stem from factors such as motion artifacts or the device's limitations in capturing precise measurements. As a result, the heart rate data obtained become fuzzy, with imprecise boundaries and uncertain values.

When these fuzzy heart rate data are shared or integrated with other healthcare systems, interoperability challenges emerge. Different systems or platforms may have varying definitions or ranges for heart rate categories (e.g., normal, elevated, or tachycardia). For example, one system may define a heart rate of 90–100 beats per minute as elevated, while another system may define it as normal. This lack of standardized interpretation can hinder effective data exchange, analysis, and decision-making between different devices and healthcare providers.

To overcome these interoperability challenges, a fuzzy ontology tailored to the context of medical IoT can be employed. The fuzzy ontology can capture and represent imprecise data, incorporating fuzzy logic to handle uncertainty and variations [20]. It can define fuzzy sets and membership functions to categorize health data, considering factors such as age, activity level, and individual differences. Using a fuzzy ontology, healthcare systems can achieve a common understanding of fuzzy health data, allowing seamless integration, analysis, and decision support across diverse devices and IoT platforms.

From this perspective, the main objective of this work is summarized as follows.

- We extend the HealthIoT ontology to enable the representation of fuzzy data using the expressive Fuzzy OWL2 language.
- To showcase the practical applicability of our approach, we present a use case study focusing on the comorbidity management. This use case focuses

on predicting heart failure and stroke in patients with multiple diseases, such as cholesterol, hypertension, and diabetes.
- To address this use case, we utilize two distinct datasets obtained from Kaggle. The first data set pertains to heart failure detection[1], which contains 918 samples, while the second data set relates to stroke prediction[2] and contains 1200 samples. The main attributes of these data are summarized as follows.
  - Personal data: this data describes data related to the patient namely: age, sex, work type, residence type, civil situation, smoking_status.
  - Health Properties: contains the different properties specific for each predefined disease such as Resting blood pressure, Cholesterol, Fasting blood sugar, resting electrocardiogram, maximum heart rate, Oldpeak, Body Index Mass, etc.

In fact, comorbidity is characterized by the coexistence of two or more pathological conditions or diseases within the same individual [23]. For example, Long et al. [13] investigated the comorbidity between diabetes and hypertension. Various research studies have also examined the relationship between the comorbidity of COVID and chronic diseases. Kamyshnyi et al. [12] examined the comorbidity of COVID and hypertension, while Guan et al. [10] focused on a patient with cardiovascular disease.

The remaining sections of this paper are structured as follows. In Sect. 2, we provide a motivational scenario to contextualize our work. Section 2.1 outlines the key concepts utilized in our research. Section 2.2 delves into the state-of-the-art in the field. Section 3 presents the extension of the HealthIoT ontology. Subsequently, in Sect. 4, we describe the validation of the Fuzzy Health-IoT ontology. Finally, Sect. 5 concludes the paper.

2 Related Work

In this section, we focus mainly on the state of the art in ontology development in the context of the medical domain. We will first look at deterministic ontologies, and then we will describe the development of fuzzy ontologies.

2.1 Classic Ontologies in the IoMT

Ensuring the semantic interoperability in the IoT and healthcare fields led to various research issues. Elsapagh et al. [5] proposed to extend the SSN ontology in the health domain to present data from mobile objects. They developed an ontological model called FASTO designed for real-time insulin management in patients with Type 1 diabetes. FASTO is based on the high-level BFO ontology, the SSN sensor ontology, and the HL7 FHIR standard. The European Telecommunications Standards Institute (ETSI) has developed an extension of the SAREF ontology for the eHealth Ageing Well domain, known as

[1] https://www.kaggle.com/datasets/fedesoriano/heart-failure-prediction.
[2] https://www.kaggle.com/datasets/fedesoriano/stroke-prediction-dataset.

SAREF4EHAW[3]. This ontology explicitly defines and describes the coupling between components of the IoT domain and the healthcare domain. Its main concepts concern the devices used, the types of communication, the main actors, the measures and the services for the healthcare domain. The authors in [6] have proposed an ontology-based healthcare monitoring system called Do-Care that supports the supervision and follow-up of outdoor and indoor patients suffering from chronic diseases. The developed ontology is a modular and dynamic ontology composed of FOAF, SSN/SOSA and ICNP ontologies with a scalable set of inference rules. The rule bases are dynamic and adjustable to reflect changes in the drug market, medical discoveries, and personal user profiles.

Rhayem et al. [17] have proposed a semantic-enabled and context-aware monitoring system for IoMT. The developed ontology entitled "HealthIoT" represents core domain concepts of the IoT and the healthcare domain. It is based on diverse ontologies such as the SSN ontology, the IoT-lite ontology, time ontology, and so one. Then, this ontology was instantiated with vital signs obtained from medical objects. To exploit and analyze these data 65 SWRL rules are implemented to propose services related to object configuration, disease diagnosis, and notifications proposing.

2.2 Fuzzy Ontologies in the Healthcare Domain

Treating and representing the uncertainty of health data through ontology was taken into account.

Gayathri et al. [8] proposed a fuzzy ontology for activity recognition and temporal information representation(FOAR) using fuzzy logic. This ontology offers enhanced activity recognition through its semantically clear representation and reasoning via fuzzy SWRL rules for aiding activity recognition and abnormality detection for health care.

The authors in [24] proposed a decision support system that allows personalization of imprecise medical knowledge according to the progressive phases of the disease and pathological cases. A rule management process first personalizes the rules according to the specificities of each disease phase and then associates a private knowledge base to each registered patient. This base contains only the patient's personalized knowledge. After reasoning, another customization process is performed by the component, Result Manager, which ensures the validation of the system's results by the experts in case of pathology, before being recommended. The authors in [9] have proposed a type-2 fuzzy ontology for the treatment of depression. This ontology presents data about the medical devices used through the reuse of the sensor ontology and the data related to depression like the mood, the sleep length, the patient's social history, etc. through the depression ontology. A recent study has presented [7] a system that uses a probabilistic ontology to predict the diagnosis of COVID-19, considering the inherent aspects of randomness and incompleteness in knowledge. To incorporate the probabilistic elements into the COVID-19 ontology, a Multi-Entity Bayesian Network is

[3] https://saref.etsi.org/saref4ehaw/v1.1.1/.

employed, enabling robust modeling of uncertainty. The system utilizes probabilistic inference techniques, specifically the Situation-Specific Bayesian Network (SSBN), to facilitate accurate predictions of the diagnosis of COVID-19.

Another study was proposed by Riali et al. [21], have developed a system that integrates fuzzy ontologies and Bayesian networks for the diagnosis of Hepatitis C. The system uses a fuzzy ontology to effectively represent sequences of uncertain and fuzzy patient data. In fact, the system introduces a novel semantic diagnosis process that relies on a fuzzy Bayesian network as its inference engine.

The proposed study in [19] introduced a new approach that integrates hybrid models combining fuzzy logic and Bayesian networks. As part of this approach, the article proposed a language to overcome the limitations of the Probabilistic Ontology Web Language (PR-OWL) when dealing with vague and probabilistic knowledge within ontologies. To validate this proposal, a case study in the medical field, focusing on diabetes diseases, is conducted. In summary, the article offers a solution to enhance ontology-based representation and reasoning in the face of uncertainty using fuzzy multi-entity bayesian networks [11] and demonstrates its effectiveness through a medical case study.

2.3 Synthesis

The aforementioned work has shown promising results in the representation and management of uncertain healthcare data. However, certain shortcomings have been identified in these approaches, such as:

- The fuzzification was mainly focused on health data. There is no work that focuses on data related to medical connected objects.
- All works that use a fuzzy ontology in the medical domain have defined membership functions with the help of domain experts, which is difficult in most cases.

To address these shortcomings, we will focus on fuzzifying the HealthIoT ontology, which combines information related to the medical sector and the IoT domain. The main novelties of the Fuzzy HealthIoT ontology are as follows:

- Fuzzification of concepts related to the medical connected objects.
- Fuzzification of concepts related to the health domain for the management of comorbidities.
- Fuzzification of contextual concepts.
- The use of the DATIL framework and c-means to manage uncertainty in data captured by connected medical objects that are linked to comorbidity management.

To the best of our knowledge, there is no work that addresses the above-mentioned aspects in the IoMT.

3 HealthIoT-Ontology Overview

In our previous work [18], a HealthIoT-Ontology was proposed and described. Concepts in the HealthIoT ontology were classified into three categories, namely Medical-Objects Knowledge, Health care knowledge and Context Knowledge as shown in Fig. 1.

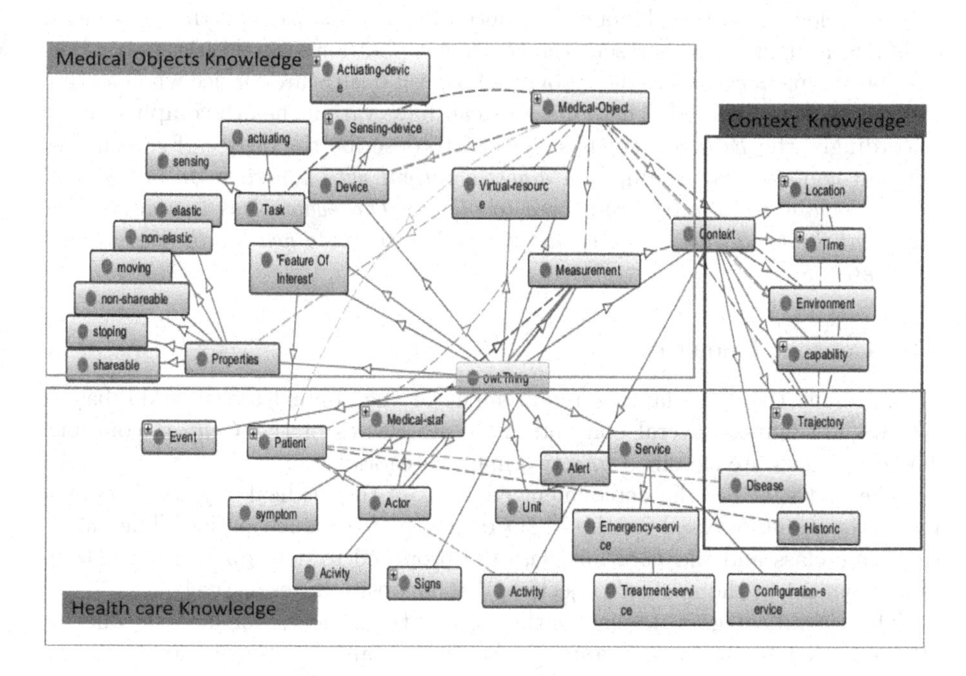

Fig. 1. HealthIoT Ontology Overview.

3.1 Medical Objects Knowledge

In this category, several concepts were proposed to represent the dynamic resources of health data and their relationships. The *Medical Object* concept designed the physical objects that healthcare professionals use to monitor their patients. Diverse objects, defined as instances of this class, such as Withings, Fitbit, scanner, etc. The *Device* class is reused from the SSN ontology [4] to represent the sensors through the sub-class *sensing-device* and the actuators through the subclass *actuating-device*, which is reused from the IoT-lite ontology [3]. In order to define the main task of the Medical object, we proposed the *Task* class that has two subclasses (sensing, actuating). Furthermore, to describe the properties of medical objects such as elastic, non-elastic, shareable, stopping, moving), we suggested the *Property* class. Moreover, to highlight our monitored phenomena, we reused the class *Feature of Interest* from the SSN ontology [4].

3.2 Health Care Knowledge

In this category, we are interested in describing the health care domain by suggesting some concepts as follows.

The *medical staff* class determines the health care professionals (doctor, nurse, surgeon) who maintain continuous monitoring of the patient. This *patient* is a subclass of the feature interest class that refers to the principal and the observed element in this domain. Furthermore, the *Disease*, and the *Treatment* classes were defined to represent the treatment plan for several diseases.

The *emergency service* class provides hospital healthcare service when a critical situation is detected. These diseases can have various health complications. Accordingly, the *Risk* concept is suggested to describe the degree of severity in different contexts. For example, *a diabetic patient with hypertension has a high risk of having heart failure compared to others. The event class was defined to represent the health events that will occur when the obtained health data exceed its threshold.*

3.3 Context Knowledge

To represent the self-adapting requirement of the context-aware IoMT-based system, we defined several concepts as a sub-classes of the *Context* concept. These concepts are classified into two main categories.

The first specifies the deployment contexts of the medical objects. It determines the points crossed by the MCO during a determined period. Therefore, the *Time* class and the *Location* class are proposed to determine the duration of deployment of the medical objects and their position, respectively.

The second category designates the state of the patient. The *disease* concept was proposed to distinguish appropriate actions and treatments. In fact, the treatment plan used by a diabetic patient is different from which is used by a diabetic patient with kidney failure. In addition to that, the *Activity* class is performed to detail the possible activities of the patient and their changes that affect the diagnosis of the disease. For example, an elevated heart rate that is an abnormal event, but it is considered a normal one with a patient in running.

HealthIoT ontology is a crisp ontology, which is limited to represent and resonate precise medical data. Indeed, the membership degrees of all concepts and properties are equal to 1. However, most patient data is vague, especially symptoms, tests, activities, signs, etc.

To meet this challenge, we propose an extension of the HealthIoT ontology. The following section describes the followed process of the fuzzification.

4 Fuzzy HealthIoT-Ontology

The main steps of our development process, as depicted in Fig. 2, will be detailed. Firstly, we will present the proposed fuzzy extension of the HealthIoT ontology, followed by an explanation of how the HealthIoT ontology was adapted to effectively represent comorbidity.

4.1 Step 1: Extending HealthIoT with Stroke and Heart Diseases

This step aims at extending the HealthIoT ontology in order to allow the representation of data on heart failure and stroke. This step is guided by a domain expert and goes through three points:

- **Identify relevant concepts:** In this step, we identify the key concepts that are relevant to representing the heart disease and the stroke. These concepts include specific cardiovascular conditions such as blood glucose level, cholesterol level, body mass index, blood pressure, heart rate.
- **Define Data properties:** Once the relevant concepts have been identified, the next step is to define the data properties within the ontology. The main identified properties from the used data sets are related to the patient and are the age, the gender, the civil situation, etc.
- **Define Object properties:** Object properties establish relationships between different concepts in the ontology.

4.2 Step 2: Fuzzification of HealthIoT Ontology

This step is divided into three main phases as shown in Fig. 2 namely the identification of fuzzy concepts, the fuzzification of the HealthIoT ontology through the DATIL framework, and representing the fuzzy part of the ontology based on additional annotations properties using the Fuzzy OWL 2 plugin[4].

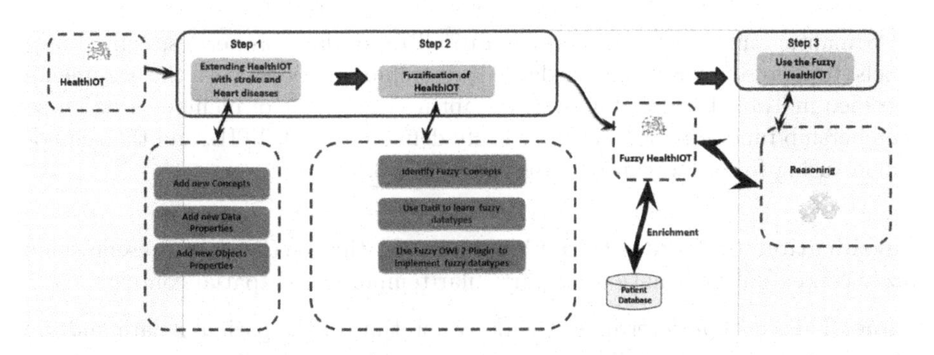

Fig. 2. Proposed process of development

Fuzzification of Patient Data: Diverse information about the patient are fuzzy like the age, and its vital signs such as the glucose level, the temperature level, the blood pressure level among others. The main fuzzy concept that describe this information is the Measurement concept.

[4] https://protegewiki.stanford.edu/wiki/FuzzyOWL2.

Property Concept is about raw data collected from connected medical objects (MCO). It designs the blood pressure (systolic and diastolic), the glycemia, the heart rate, and other vital signs, which are presented as subconcepts of the property concept.

For example, the fuzzy class **BodyMassIndex** can be defined by several subclasses: **LowBMI, NormalBMI and HighBMI**. The class definitions in Description Logic (DL) syntax are as follows:

- **LowBMI**: Represents a collection of BodyMassIndex instances with assigned values of the LowBMI data type, given by the intersection of BodyMassIndex and the existence of the property hasBMI.LowBMI.

$$LowBMI \equiv BodyMassIndex \cap \exists hasBMI.LowBMI$$

- **MediumBMI**: Represents a collection of BodyMassIndex instances with assigned values of the MediumBMI data type, given by the intersection of BodyMassIndex and the existence of the property hasBMI.MediumBMI.

$$MediumBMI \equiv BodyMassIndex \cap \exists hasBMI.MediumBMI$$

- **HighBMI**: Represents a collection of BodyMassIndex instances with assigned values of the HighBMI data type, given by the intersection of Body-MassIndex and the existence of the property hasBMI.HighBMI.

$$HighBMI \equiv BodyMassIndex \cap \exists hasBMI.HighBMI$$

Semantic annotations were applied to all fuzzy data properties, using fuzzy labels, to represent vague knowledge. Membership functions, with the arguments specified in Table 1, were employed to capture the degree of membership. These membership functions were automatically defined using DATIL, and their annotations using fuzzy DL can be observed in Fig. 3.

Fuzzification of Contextual Concepts: In this section, we present some fuzzy contextual information, in particular temporal and spatial contexts.

Time: This concept determines the detection time of the patient's data and the treatment time. For example, doctors prescribe treatment to patients three times a day after each meal. This information is uncertain and imprecise. To represent fuzzy temporal knowledge, we reused the ontology proposed by Nassira et al. [1]. In this work, the authors proposed an ontology called UncertTimeOnto that extends Allen's relations by instantiating them with 13 properties "RelationIntervals, RelationIntervalsCertainty, RelationIntevalPoint, RelationIntervalPointCertainty, RelationPointInterval, RelationPointIntervalCertainty, RelationPoints, RelationPointsCertainty" to present the degree of certainty of temporal indications.

Location: This concept is proposed to represent information about the location of the monitored patient or/and to represent the locations closest to him.

Table 1. Fuzzy Concepts

Fuzzy Concept	Fuzzy Data Property	Sub-concepts	Membership Functions
Patient	Age	YoungPatient OldPatient VeryOldPatient	leftshoulder a = 40.72 b = 53.62 triangular a = 40.72 b = 53.62 c = 64.12 rightshoulder a = 53.62 b = 64.12
Blood-Sugar	hasAvg-glucose-level	LowAvgGlucoseLevel MediumAvgGlucose-Level HighAvgGlucoseLevel	left-shoulder a = 76.37 b = 111.93 triangular a = 76.37 b = 111.93 c = 210.75 right-shoulder a = 111.93 b = 210.75
BodyMass-Index	hasBMI	LowBMI MediumBMI HighBMI	left-shoulder a = 17.11 b = 27.58 triangular a = 17.11 b = 27.58 c = 38.99 right-shoulder a = 27.58 b = 38.99
RestingBlo-odPressure	hasRestingBP	LowRestingBP MediumRestingBP HighRestingBP	leftshoulder a = 115.0 b = 135.46 triangular a = 115.0 b = 135.46 c = 160.96 rightshoulder a = 135.46 b = 160.96
Cholesterol	hasCholesterol	LowCholesterol HighCholesterol MediumCholestterol	leftshoulder a = 1.83 b = 211.3 rightshoulder a = 211.3 b = 297.63 triangular a = 1.83 b = 211.3 c = 297.63
HeartRate	hasMaxHeart-Rate	LowMaxHeartRate MediumMaxHeartRate HighMaxHeartRate	leftshoulder a = 105.19 b = 135.7 triangular a = 105.19 b = 135.7 c = 166.74 rightshoulder a = 135.7 b = 166.74

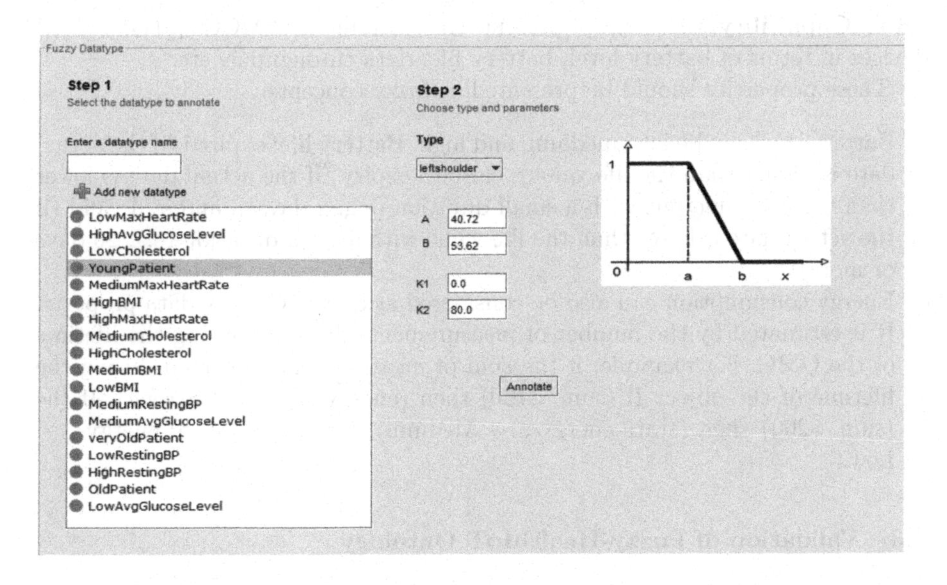

Fig. 3. Fuzzy data types

For example, when a patient has a drop in blood pressure and feels dizzy, his MCO object should send an alert to the nearest hospital. This alert contains detailed information about the location of the patient, as for example "the X patient is actually in the central park that is behind the Carrefour market and he is sitting near the lotus tree".

In order to model and present this fuzzy and uncertain knowledge, we suggest diverse fuzzy relations that will be assigned between places. In the following table we detail these relations.

Table 2. Fuzzy Location relations.

Approaches	Semantic languages	Domains
Before (l1, l2)	place L1 is before place l2	After (l1, L2)
Behind (l1, l2)	place l1 is behind the place l2	in front-of (l1, l2)
under (l)	under place l	up(l)
near (l1, l2)	place l1 is near to l2	far-away(l1, l2)
at-right(l1, l2)	place l1 is at right of place l2	at left (l1, l2)

Fuzzification of MCO Concepts. This section focuses on the fuzzification of information about connected objects. Indeed, the uncertain concepts are the following:

HIoT:Capability: this concept presents the capabilities of MCO and embedded devices in terms of battery level, battery life, data throughput, etc.

These properties should be presented as fuzzy concepts.

- Battery level can be low, medium, and high. Battery life: expired (if the actual date is higher than the life value), close to expiry (if the actual date is lower than the life value but with a small duration (e.g. 2 days)), and still valid (if the actual date is lower than the life value with a good duration (e.g. 20 days or more)).
- Energy consumption can also be considered as an OMC fuzzy data property. It is estimated by the number of measurements detected during the lifetime of the OMC. For example, if the sum of measurements detected during the lifetime of the object If (sum <160) then (energy-state $< -$Strong). If the (sum >200) then (state-energy$< -$ Medium). Otherwise, state energy$< -$ Low.

4.3 Validation of Fuzzy-HealthIoT Ontology

To validate our ontology, we propose a process as shown in the following figure. This process consists of three main phases:

- The first step is to collect the data from the medical objects and pre-process them. At this stage, we use OpenRefine[5] software. The main operations are: deleting missing data, modifying the time format, dividing health data into different levels (high, medium, low).
- The second step is the construction of a fuzzy knowledge graph. In this phase, we plan to use the fuzzy HealthIoT ontology and the fuzzy Bayesian network. After that, various rules will be defined for the treatment of comorbidity.
- The final step is to implement a fuzzy medical system that helps doctors and patients obtain appropriate services using SPARQL rules (Fig. 4).

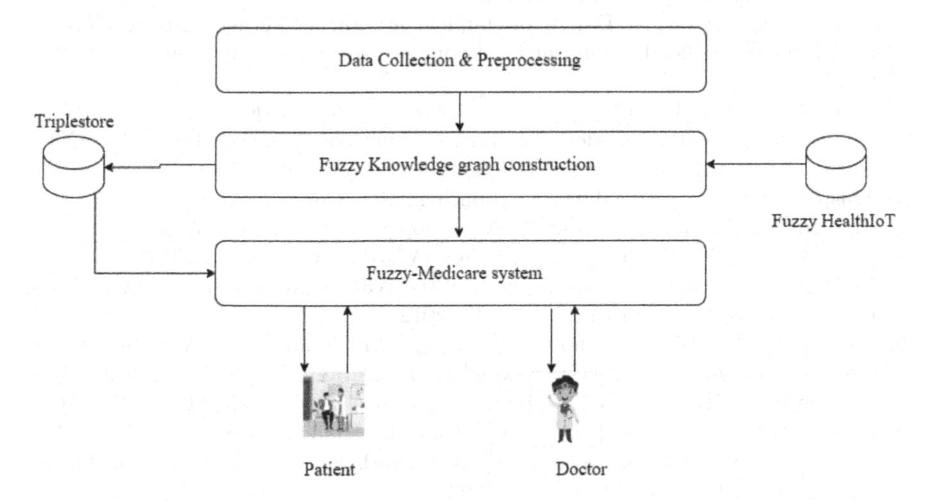

Fig. 4. Validation Process

5 Conclusion

In this article, we have proposed a solution to the challenge of data uncertainty in the context of the Internet of Medical Things. We focus mainly on comorbidity management. In fact, we extend the HealthIoT ontology to represent the complex relationship between heart disease and stroke. Through the use of the DATIL framework, we have proposed a fuzzy extension to the ontology, allowing the incorporation of fuzzy data types learned from real-world data. This fuzzy ontology was implemented using the Fuzzy OWL 2 language.

In future work, we plan to implement the validation process by fuzzy medical system that combines the HealthIoT fuzzy ontology and the fuzzy Bayesian network for the treatment of comorbidities.

[5] https://openrefine.org/.

Acknowledgements. This work is partially funded by the Madrid Government (Comunidad de Madrid-Spain) under the Multiannual Agreement with the Universidad Politécnica de Madrid in the Excellence Programme for University Teaching Staf, in the context of the V PRICIT (Regional Programme of Research and Technological Innovation).

References

1. Achich, N., Ghorbel, F., Hamdi, F., Métais, E., Gargouri, F.: Certain and uncertain temporal data representation and reasoning in OWL 2. Int. J. Semant. Web Inf. Syst. (IJSWIS) **17**(3), 51–72 (2021)
2. Aljabr, A.A., Kumar, K.: Design and implementation of Internet of Medical Things (IoMT) using artificial intelligent for mobile-healthcare. Measur. Sens. **24**, 100499 (2022)
3. Bermudez-Edo, M., Elsaleh, T., Barnaghi, P., Taylor, K.: IoT-lite: a lightweight semantic model for the Internet of Things. In: 2016 INTL IEEE Conferences on Ubiquitous Intelligence & Computing, Advanced and Trusted Computing, Scalable Computing and Communications, Cloud and Big Data Computing, Internet of People, and Smart World Congress (UIC/ATC/ScalCom/CBDCom/IoP/SmartWorld), pp. 90–97. IEEE (2016)
4. Compton, M., et al.: The SSN ontology of the W3C semantic sensor network incubator group. J. Web Semant. **17**, 25–32 (2012)
5. El-Sappagh, S., Ali, F., Hendawi, A., Jang, J.H., Kwak, K.S.: A mobile health monitoring-and-treatment system based on integration of the SSN sensor ontology and the HL7 FHIR standard. BMC Med. Inform. Decis. Making **19**(1), 97 (2019)
6. Elhadj, H.B., Sallabi, F., Henaien, A., Chaari, L., Shuaib, K., Al Thawadi, M.: Do-Care: a dynamic ontology reasoning based healthcare monitoring system. Future Gener. Comput. Syst. **118**, 417–431 (2021)
7. Fareh, M., Riali, I., Kherbache, H., Guemmouz, M.: Probabilistic reasoning for diagnosis prediction of Coronavirus disease based on probabilistic ontology. Comput. Sci. Inf. Syst. **20**(3), 1109–1132 (2023)
8. Gayathri, K., Easwarakumar, K., Elias, S.: Fuzzy ontology based activity recognition for assistive health care using smart home. Int. J. Intell. Inf. Technol. (IJIIT) **16**(1), 17–31 (2020)
9. Ghorbani, A., Davoodi, F., Zamanifar, K.: Using type-2 fuzzy ontology to improve semantic interoperability for healthcare and diagnosis of depression. Artif. Intell. Med. **135**, 102452 (2023)
10. Guan, W.J., Liang, W.H., He, J.X., Zhong, N.S.: Cardiovascular comorbidity and its impact on patients with COVID-19. Eur. Respir. J. **55**(6), 2001227 (2020)
11. Ishak, R., Messaouda, F., Hafida, B.: FzMEBN: toward a general formalism of fuzzy multi-entity Bayesian networks for representing and reasoning with uncertain knowledge. In: International Conference on Enterprise Information Systems, vol. 2, pp. 520–528. SCITEPRESS (2017)
12. Kamyshnyi, A., Krynytska, I., Matskevych, V., Marushchak, M., Lushchak, O., et al.: Arterial hypertension as a risk comorbidity associated with COVID-19 pathology. Int. J. Hypertens. **2020**, 8019360 (2020)
13. Long, A.N., Dagogo-Jack, S.: Comorbidities of diabetes and hypertension: mechanisms and approach to target organ protection. J. Clin. Hypertens. **13**(4), 244–251 (2011)

14. Ma, Y., Wang, Y., Yang, J., Miao, Y., Li, W.: Big health application system based on health Internet of Things and big data. IEEE Access **5**, 7885–7897 (2016)
15. Mbengue, S.M., Diallo, O., El Hadji, M.N., Rodrigues, J.J., Neto, A., Al-Muhtadi, J.: Internet of medical things: remote diagnosis and monitoring application for diabetics. In: 2020 International Wireless Communications and Mobile Computing (IWCMC), pp. 583–588. IEEE (2020)
16. Razdan, S., Sharma, S.: Internet of medical things (IoMT): overview, emerging technologies, and case studies. IETE Tech. Rev. **39**(4), 775–788 (2022)
17. Rhayem, A., Mhiri, M.B.A., Drira, K., Tazi, S., Gargouri, F.: A semantic-enabled and context-aware monitoring system for the internet of medical things. Expert. Syst. **38**(2), e12629 (2021)
18. Rhayem, A., Mhiri, M.B.A., Gargouri, F.: HealthIoT ontology for data semantic representation and interpretation obtained from medical connected objects. In: 2017 IEEE/ACS 14th International Conference on Computer Systems and Applications (AICCSA), pp. 1470–1477. IEEE (2017)
19. Riali, I., Fareh, M., Bouarfa, H.: Fuzzy probabilistic ontology approach: a hybrid model for handling uncertain knowledge in ontologies. Int. J. Semant. Web Inf. Syst. (IJSWIS) **15**(4), 1–20 (2019)
20. Riali, I., Fareh, M., Bouarfa, H.: A semantic approach for handling probabilistic knowledge of fuzzy ontologies. In: ICEIS (1), pp. 407–414 (2019)
21. Riali, I., Fareh, M., Ibnaissa, M.C., Bellil, M.: A semantic-based approach for hepatitis C virus prediction and diagnosis using a fuzzy ontology and a fuzzy Bayesian network. J. Intell. Fuzzy Syst. **44**(2), 2381–2395 (2023)
22. Rose, K., Eldridge, S., Chapin, L.: The Internet of Things: an overview. Internet Soc. (ISOC) **80**, 1–50 (2015)
23. Valderas, J.M., Starfield, B., Sibbald, B., Salisbury, C., Roland, M.: Defining comorbidity: implications for understanding health and health services. Ann. Family Med. **7**(4), 357–363 (2009)
24. Zekri, F., Ellouze, A.S., Bouaziz, R.: A fuzzy-based customisation of healthcare knowledge to support clinical domestic decisions for chronically ill patients. J. Inf. Knowl. Manag. **19**(04), 2050029 (2020)

Healthcare Applications

Advancing Brain Tumor Segmentation via Attention-Based 3D U-Net Architecture and Digital Image Processing

Eyad Gad[1], Seif Soliman[1], and M. Saeed Darweesh[1,2](✉)

[1] School of Engineering and Applied Sciences, Nile University, Giza 12677, Egypt
{e.gad,s.soliman,mdarweesh}@nu.edu.eg
[2] Wireless Intelligent Networks Center (WINC), Nile University, Giza 12677, Egypt

Abstract. In the realm of medical diagnostics, rapid advancements in Artificial Intelligence (AI) have significantly yielded remarkable improvements in brain tumor segmentation. Encoder-Decoder architectures, such as U-Net, have played a transformative role by effectively extracting meaningful representations in 3D brain tumor segmentation from Magnetic resonance imaging (MRI) scans. However, standard U-Net models encounter challenges in accurately delineating tumor regions, especially when dealing with irregular shapes and ambiguous boundaries. Additionally, training robust segmentation models on high-resolution MRI data, such as the BraTS datasets, necessitates high computational resources and often faces challenges associated with class imbalance. This study proposes the integration of the attention mechanism into the 3D U-Net model, enabling the model to capture intricate details and prioritize informative regions during the segmentation process. Additionally, a tumor detection algorithm based on digital image processing techniques is utilized to address the issue of imbalanced training data and mitigate bias. This study aims to enhance the performance of brain tumor segmentation, ultimately improving the reliability of diagnosis. The proposed model is thoroughly evaluated and assessed on the BraTS 2020 dataset using various performance metrics to accomplish this goal. The obtained results indicate that the model outperformed related studies, exhibiting dice of 0.975, specificity of 0.988, and sensitivity of 0.995, indicating the efficacy of the proposed model in improving brain tumor segmentation, offering valuable insights for reliable diagnosis in clinical settings.

Keywords: Brain Tumor Segmentation (BraTS) · Artificial Intelligence (AI) · Convolutional Neural Networks (CNNs) · Attention U-Net · 3D MRI Segmentation · Digital Image Processing

1 Introduction

A brain tumor is a collection of abnormal cells that grows in the brain or central spine canal. These cells undergo a process called mitosis, where a single

M. Mosbah et al. (Eds.): MEDI 2023, LNCS 14396, pp. 245–258, 2024.
https://doi.org/10.1007/978-3-031-49333-1_18

cell duplicates its entire contents, including chromosomes, and divides into two identical daughter cells. Consequently, an abnormal mass, which can be either non-cancerous (benign) or cancerous (malignant), forms. Depending on their location and size, brain tumors can manifest a range of symptoms, such as headaches, seizures, cognitive decline, mood alterations, and impaired movement or coordination. Brain tumors can affect people of all ages, but they are more common in older adults. According to the Global Cancer Statistics, there are approximately 300,000 new cases of brain tumors diagnosed globally each year [1]. Thus, early detection and treatment play a crucial role in managing brain tumors and enhancing patients' quality of life.

Recent advancements in AI have revolutionized the field of medical imaging, particularly in the domain of brain malignancies. AI technologies have significantly improved the segmentation, identification, and survival prediction of tumors and other diseases [2–4]. By accurately delineating brain tumors from medical images, AI facilitates early diagnosis, leading to better prognostic capabilities and informed decision-making by medical professionals. Furthermore, AI algorithms enable the tracking of tumor changes over time, aiding in disease monitoring and providing real-time support during surgical interventions [4,5]. The automation of image analysis through AI not only reduces the risk of errors but also enhances efficiency within the healthcare system. Additionally, AI has the potential to facilitate remote consultations, reducing the need for in-person visits and lowering associated costs. As AI continues to advance, its transformative impact on brain tumor segmentation and patient care will continue to expand.

In the past, classical machine learning techniques were commonly employed for segmentation tasks until the advent of deep learning techniques, such as CNNs, revolutionized the field. CNNs, specifically, have been widely adopted for MRI segmentation, with a focus on brain tumor segmentation. Architectures like U-Net have demonstrated high performance by effectively capturing local features and extracting meaningful representations from MRI scans [6]. While traditional U-Net models have achieved considerable success in numerous studies, they may encounter challenges when accurately distinguishing between tumor regions and healthy brain tissue, especially in cases where tumors have irregular shapes or indistinct boundaries. This can lead to imprecise segmentation outcomes, making it difficult to precisely locate tumors. Furthermore, the computational requirements for brain tumor segmentation are noteworthy, as achieving accurate segmentation heavily relies on substantial computational power for training U-Net models on high-resolution MRI scans. Scaling up these models to handle large datasets and effectively testing novel concepts and adjustments exacerbates these challenges [7–9].

Another challenge arises from the imbalanced distribution of pixels across the different classes, with some classes having significantly fewer samples compared to others. This class imbalance complicates the training process as the model tends to prioritize the majority class, leading to biased results. The limited representation of minority classes can result in their underestimation or misclassification during the segmentation process.

The study proposes the integration of the attention mechanism into the U-Net model to mitigate the challenges in brain tumor segmentation. By incorporating the attention mechanism, the model can selectively focus on relevant regions and prioritize important features during segmentation [10], resulting in improved differentiation between tumor regions and healthy brain tissue, even in scenarios involving irregular tumor shapes or unclear boundaries. To address the class imbalance, a tumor detection method based on digital image processing is employed as a data preparation step to detect tumors in scans, which balances the classes prior to training and reduces bias in the segmentation results.

To sum up, this study makes two key contributions. Firstly, it integrates the attention mechanism into the U-Net model. Secondly, it employs a tumor detection method based on digital image processing techniques to address class imbalance and reduce bias in the segmentation results. These contributions aim to enhance the accuracy and reliability of brain tumor segmentation, leading to improved diagnosis and treatment planning.

2 Literature Review

Over the past few years, several approaches have been proposed for brain tumor segmentation, ranging from traditional image-processing techniques to deep learning-based methods. Throughout the upcoming literature review, an overview is provided containing recent advances in brain tumor segmentation methods and techniques, highlighting their strengths and limitations.

Starting with Montaha et al. [11], which proposed a 2D U-Net implementation for brain tumor segmentation that uses a single slice of 3D MRI to minimize computational cost while achieving high performance. MRI intensity normalization is employed as a pre-processing technique, converting minimum values to 0 and maximum values to 1. Rescaling is applied to the middle single slices instead of all 155 slices, reducing computational complexity. The segmentation pipeline uses a 2D U-Net architecture with a compact encoder for feature extraction and a decoder for image reconstruction, incorporating skip connections to reduce information loss and address the vanishing gradient problem. The model is separately trained and validated on four MRI modalities and labeled segmented ROI images. The proposed model achieved a dice score of 0.9386, accuracy of 0.9941, and sensitivity of 0.9897 on the BraTS2020 dataset, demonstrating its potential for clinical application.

Another recently proposed approach from Ilhan et al. [12] introduces a brain tumor segmentation system that uses non-parametric tumor localization and enhancement methods with a U-Net implementation. The study heavily relies on pre-processing techniques and a U-Net architecture, and the data preparation modules used for this study comprise 2D axial images obtained from patients' FLAIR modalities. To determine the background and tumorous regions in brain MRI scans, image histograms are obtained, and a non-parametric threshold value is calculated using the frequency of intensity values. In low-contrast MRI scans, the standard deviation is used to enhance the distinguishability and contrast between the background and tumorous regions, allowing easy localization of tumors. The enhanced images are fed into the U-Net architecture for segmentation, where the contraction path (encoder) and expansion path (decoder) enable the architecture to learn fine-grained details. The system achieved outstanding segmentation performance on three benchmark datasets, with dice scores of 0.94, 0.87, and 0.88 for BraTS2012 HGG-LGG, BraTS2019, and BraTS2020, respectively.

Moreover, N. Cinar et al. [13] developed a novel hybrid architecture with a pre-trained DenseNet121 and U-Net architecture to identify multiple tumor subregions for the segmentation process. In terms of data preparation, MRI images were enhanced using clipping and cropping into a (64×64) size whilst centering the tumor in the image. The images were also divided into smaller parts to shorten the training period. The Otsu threshold technique was also applied as it's tasked to differentiate between-class variance of foreground (tumor) and background pixels, and z-score normalization was used to ensure homogeneity. The architecture of the proposed model combines the DenseNet121 architecture, which is pre-trained on ImageNet, and a U-Net architecture implementation. DenseNet121 network was used as the encoder which is responsible for feature extraction, and the fully connected layer was removed and replaced with a decoder. Skip connections were used to transmit the attributes of the input image to the decoder layers. The segmented tumor areas are concatenated together in order to obtain the fully segmented image. This model was validated on the BraTS2019 dataset whilst disregarding all T1 scans due to lack of resolution. Dice coefficient obtained for the sub-regions are as follows WT 0.959, CT 0.943 and ET 0.892.

Lastly, Raza et al. [14] proposed architecture, referred to as the Deep Residual U-Net (dResU-Net), which is an end-to-end encoder-decoder-based model that utilizes residual blocks with shortcut connections in the encoder part of the U-Net model. The authors utilized image standardization and normalization on all MRI images, resizing them to $128 \times 128 \times 128$ dimensions and stacking them together to form a $128 \times 128 \times 128 \times 4$ input. The proposed architecture utilized a U-Net with residual blocks in the encoder part to extract low and high-level features, overcoming the vanishing gradient problem whilst transferring feature maps of each encoder level to its corresponding decoder level using the skip connection. The bottleneck and decoder parts contain plain convolutional blocks for predicting segmentation masks, while the expanding path recovers the image into its original shape using the traditional up-sampling functions. The proposed model was cross-validated on an external dataset to evaluate its robustness, achieving a Dice Score of 0.8357, 0.8660, and 0.8004 for BraTS 2020 CT, WT, and ET, respectively.

The recent advancements in brain tumor segmentation techniques, particularly in U-Net variants, have become an increasingly popular choice for brain tumor segmentation due to their ability to effectively learn and extract features from medical images. The various modifications and improvements made to the original U-Net architecture have addressed some of its limitations, such as the vanishing gradient problem, and have resulted in improved segmentation performance.

3 Materials and Methods

3.1 Proposed Approach

This paper presents a novel approach for brain tumor segmentation, as illustrated in Fig. 1. The proposed method involves several steps, starting with the partitioning of MRI scans from the BraTS dataset into training and validation sets. Preprocessing techniques, such as cropping, resizing, and normalization, are applied, along with a tumor detection algorithm based on digital image processing, which focuses on the tumor region to improve class balancing within the training set. Subsequently, the attention U-Net model is then trained on the preprocessed data and evaluated using performance metrics. This comprehensive approach aims to achieve precise and efficient brain tumor segmentation.

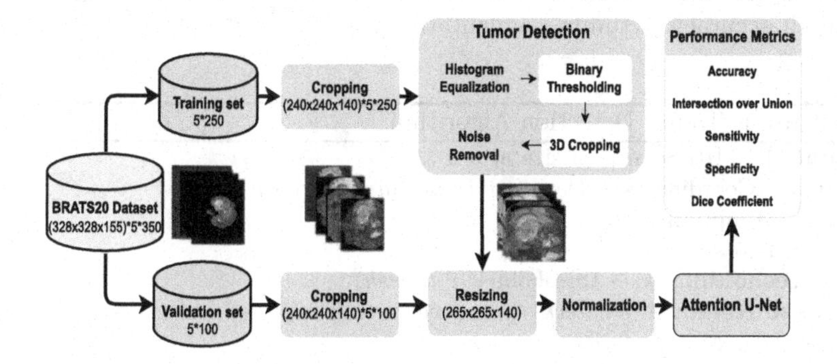

Fig. 1. The pipeline of the proposed approach for 3D BraTS

3.2 Dataset Sources and Preparation

In the BraTS dataset, brain tumor segmentation involves classifying pixels into distinct classes representing three non-overlapping subregions: edema, enhancing tumor, and necrotic core or non-enhancing tumor (NCR/NET). These subregions have different biological properties and are crucial for tumor characterization. In this study, three commonly used regions of interest (ROIs) or classes based on these subregions are utilized.

The first ROI of interest is the whole tumor (WT), which includes all three subregions and provides a comprehensive representation of the tumor's extent.

The second ROI is the tumor core (TC), focusing on the NCR/NET and enhancing tumor subregions for characterization and treatment planning. Lastly, the third ROI is the enhancing tumor (ET), targeting the area that shows enhancement and providing important information for clinical decisions.

In this study, the BraTS2020 dataset was examined, which consists of 350 MRI scans specifically focusing on glioma brain tumors. These scans encompass four different modalities, namely T1-weighted (T1), T1-weighted with contrast enhancement (T1ce), T2-weighted (T2), and Fluid-Attenuated Inversion Recovery (FLAIR). Brain tumors can emerge in various regions of the brain and can vary significantly in terms of their size and shape. Additionally, the intensity of tumor tissue can overlap with that of healthy brain tissue, presenting a significant challenge in distinguishing between the two. For instance, in T1 MRI scans, a bright tumor border might be visible, while the tumor area itself might be highlighted in T2 MRI scans. Moreover, FLAIR MRI scans assist in differentiating edema from cerebrospinal fluid. To tackle this challenge, a fundamental approach involves integrating data from multiple MRI modalities, such as T1, T1ce, T2, and FLAIR. By combining information from these different modalities, it becomes possible to obtain a more comprehensive and accurate representation of the tumor and its surrounding tissues. This multi-modal approach improves the ability to differentiate between tumor regions and healthy brain tissue, enhancing the overall accuracy of tumor segmentation.

Algorithm 1. Tumor Detection Algorithm

1: **Input**: 3D MRI Scan V of size $m \times n \times s$
2: **Output**: Coordinates of Detected Tumor *largest_coor*
3: *coors* $\leftarrow$ Empty list
4: **for** $i = 1$ to s **do**
5:     **Thresholding:** $k \leftarrow$ threshold($V[i]$,*thresh*)
6:     **Noise Removal:** Dilate(k) to connect nearby objects
7:     **for** each *object* in k **do**
8:         **if** area of *object* $<$ area_thresh **then**
9:             Eliminate *object*
10:         **end if**
11:     **end for**
12:     *coors* $\leftarrow$ coordinates of square points containing large objects in k
13: **end for**
14: **Tumor Cropping:**
15: *largest_area* $\leftarrow$ area of *coors*[0]
16: *largest_coor* $\leftarrow$ *coors*[0]
17: **for** *cur_coor* $= 1$ to len(*coors*) **do**
18:     **if** area of *coors*[*cur_coor*] $>$ *largest_area* **and** region of *coors*[*cur_coor*] includes region of *largest_coor* **then**
19:         *largest_area* $\leftarrow$ area of *coors*[*cur_coor*]
20:         *largest_coor* $\leftarrow$ *coors*[*cur_coor*]
21:     **end if**
22: **end for**

Training on a dataset of 250 scans, each consisting of 5 dimensions (4 modalities and a mask), with each dimension having 328 × 328 × 155 pixels, would be inefficient and impractical. To address this, preprocessing procedures are undertaken to eliminate irrelevant areas such as the background and unaffected healthy tissues, and enhance tumor detection. The first step involves cropping the scans and their corresponding masks to constrain them to the brain region defined by assigned coordinates, ensuring that subsequent analysis focuses only on the relevant areas for tumor detection. Next, a tumor detection algorithm is applied, utilizing multiple image processing techniques to enhance accuracy. The algorithm begins with histogram equalization, which improves the contrast and visibility of tumor regions by making darker and brighter regions more distinguishable, making them more prominent and easier to detect. The algorithm, as illustrated in Algorithm 1, employs thresholding to differentiate between tumor and non-tumor regions based on a specific intensity threshold. Pixels with intensities above the threshold are classified as part of the tumor region and vice versa. However, it is important to note that fully distinguishing the tumor region can be challenging due to healthy regions with tumor-like intensities, leading to false positives. As shown in Fig. 2, small objects considered noise in comparison to the large tumor object.

To address noise and refine tumor detection further, noise removal techniques are applied. Initially, dilation is employed to connect nearby large objects to the tumor region and create thickness in the boundary. This ensures that when the detected tumor is cropped, the entire tumor is included within the cropped object. Subsequently, the algorithm iterates through the objects and eliminates those with small areas, effectively removing small false positives or noise that may have been detected.

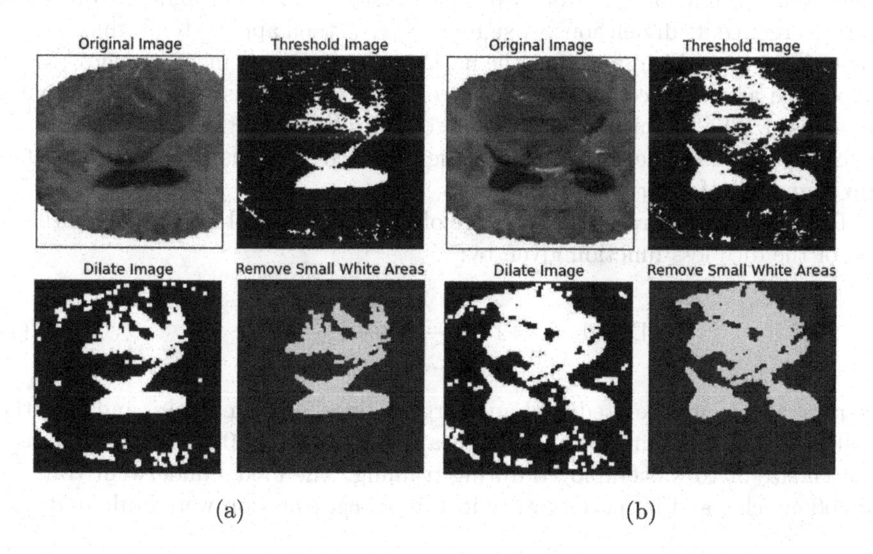

Fig. 2. Tumor detection results

After applying these steps to each 2D scan in the MRI volume and obtaining the detected tumor for each scan, the next stage involves cropping the tumor to encompass the entire volume. This is achieved by determining the coordinates that encapsulate the tumor throughout the entire MRI volume. The algorithm iterates through the detected tumor coordinates obtained from each 2D scan and identifies the largest area. By examining the coordinates of this largest area, it can be determined whether they encompass all the other detected tumor coordinates. This ensures that the resulting cropping includes the entire tumor within the MRI volume, providing a comprehensive representation of the tumor across multiple scans. Lastly, all the cropped scans are resized and normalized, preparing them for training.

3.3 Attention U-Net

In this study, the attention U-Net architecture was utilized for segmenting brain tumor. The attention U-Net consists of a contracting path, which includes convolutional and max pooling layers, followed by an expanding path with convolutional and up-sampling layers. What sets the attention U-Net apart from a standard U-Net is the incorporation of an attention mechanism or attention gate in the skip connections, as it improves the network's ability to focus on relevant features while disregarding irrelevant ones, thereby enhancing the accuracy of tumor segmentation. As illustrated in Fig. 3, the attention gate involves two input vectors: X and G. Vector G is derived from the lower layer of the network and has smaller dimensions but better feature representation. Vector X undergoes a strided convolution, while vector G undergoes a 1×1 convolution. These vectors are then combined through element-wise summation, resulting in a new vector. This new vector is passed through activation and convolutional layers to reduce its dimensions. A sigmoid layer is then applied to produce attention coefficients, which represent the importance of each element in vector X. To restore the attention coefficients to the original dimensions, trilinear interpolation is used. The attention coefficients are multiplied with the original vector X, scaling it based on relevance. Finally, the modified vector is passed through the skip connection for further processing.

The brain tumor training procedure of the U-Net model included the utilization of the dice loss function given by:

$$\text{Dice Loss} = 1 - \frac{2 \cdot \sum_{i=1}^{N} p_i \cdot t_i}{\sum_{i=1}^{N} p_i^2 + \sum_{i=1}^{N} t_i^2}, \quad (1)$$

where p_i and t_i are the predicted and target values for each pixel i, and N is the total pixels. The Adam optimizer with a learning rate of 0.0001 was used, and a batch size of 16 was employed during training. The model underwent training for 100 epochs, and its performance in tumor segmentation were evaluated.

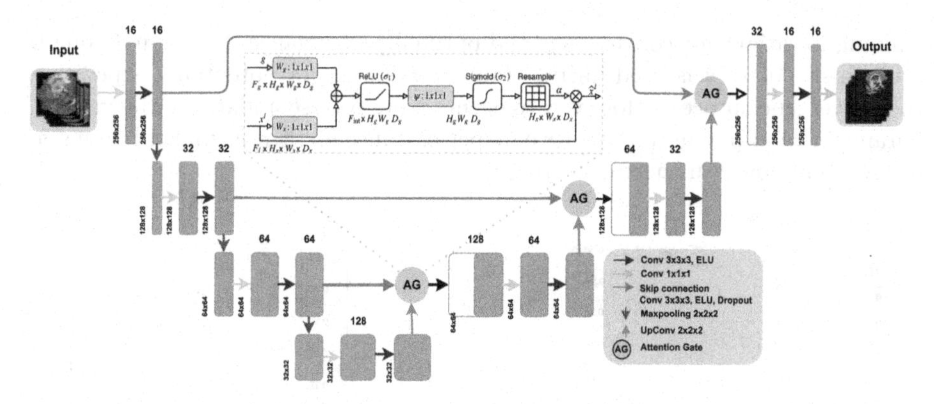

Fig. 3. The architecture of 3D U-Net model with attention gate

3.4 Performance Metrics

To evaluate the model's performance, various metrics, including Accuracy, Dice Coefficient, Intersection over Union (IoU), Sensitivity, and Specificity were employed. These metrics were assessed during both the training and validation phases using a confusion matrix that includes true positives (TP), false positives (FP), false negatives (FN), and true negatives (TN).

The Dice Coefficient is a widely-used metric to measure pixel-level agreement between predicted segmentation and ground truth. It is computed by subtracting the Dice Loss from 1. Additionally, the IoU metric, also known as the Jaccard index, quantifies the overlap between predicted and true positive regions [15]. It is calculated as:

$$\mathrm{IoU} = \frac{\mathrm{TP}}{\mathrm{TP} + \mathrm{FP} + \mathrm{FN}} \tag{2}$$

Additionally, Sensitivity, also known as recall or true positive rate, measures the proportion of actual positives correctly identified by the model. It can be calculated using the following formula:

$$\mathrm{Sensitivity} = \frac{\mathrm{TP}}{\mathrm{TP} + \mathrm{FN}} \tag{3}$$

Furthermore, Specificity evaluates the proportion of actual negatives correctly identified by the model and is computed as:

$$\mathrm{Specificity} = \frac{\mathrm{TN}}{\mathrm{TN} + \mathrm{FP}} \tag{4}$$

These performance metrics provide valuable insights into the effectiveness and accuracy of the model in accurately capturing true positive regions.

4 Experimental Results

The training process was carried out in two rounds, each consisting of 50 epochs, determined through empirical evaluation and prior research, ensuring sufficient training iterations for convergence and optimal performance. In the first round, a batch size of 16 was used during training, allowing for potentially expediting the convergence process. However, to ensure more precise updates to the model's parameters and potentially improve its performance further, a smaller batch size of 8 was employed in the second round.

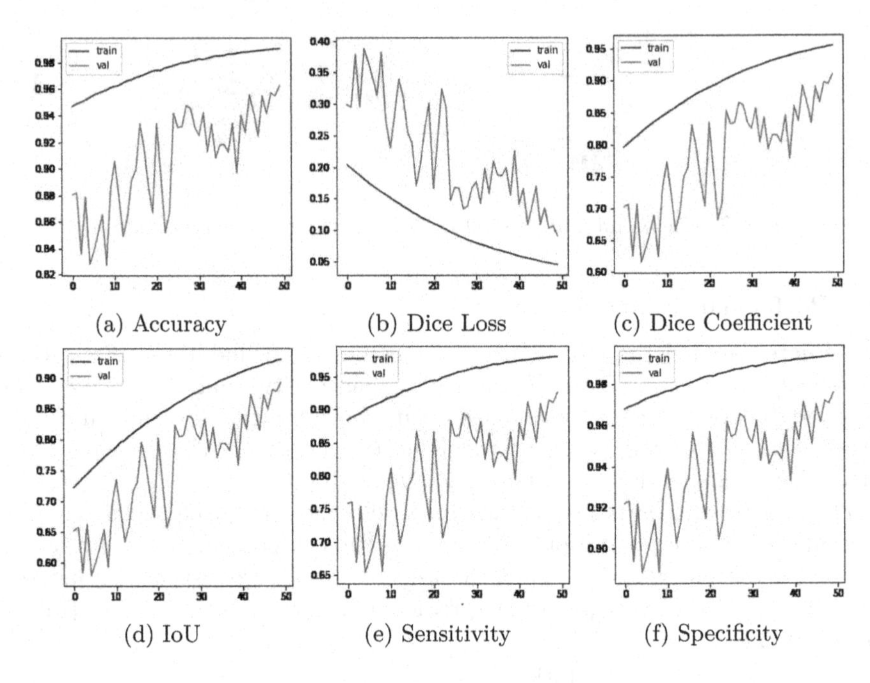

(a) Accuracy (b) Dice Loss (c) Dice Coefficient

(d) IoU (e) Sensitivity (f) Specificity

Fig. 4. Performance metrics vs. epochs for the first round

During the initial round of training with a batch size of 16, the attention U-Net model demonstrated steady and promising improvement in performance, as depicted in Fig. 4. The accuracy of the model started at 0.88 and consistently increased, reaching an impressive value of 0.96 at the end of the 50 epochs. Similarly, other evaluation metrics showed continuous progress as well. The dice coefficient and IoU, which assess the agreement between the model's predictions and the ground truth segmentations, also exhibited a positive trend, achieving values of 0.9 and 0.875, respectively, by the end of this round. Additionally, sensitivity and specificity metrics, which measure the model's ability to correctly identify positive and negative cases, also displayed favorable trends throughout the training process. Sensitivity increased from 0.75 to 0.9, while specificity improved from 0.92 to 0.97.

Building upon the progress achieved in the first round, the attention U-Net model's performance further excelled during the second round of training with a batch size of 8. As illustrated in Fig. 5, the model continued to demonstrate exceptional performance across various metrics after an additional 50 epochs.

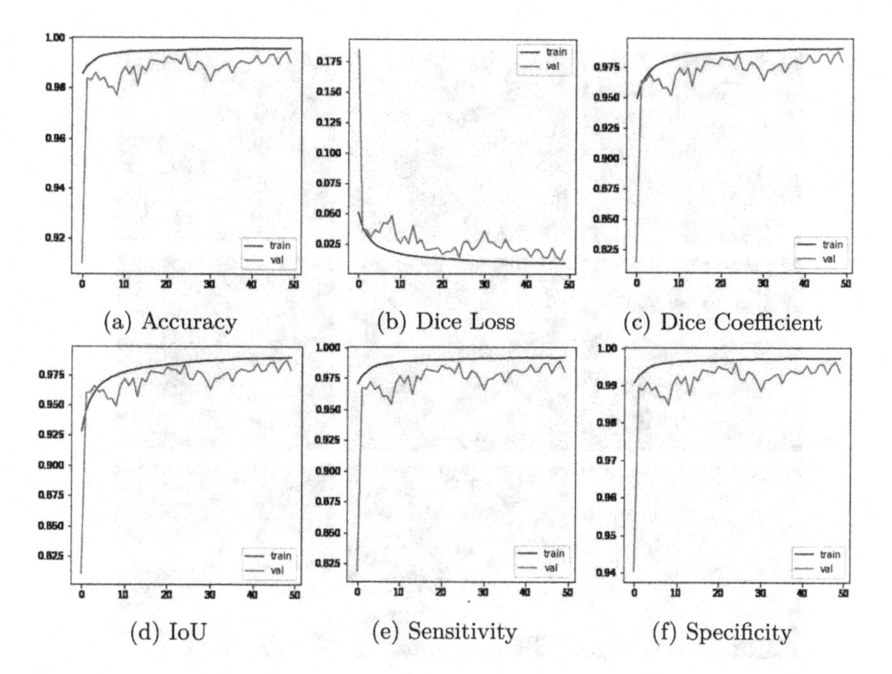

(a) Accuracy (b) Dice Loss (c) Dice Coefficient

(d) IoU (e) Sensitivity (f) Specificity

Fig. 5. Performance metrics vs. epochs for the second round

The accuracy remained consistently high at 0.984, showcasing the model's overall proficiency in segmentation tasks. The dice coefficient and IoU both reached nearly 0.98, indicating a significant improvement over the initial scores and suggesting that the model successfully captured tumor regions with higher precision and improved overlap with the ground truth. The loss function also remained consistently low at 0.025, signifying the model's stability and efficiency in minimizing errors. Lastly, sensitivity and specificity scores remained impressively high at 0.975 and 0.996, respectively, further affirming the model's ability to accurately detect tumor and non-tumor regions.

5 Discussion

By incorporating attention mechanisms, the model effectively focused on relevant features while disregarding irrelevant ones, leading to improved segmentation performance. For a qualitative assessment of the model's performance, a further analysis was conducted by visually comparing the ground truth with the predicted segmentation of three scans, as depicted in Fig. 6. The visualization of the segmentation results aligns well with the results of the performance metrics discussed earlier, providing evidence of the model's proficiency in accurately capturing true positive regions. Of particular note is the second test (second row), which illustrates the model's effectiveness in correctly identifying true negatives.

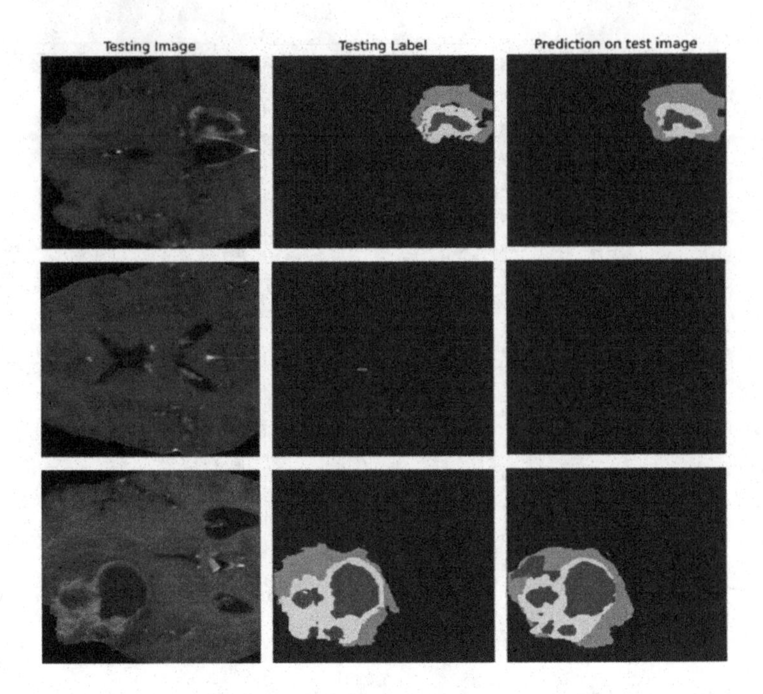

Fig. 6. Segmentation results on three test scans

This aligns with the high specificity observed in the performance metrics, indicating the model's capability to correctly identify negative cases. However, it is worth noting that in the third test (third row), there are some misclassified pixels in the ET region. This can be attributed to the fact that it has the lowest class distribution, making it more challenging for the model to accurately identify and segment these regions.

The proposed model was compared to recent studies, and the results are summarized in Table 1. While this study achieved slightly lower accuracy, sensitivity, and specificity compared to Montaha et al. and Ilhan et al., it demonstrated a

Table 1. Comparison of performance metrics with recent studies

Study	Accuracy	Sensitivity	Specificity	Dice Coefficient
Montaha et al. [11]	0.994	0.989	0.997	0.939
Ilhan et al. [12]	0.994	0.836	0.998	0.880
Cinar et al. [13]	N/A	0.931	0.995	0.931
Raza et al. [14]	N/A	0.971	0.986	0.834
Gab Allah et al. [16]	N/A	0.912	0.996	0.893
Cao et al. [17]	N/A	0.870	0.996	0.852
This study	**0.992**	**0.988**	**0.995**	**0.975**

higher Dice coefficient, indicating better overlap between predicted and ground truth segmentations. Cinar et al. achieved high sensitivity and specificity but did not report accuracy. Raza et al., Gab Allah et al., and Cao et al. reported good sensitivity but did not provide accuracy or specificity values. Overall, this study showed competitive performance in brain tumor segmentation, particularly with a notable improvement in the Dice coefficient, which reflects overall segmentation accuracy.

6 Conclusion

This study explores the integration of the attention mechanism into the 3D U-Net model and a tumor detection algorithm based on digital image processing techniques for brain tumor segmentation. By incorporating the attention mechanism, the model effectively focuses on relevant regions and important features, capturing intricate details and prioritizing informative areas during segmentation. Additionally, the tumor detection algorithm addresses the challenge of imbalanced training data. The evaluation of the proposed model on the BraTS2020 dataset demonstrates its effectiveness, achieving an accuracy of 0.992 and a dice coefficient of 0.975. These results indicate precise tumor boundary delineation with minimal error and reliable localization of tumor regions. This study contributes to brain tumor segmentation research and highlights the potential of advanced deep-learning techniques, such as the attention mechanism, to enhance accuracy and reliability in medical imaging. Further exploration of these techniques promises significant advancements in medical diagnostics.

References

1. Cancer.Net. Brain tumor: Statistics. Accessed February 2023
2. Saleh, A., Sukaik, R., Abu-Naser, S.S.: Brain tumor classification using deep learning. In: 2020 International Conference on Assistive and Rehabilitation Technologies (iCareTech) (2020)
3. Alqazzaz, S., Sun, X., Yang, X., Nokes, L.: Automated brain tumor segmentation on multi-modal MR image using SegNet. Comput. Visual Media **5**(2), 209–219 (2019)
4. Hollon, T.C., et al.: Near real-time intraoperative brain tumor diagnosis using stimulated Raman histology and deep neural networks. Nat. Med. **26**(1), 52–58 (2020)
5. Ronneberger, O., Fischer, P., Brox, T.: U-net: convolutional networks for biomedical image segmentation. In: Navab, N., Hornegger, J., Wells, W.M., Frangi, A.F. (eds.) MICCAI 2015. LNCS, vol. 9351, pp. 234–241. Springer, Cham (2015). https://doi.org/10.1007/978-3-319-24574-4_28
6. Wang, F., Jiang, R., Zheng, L., Meng, C., Biswal, B.: 3D U-net based brain tumor segmentation and survival days prediction. In: Crimi, A., Bakas, S. (eds.) BrainLes 2019. LNCS, vol. 11992, pp. 131–141. Springer, Cham (2020). https://doi.org/10.1007/978-3-030-46640-4_13

7. Aboelenein, N.M., Songhao, P., Koubaa, A., Noor, A., Afifi, A.: HTTU-Net: hybrid two track U-net for automatic brain tumor segmentation. IEEE Access **8**, 101406–101415 (2020)
8. Al Nasim, M.A., et al.: Brain tumor segmentation using enhanced U-net model with empirical analysis. In: 2022 25th International Conference on Computer and Information Technology (ICCIT) (2022)
9. Henry, T., et al.: Brain tumor segmentation with self-ensembled, deeply-supervised 3D U-net neural networks: a BraTS 2020 challenge solution. In: Crimi, A., Bakas, S. (eds.) BrainLes 2020. LNCS, vol. 12658, pp. 327–339. Springer, Cham (2021). https://doi.org/10.1007/978-3-030-72084-1_30
10. Vaswani, A., et al.: Attention is all you need. In: Proceedings of the 31st International Conference on Neural Information Processing Systems, pp. 6000–6010 (2017)
11. Montaha, S., et al.: Brain tumor segmentation from 3D MRI scans using U-net. SN Comput. Sci. **4**(4), 386 (2023)
12. Ilhan, A., Sekeroglu, B., Abiyev, R.: Brain tumor segmentation in MRI images using nonparametric localization and enhancement methods with U-net. Int. J. Comput. Assist. Radiol. Surg. **17**(3), 589–600 (2022)
13. Cinar, N., Ozcan, A., Kaya, M.: A hybrid DenseNet121-UNet model for brain tumor segmentation from MR images. Biomed. Signal Process. Control **76**, 103647 (2022)
14. Raza, R., Ijaz Bajwa, U., Mehmood, Y., Waqas Anwar, M., Hassan Jamal, M.: DResU-Net: 3D deep residual U-net based brain tumor segmentation from multimodal MRI. Biomed. Signal Process. Control **79**, 103861 (2023)
15. Everingham, M., Van Gool, L., Williams, C.K., Winn, J., Zisserman, A.: The pascal visual object classes (VOC) challenge. Int. J. Comput. Vision **88**(2), 303–338 (2009)
16. Gab Allah, A.M., Sarhan, A.M., Elshennawy, N.M.: Edge U-Net: brain tumor segmentation using MRI based on deep u-net model with boundary information. Expert Syst. Appl. **213**, 118833 (2023)
17. Cao, Y., et al.: MBANet: a 3D convolutional neural network with multi-branch attention for brain tumor segmentation from MRI images. Biomed. Signal Process. Control **80**, 104296 (2023)

Breast Cancer Detection Based DenseNet with Attention Model in Mammogram Images

Tawfik Ezat Mousa(✉), Ramzi Zouari, and Mouna Baklouti

National School of Engineering of Sfax, University of Sfax, Sfax, Tunisia
tawfikezat@yahoo.com, mouna.baklouti@enis.tn

Abstract. Breast cancer has become a very interesting topic due to the massive number of deaths among women across the world. Radiologists can diagnose breast cancer faster and more accurately because of advances in the computer-aided diagnosis (CAD) system. In this paper, we presented a new breast cancer detection system based on the integration of self attention model in the pre-trained deep neural networks DenseNet. First, we extracted automatic high-level features from breast images using DenseNet extraction layers, and thereafter attention model was applied to focus the treatment on the relevant parts of the region of interest. The experiments were conducted on a multi-class Mammographic Image Analysis Society (MIAS) database, including three classes of breast cancer images. We achieved the accuracy of 0.9939 when applying both transfer learning, data augmentation, and self attention mechanism.

Keywords: Breast cancer · DenseNet · Self Attention · Detection · Mammogram

1 Introduction

Cancer is a disease that affects cells and its detection at an early stage increases the chances of recovery. In fact, breast cancer is one of the most common types among women, and breast cancer has always shown a very high incidence and mortality rate of about 10% of women at some point in their lives. It is the second-largest cause of death among females after lung cancer [1]. The World Health Organization's International Agency for Research on Cancer (IARC) reported an anticipated increase in the number of breast cancer cases to 1.1 million by 2030, with the gap between developed and developing nations expected to widen [2]. Cancer can be described as the uncontrolled proliferation of abnormal cells that form masses. These tumors can be benign or malignant. Benign tumors remain localized and grow slowly. Malignant tumors invade nearby structures and may destroy other parts of the body [3]. Therefore, mammography is an effective imaging technique used in the detection of breast cancer. It is the

M. Mosbah et al. (Eds.): MEDI 2023, LNCS 14396, pp. 259–271, 2024.
https://doi.org/10.1007/978-3-031-49333-1_19

most used imaging technique in screening programs [4]. It helps in detecting suspicious lesions like masses and micro calcification. However, mammography in a 2D image results in tissue overlap, which can mask the lesion, or create a false lesion, thus producing false-positive and false-negative results. In addition, mammography is known to be less sensitive to breast density (30–64%) compared to fatty breasts (76–98%), as it has been shown that women with dense breasts are more susceptible to breast cancer.

In the last decade, several works based artificial intelligence tools were developed to enhance computer-assisted breast cancer (CAD) diagnosis. These approaches have shown their ability to treat the problem of abnormal lumps and calcification in the breast and predict their growth. They help radiologists and oncologists diagnose breast cancer by providing a second opinion. In this work, we propose a new system of breast cancer detection using a pre-trained DenseNet, with the integration of attention mechanism. The combination of these two models has demonstrated their effectiveness in improving the detection performance. In fact, attention mechanism allows to increase the weights of the relevant features of the model and decreasing the others, to make a better decision. Furthermore, we applied data augmentation technique to increase the number of training images and to improve the model generalization.

The rest of this paper is organized as follows. Section 2 presents some related works in breast cancer detection field. Section 3 illustrates the different parts of the proposed methodology. Section 4 presents the experimental results and a comparison with similar works. We finish this paper by a conclusion and some prospects.

2 Related Works

In the literature, numerous approaches were proposed for breast cancer detection based on mammography images. Samee et al. [5] proposed a breast cancer detection system based on several deep learning architectures. In fact, they extracted automatic features from both AlexNet, VGG, and GoogleNet models. These features are fused to make the prediction task. This system was evaluated on INbreast database and achieved the accuracy of 98.50%. In other work, Jiang et al. [6] integrated a new dataset of breast mammograms named Film Mammography dataset number 3 (BCDR-F03). They applied both GoogLeNet and AlexNet models to classify segmented tumors found on mammograms, and obtained the accuracy of 88% and 83% for GoogleNet and AlexNet respectively. Ribli et al. [7] used Regional based CNN(R-CNN) model to detect and classify breast lesions using mammograms. They obtained the accuracy of 95% on INbreast dataset. Alruwaili et al. [8] used ResNet50 model to distinguish between malignant and benign breast cancer. Data augmentation technique was applied to increase the number of training images and prevent the model from over-fitting. The proposed model was assessed on MIAS dataset and achieved the accuracy of 89.5%. Kaur et al. [9] proposed an hybrid model where both deep learning neural networks (DNN) and Support Vector Machines (SVM) were used. SVM was implemented after the DNN classification part instead of regular dense layers. The results showed that SVM allows improving the recognition rate from 70% to 96.9% on multiclass MIAS dataset. Mohapatra et al. [10] evaluated several pre-trained deep learning models such as AlexNet, VGG16, and ResNet50 on mammogram images. Due to the limited number of training images, they applied data augmentation to address the problem of over-fitting. The experiments were done on Mini-DDSM dataset and reached the accuracy of 65% when using ResNet50. Muduli et al. [11] proposed a deep convolution neural network (CNN) model for breast cancer classification using mammograms and ultrasound images. This model facilitates the extraction of prominent features from the images with only few tune parameters. They applied data augmentation to increase the number of training images and prevent the model from over-fitting. The proposed model was evaluated on MIAS and INbreast datasets, and achieved the accuracy of 90.68% and 91.28% respectively. Rouhi et al. [12] proposed a new model of primary breast cancer detection using region growing method. Their model is based on the hybridization of cellular neural network with genetic algorithm, and achieved the accuracy of 96.47% and 95.13% on MIAS and DDSM databases respectively. In [13], transfer learning technique was applied with the pre-trained deep neural networks Inception V3, ResNet50, VGG16, and Inception-ResNet. The best result was obtained using VGG16 model which achieved the accuracy of 98.96% on MIAS database. Punithavathi et al. [14] proposed an hybrid model based on SVM and KNN classifiers. They introduced multiple categories of images to the SVM, and the final decision was done by KNN algorithm. This model produces higher diagnostic accuracy on MIAS dataset and achieved the accuracy of 99.34%. Pillai et al. [15] evaluated several pre-trained deep learning models such as EfficientNet, AlexNet, VGG16, and GoogleNet on MIAS database. They applied data augmentation to increase the number of training images and prevent the model from over-fitting. The best performance was obtained with VGG16 model and achieved the average accuracy of 75.46%. Chougrad et al. [16] applied fine-tuned Inception-v3 model on MIAS database to classify breast lesions and obtained the accuracy of 98.23%. Selvathi et al. [17] proposed a new system for breast cancer detection. Their approach consists of using a stack autoencoder architectures with softmax classifier. Moreover, they applied some processing on MIAS database images to remove noise, background, and pectoral muscle, and obtained the accuracy of 98.5%.

3 Proposed Methodology

In this section, we present our system for multi-class breast cancer detection based on mammogram images. The proposed methodology employs the pre-trained DenseNet121 model truncated at the feature extraction part, followed by an attention model to give more importance to relevant features of the Region of Interest (ROI). Thereafter, convolutions and attention modules are combined to fuse both the high-level information and the interesting semantic information. The obtained features are fed into a Global Average Pooling (GAP) to reduce the feature maps dimensions and preserve pertinent features for the classification part.

3.1 DenseNet121 Architecture

Dense Convolution Network (DenseNet) is a modern CNN architecture designed for visual object recognition with only few parameters [18]. It achieved the state-of-the-art results on several image classification datasets, such as CIFAR-10, SVHN and ImageNet [19]. The basic structure of the network mainly includes two-component modules: Dense and Transition blocks (Fig. 1). In DenseNet-121, there are a total of 4 dense blocks and 3 transition blocks. Each layer in the Dense Block is connected to all subsequent layers in a densely manner [20]. Moreover, each dense block is composed of a stack of two convolution layers with a kernel size of (1×1) and (3×3) respectively. In each transition block, (1×1) convolution and (2×2) max pooling operations are done. Table 1 shows the overall architecture of DenseNet121 model. We notice that DenseNet121 alternates dense and transition blocks. At each pass, the convolution layers of the dense block are reproduced 6, 12, 24, and 16 times respectively.

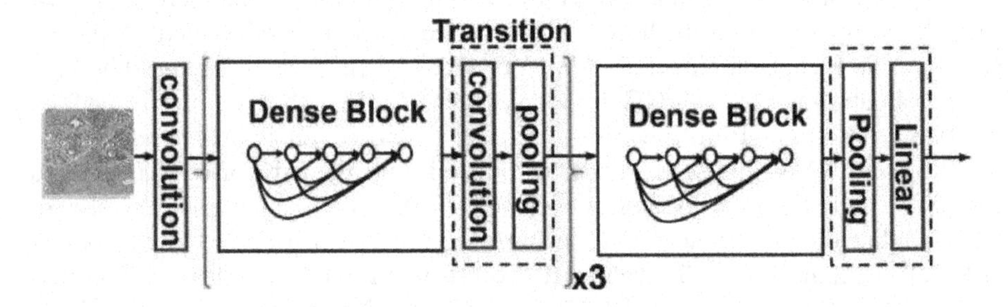

Fig. 1. DenseNet121 model concept [21]

Table 1. DenseNet121 structure

DenseNet121 parts	Layers	Parameters
Input	–	$(224 \times 224 \times 3)$
Extraction	convolution	kernel size = (7×7)
	Max Pooling	pool size = (2×2)
Dense Block $(\times 4)$	convolution	kernel size = (1×1)
	convolution	kernel size = (3×3)
Transition Block $(\times 3)$	convolution	kernel size = (1×1)
	Average pooling	pool size = (2×2)
Classification	Global Average Pooling	pool size = (7×7)

3.2 Self Attention Model

After the global average pooling layer, we implemented a Multi-Head Self Attention (MHSA) model to improve the model effectiveness (Fig. 2). In fact, MHSA

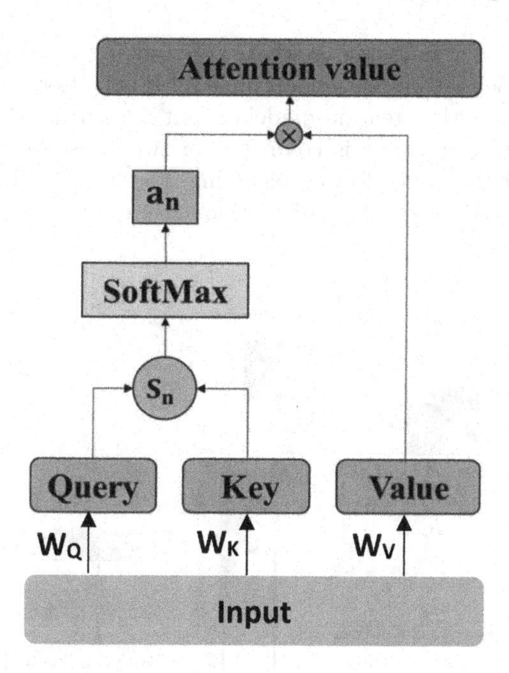

Fig. 2. Attention model architecture

is a mechanism used to provide an additional focus on a specific component in the data. It enables the network to concentrate on a few aspects at a time and ignore the rest [22]. MHSA consists of several attention layers running in parallel, instead of performing one single attention function. In particular, the input consists of queries and keys of dimension d_k (Q and K respectively), and values of dimension d_v (V). The output of the attention model is done by computing the scaled dot product of the queries with all keys and applying a SoftMax function to obtain the weights on the values V (Eq. 1). The attention mechanism is linearly projected h times with different learned weights (W_Q, W_K, W_V). These different representation sub spaces are concatenated into one single attention head to form the final output result (Eq. 2). We applied a particular version of attention model called self-attention, in which query, key and value inputs are the same. The calculation process follows these steps: First, we made the dot product (MatMul) of query and keys tensors and scale the obtained scores. Next, we apply a SoftMax function on these scores to obtain attention probabilities. Finally, we take a linear combination of these distributions with the value input tensors and concatenate them into one channel.

$$Attention(Q, K, V) = Softmax\left(\frac{Q \times K^T}{\sqrt{d_k}}\right) \times V \tag{1}$$

$$\begin{cases} \text{MHA (Q, K, V)} = \text{concat}(\text{head}_1, ..., \text{head}_h) \\ \text{head}_{i=1..h} = \text{Attention}(QW_Q, KW_k, VW_V) \end{cases} \tag{2}$$

The proposed methodology consists of making the dot product of DenseNet121 and Self Attention models outputs. Thereafter, we applied Global Average Pooling on both attention model output and the resulting dot product tensors. The classification part is composed of two dense layers with dropout function to prevent the model from over-fitting. Figure 3 illustrates the different parts of the proposed breast cancer detection system.

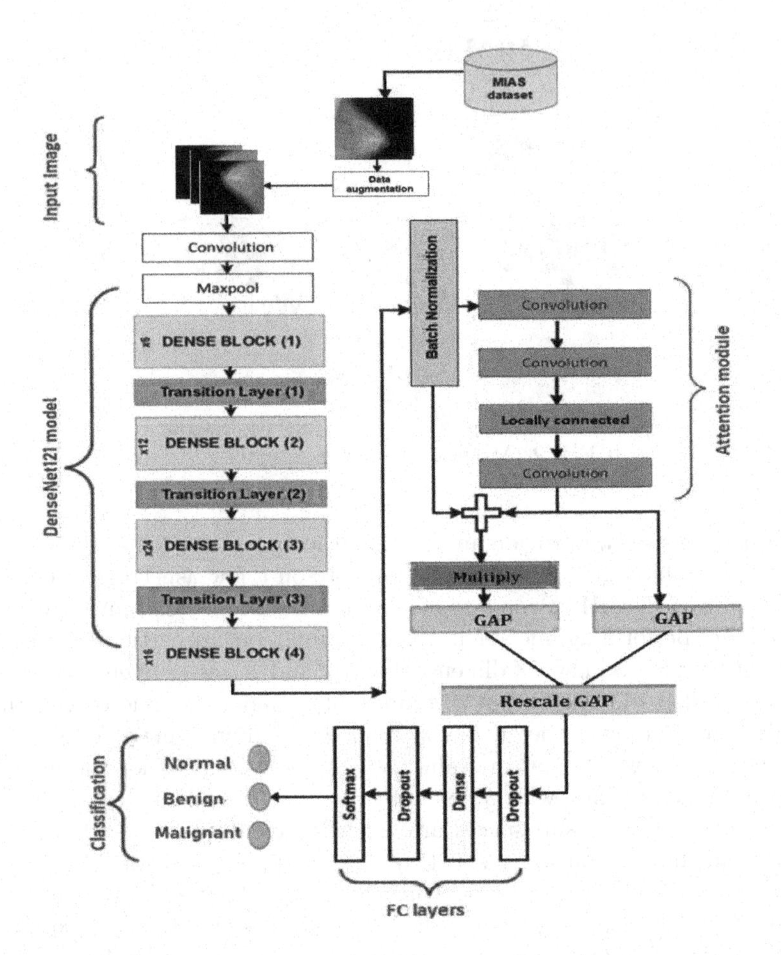

Fig. 3. Proposed methodology architecture

4 Experimentation and Results

4.1 Database Description

The proposed methodology was evaluated on MIAS multi-class database containing images of normal, benign, and malign breast cancer [23]. This database consists of 322 mammogram images of size (1024×1024) pixels and stored according to Portable Gray Map (PGM) format. These images belong to three types: glandular dense, fatty, and fatty glandular. Each type is divided into three categories: normal, benign, and malignant. The dataset also contains radiologists' actual estimations of the location of abnormalities (benign, malignant), with an approximate determination of the radius surrounding the center of the anomaly. In this work, we use all the images in the dataset, which consists of 207 normal images, 64 benign images, and 51 malignant images. Figure 4 shows three images from MIAS database representing three categories (Normal, Benign, and Malign).

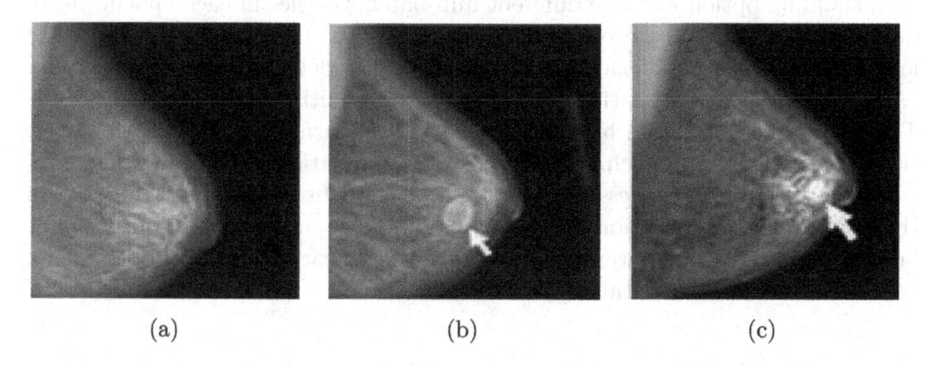

(a) (b) (c)

Fig. 4. MIAS database samples. (a) Normal (b) Benign (c) Malign

4.2 Data Augmentation

Since MIAS dataset contains only 322 images, the proposed model may not be generalized. For this purpose, we applied data augmentation operation to increase the number of training samples in each class and prevent the model from overfitting. In this work, data augmentation is mainly based on geometric transformations including rotation, flipping, and shifting. Thus, we obtained a new database of 1836 breast cancer images evenly distributed over the three classes (612 images per class). Figure 5 shows an example of data augmentation where vertical and horizontal flip were applied on the original image.

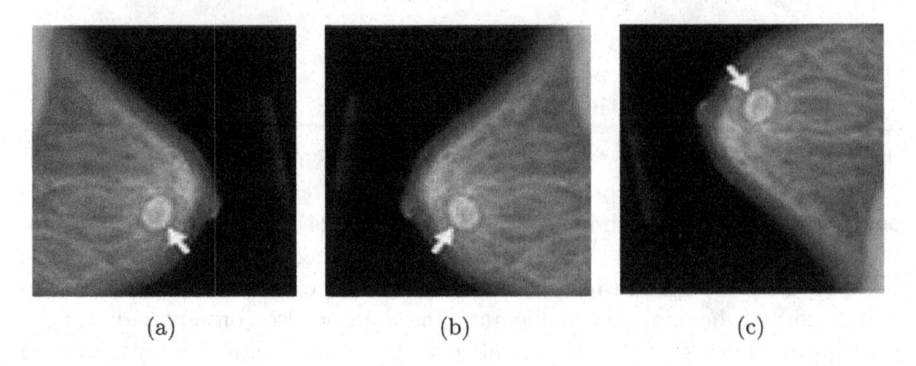

Fig. 5. Data augmentation samples. (a) original (b) horizontal flip (c) vertical flip

4.3 Experimental Setup

During the experiments, the training database was divided into batches of size 32, with shuffling option to make different min-batch samples in each epoch. Moreover, in each iteration categorical cross entropy method was used to compute the loss between desired and calculated outputs. The model was trained using Adam (Adaptive Moment Estimation) optimizer with an initial learning rate of 0, 001. This value can be reduced by a factor of 0.5 once learning stagnates. Moreover, the early stopping approach is applied as a regularization method. It consists of stopping the training process early before it has over-fit the training database. In the multi-head self attention model, we employed 8 parallel attention layers or heads. For each of these, we used 64 units in the linear projector of both query, key, and value matrices (Table 2).

Table 2. Hyperparameters setting

Hyper parameter	Value
learning rate (lr)	0.001
lr decrease rate	0.5
optimizer	Adam
batch size	32
epochs	100
loss function	cross entropy

4.4 Evaluation Metrics

To illustrate the performance of the proposed model, the confusion matrix and other metrics were calculated like Accuracy, Recall, Precision, and F1-score (Eq.

3–6). They are all based on the calculation of True positive (TP), False positive (FP), False negative (FN) and True negative (TN) values. TP denotes images predicted with breast cancer when they were. TN relates to normal images predicted as healthy. FP concerns normal images which are predicted as breast cancer, and FN refers to images predicted as normal, but they were not.

$$Accuracy = \frac{TP + TN}{TP + TN + FP + FN} \tag{3}$$

$$Recall = \frac{TP}{TP + FN} \tag{4}$$

$$Precision = \frac{TP}{TP + FP} \tag{5}$$

$$\text{F1-score} = 2 \times \frac{Precision \times Recall}{Precision + Recall} \tag{6}$$

4.5 Experimental Results

In the experiments, the images shape was fixed to ($256 \times 256 \times 3$). Moreover, several models were studied with different values of splitting and optimizers. All of these models have been used with pretrained weights. First, we evaluated the model's performance without the use of self attention mechanism. The best result was obtained with DenseNet-121 (Table 3). When applying multi-head self attention mechanism, the DenseNet-121 accuracy was improved by 6%, and we reached the accuracy of 0.9939 for 90% of database split. On the other hand, several other metrics were evaluated such recall, precision, and AUC (Table 4). In all of these metrics, the best results have been obtained using DenseNet-121 model with Adam optimizer. Figures 6 and 7 represent the confusion matrices related to the classification report for different split ratios. We notice that the model performances was improved when using multi-head self attention mechanism. Moreoever, the proposed model allows a good discrimination between benign and malign image samples, but it confuses between normal and benign classes (Table 5).

Table 3. Models accuracies without and with attention

Model	Accuracy (split ratio 90:10)	
	without Attention	with Attention
VGG-16	0.8712	0.9387
MobileNet	0.9027	0.9675
Xception	0.9029	0.9600
InceptionResNetv2	0.9363	0.9714
DenseNet-121	0.9405	0.9939

Table 4. DenseNet model performance with different split ratio

	Without Attention			With Attention		
Split ratio (Train:Test)	(70:30)	(80:20)	(90:10)	(70:30)	(80:20)	(90:10)
Accuracy	0.6363	0.8727	0.9405	0.7490	0.9264	0.9939
Recall	0.6363	0.8545	0.9297	0.7436	0.9202	0.9939
Precision	0.6375	0.8935	0.9398	0.7490	0.9259	0.9939
F1-score	0.6368	0.8735	0.9347	0.7462	0.9230	0.9939
AUC	0.7807	0.9539	0.9749	0.8964	0.9882	0.9978

Table 5. Performance results with optimizers

Optimizer	Accuracy	Recall	Precision	F1-score	AUC
Rmsprop	0.9693	0.9632	0.9691	0.9661	0.9983
Adam	0.9939	0.9939	0.9939	0.9939	0.9978
SGD	0.9387	0.9773	0.9597	0.9166	0.9857

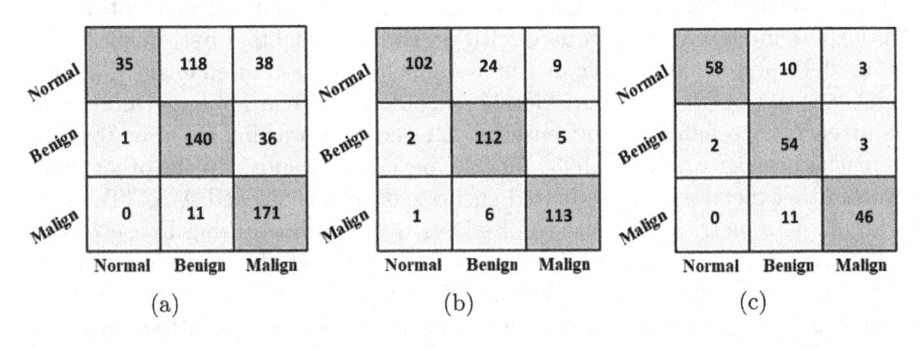

Fig. 6. Confusion matrices without self attention. (a) split ratio (70:30) (b) split ratio (80:20) (c) split ratio (90:10)

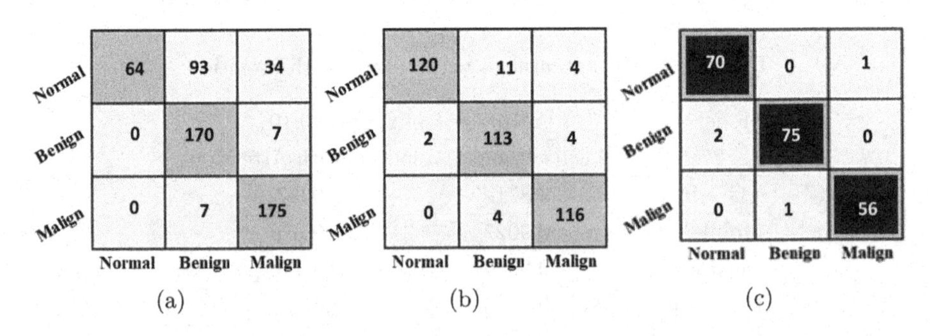

Fig. 7. Confusion matrices with self attention. (a) split ratio (70:30) (b) split ratio (80:20) (c) split ratio (90:10)

4.6 Comparative Study and Discussion

Table 6 summarizes several works evaluated on multi-class MIAS dataset. When applying the split ratio of 90% and multi-head self attention mechanism, the proposed model achieves the state-of-the-art performances on MIAS dataset, and outperforms the models based on ADL-BCD and ResNet50. However, in the case of split ratio of 80%, the proposed approach is better than DenseNet-201 model, and it is slightly less efficient than VGG16 and OMLTS-DLCN approaches. Furthermore, the proposed work is the only one to combine multi-head self attention mechanism with the pre-trained deep neural networks DenseNet-121. This combination has led to significant improvement in the classification rates. In fact, the attention model was frequently applied to sequential data. In this work, we turned it to image classification task to associate high attention weights to the parts of images with relevant features.

Table 6. Results comparison on MIAS database

Authors	Model	Split ratio	Accuracy
Alruwaili et al. [8]	ResNet50	90	89.5
Saber et al. [13]	VGG16	80	98.96
Zeng et al. [24]	DenseNet-201	80	92.73
Kvitha et al. [25]	OMLTS-DLCN	80	98.50
Maqsood et al. [26]	TTCNN	60	96.57
Gutierrez et al. [27]	ADL-BCD	90	96.07
Jebarani et al. [28]	GMM + K-means	70	95.5
Proposed work	DenseNet121 + Attention	70	74.90
		80	92.64
		90	99.39

5 Conclusion

In this paper, we proposed a deep architecture for breast cancer classification based on mammographic images to help medical doctors in breast cancer detection and diagnosis. The approach provides the breast image classification into normal, benign, and malignant. The virtue of our method is to combine pre-trained deep convolution neural networks DenseNet121 with a self-attention model. Moreover, data augmentation was applied to increase the number of images and prevent the model from overfitting. During the experiments, several hyper-parameters were tuned such as optimizer and learning rate to boost the diagnostic efficiency. The proposed methodology achieved the accuracy of 92.64% and 99.39% for a split ratio of 80% and 90% respectively. Finally, it can be concluded that by integrating the CNN using learning transfer with the attention mechanism, a clear improvement was achieved compared with other existing approaches. The results presented in this study open new windows for the use of self-attention-based architectures with vision transformer technology for breast cancer classification to obtain high-performance CAD schemes with better results.

References

1. Ara, S., Das, A., Dey, A.: Malignant and benign breast cancer classification using machine learning algorithms. In: 2021 International Conference on Artificial Intelligence (ICAI), pp. 97–101. IEEE (2021)
2. Krithiga, R., Geetha, P.: Deep learning based breast cancer detection and classification using fuzzy merging techniques. Mach. Vision Appl. **31**, 1–18 (2020)
3. Elmannai, H., Hamdi, M., AlGarni, A.: Deep learning models combining for breast cancer histopathology image classification. Int. J. Comput. Intell. Syst. **14**(1), 1003 (2021)
4. Mendes, J., Domingues, J., Aidos, H., Garcia, N., Matela, N.: AI in breast cancer imaging: a survey of different applications. J. Imaging **8**(9), 228 (2022)
5. Samee, N.A., Atteia, G., Meshoul, S., Al-antari, M.A., Kadah, Y.M.: Deep learning cascaded feature selection framework for breast cancer classification: hybrid CNN with univariate-based approach. Mathematics **10**(19), 3631 (2022)
6. Jiang, F., Liu, H., Yu, S., Xie, Y.: Breast mass lesion classification in mammograms by transfer learning. In: Proceedings of the 5th International Conference on Bioinformatics and Computational Biology, pp. 59–62 (2017)
7. Ribli, D., Horváth, A., Unger, Z., Pollner, P., Csabai, I.: Detecting and classifying lesions in mammograms with deep learning. Sci. Rep. **8**(1), 4165 (2018)
8. Alruwaili, M., Gouda, W.: Automated breast cancer detection models based on transfer learning. Sensors **22**(3), 876 (2022)
9. Kaur, P., Singh, G., Kaur, P.: Intellectual detection and validation of automated mammogram breast cancer images by multi-class SVM using deep learning classification. Inform. Med. Unlock. **16**, 100151 (2019)
10. Mohapatra, S., Muduly, S., Mohanty, S., Ravindra, J.V.R., Mohanty, S.N.: Evaluation of deep learning models for detecting breast cancer using histopathological mammograms images. Sustain. Oper. Comput. **3**, 296–302 (2022)
11. Muduli, D., Dash, R., Majhi, B.: Automated diagnosis of breast cancer using multimodal datasets: a deep convolution neural network based approach. Biomed. Signal Process. Control **71**, 102825 (2022)
12. Rouhi, R., Jafari, M., Kasaei, S., Keshavarzian, P.: Benign and malignant breast tumors classification based on region growing and CNN segmentation. Expert Syst. Appl. **42**(3), 990–1002 (2015)
13. Saber, A., Sakr, M., Abo-Seida, O.M., Keshk, A., Chen, H.: A novel deep-learning model for automatic detection and classification of breast cancer using the transfer-learning technique. IEEE Access **9**, 71194–71209 (2021)
14. Punithavathi, V., Devakumari, D.: A hybrid algorithm with modified SVM and KNN for classification of mammogram images using medical image processing with data mining techniques. Eur. J. Mol. Clin. Med. **7**(10), 2956–2965 (2020)

15. Pillai, A., Nizam, A., Joshee, M., Pinto, A., Chavan, S.: Breast cancer detection in mammograms using deep learning. In: Iyer, B., Ghosh, D., Balas, V.E. (eds.) Applied Information Processing Systems. AISC, vol. 1354, pp. 121–127. Springer, Singapore (2022). https://doi.org/10.1007/978-981-16-2008-9_11
16. Chougrad, H., Zouaki, H., Alheyane, O.: Deep convolutional neural networks for breast cancer screening. Comput. Methods Programs Biomed. **157**, 19–30 (2018)
17. Selvathi, D., Aarthy Poornila, A.: Breast cancer detection in mammogram images using deep learning technique. Middle-East J. Sci. Res. **25**(2), 417–426 (2017)
18. Hasan, N., Bao, Y., Shawon, A., Huang, Y.: DenseNet convolutional neural networks application for predicting COVID-19 using CT image. SN Comput. Sci. **2**(5), 389 (2021)
19. Albelwi, S.A.: Deep architecture based on DenseNet-121 model for weather image recognition. Int. J. Adv. Comput. Sci. Appl. **13**(10), 2022
20. Kim, T.-H.: Electricity theft detection using fusion DenseNet-RF model (2021)
21. Zeng, L., Lang, J.: Classification of breast cancer histopathological image based on lightweight network. In: CIBDA 2022; 3rd International Conference on Computer Information and Big Data Applications, pp. 1–6. VDE (2022)
22. Vaswani, A., et al.: Attention is all you need. Advances in Neural Information Processing Systems, vol. 30 (2017)
23. Yoon, W.B., Oh, J.E., Chae, E.Y., Kim, H.H., Lee, S.Y., Kim, K.G.: Automatic detection of pectoral muscle region for computer-aided diagnosis using MIAS mammograms. Biomed. Res. Int. (2016)
24. Xiang, Yu., Zeng, N., Liu, S., Zhang, Y.-D.: Utilization of DenseNet201 for diagnosis of breast abnormality. Mach. Vis. Appl. **30**, 1135–1144 (2019)
25. Kavitha, T., et al.: Deep learning based capsule neural network model for breast cancer diagnosis using mammogram images. Interdisc. Sci.: Comput. Life Sci. 1–17 (2021)
26. Maqsood, S., Damaševičius, R., Maskeliūnas, R.: TTCNN: a breast cancer detection and classification towards computer-aided diagnosis using digital mammography in early stages. Appl. Sci. **12**(7), 3273 (2022)
27. Escorcia-Gutierrez, J., et al.: Automated deep learning empowered breast cancer diagnosis using biomedical mammogram images. Comput. Mater. Continua **71**, 3–4221 (2022)
28. Jebarani, P.E., Umadevi, N., Dang, H., Pomplun, M.: A novel hybrid k-means and GMM machine learning model for breast cancer detection. IEEE Access **9**, 146153–146162 (2021)

AI-LMS: AI-Based Long-Term Monitoring System for Patients in Pandemics: COVID-19 Case Study

Nada Zendaoui[1](✉), Nardjes Bouchemal[1], and Maya Benabdelhafid[2]

[1] LIRE Laboratory of Constantine 2, University Center Abdelhafid Boussouf of Mila, Constantine, Algeria
nada.zendaoui@centre-univ-mila.dz, n.bouchemal.dz@ieee.org

[2] School of Accounting and Finance ESCF, Constantine LIRE Laboratory of Constantine 2, Constantine, Algeria
mbenabdelhafid@escf-constantine.dz

Abstract. In the context of the ongoing COVID-19 pandemic, the need for robust health monitoring systems has become increasingly evident, especially for at-risk patients. The latter refers to individuals who are more susceptible to severe illness or complications if infected with COVID-19 due to underlying health conditions, age, or other factors. To address this need, the proposed research aims to develop an intelligent health monitoring system called AI-LMS (AI-based Long-term Monitoring System) that focuses on patients in pandemics. The system will utilize IoMT (Internet of Medical Things) sensors, Machine Learning algorithms, and Mobile Cloud Computing to enable real-time identification and monitoring of at-risk patients. The suggested approach can be simply adaptable for use in various pandemic circumstances. Using COVID-19 as a case study, AI-LMS underscores the significance of robust health monitoring systems in pandemic conditions. It is separated into two phases: the first collects and processes health data using a multi-layer classifier to identify at-risk patients, whereas the second one is centered on monitoring at-risk patients with the help of IoMT sensors that provide data to a machine learning model. The model alerts healthcare professionals to any concerning trends. By making slight modifications, this research aims to design efficient health monitoring systems for pandemic situations, ultimately leading to improved patient outcomes and alleviating the burden on healthcare systems.

Keywords: Healthcare · COVID-19 pandemic · At-risk patients · telehealth · Artificial Intelligence · Machine Learning · IoMT · Monitoring

1 Introduction

The COVID-19 pandemic has caused a global disaster, prompting new innovations in accounting and viral response. The virus was formally designated by

M. Mosbah et al. (Eds.): MEDI 2023, LNCS 14396, pp. 272–285, 2024.
https://doi.org/10.1007/978-3-031-49333-1_20

the WHO on February 11, 2020, and on February 13, 2020, China reported a rise of approximately 15,000 new cases and 242 deaths in a single day in Hubei province. China's government has enacted wartime control measures, putting cities under lockdown and affecting an estimated 760 million people. Meanwhile, concerns have been raised about measures such as isolation and mass round-ups, as well as the quarantining of people in makeshift medical facilities for unspecified periods of time [10].

Currently, many countries have successfully integrated telemedicine and advanced technologies into various healthcare procedures such as diagnosis, disease prevention, treatment, and health research. The benefits of using these technologies in epidemics or pandemics, including the current COVID-19 outbreak, include assisting individuals with chronic conditions requiring medical treatment and follow-up while reducing their exposure to hospital facilities [1]. For instance, in the COVID-19 context, several nations and healthcare organizations have adopted technologies to manage this pandemic. Among them, we find Internet of Medical Things (IoMT) technology that has enormous promise in the healthcare business, particularly during a pandemic. Its devices can be used to monitor patients' health in real time, gathering data on vital signs, medication adherence, and illness management. They may also be used to increase hospital efficiency by tracking the position of equipment and decreasing patient wait times. Moreover, IoT devices can help with remote consultations and telemedicine by allowing clinicians to monitor patients from a distance, lowering the risk of infection exposure [19].

Moreover, Mobile Cloud Computing (MCC) represents a new technology that overcomes mobile devices' constraints in processing large amounts of data by offering multi-platform compatibility and dynamic provisioning. MCC might be used in a variety of fields, including education, medical science, biometry, forensics, and cars. It can also assist in overcoming the obstacles encountered during the COVID-19 pandemic by providing, on one hand, proper reach and delivery of important services using gamification, Cloud rendering, and collaborative methods, and, on the other hand, addressing security, authentication, privacy, and trust issues for the safe deployment of MCC [20].

In addition, telehealth and Artificial Intelligence (AI) are useful during a pandemic. Indeed, telehealth provides benefits such as enhanced healthcare access and reduced disease exposure, whereas AI represents a critical technology in the fourth industrial revolution that can assist in addressing global health issues and increasing pandemic preparedness. In COVID-19, for instance, AI is employed in identification and monitoring, as well as in fields like protein structure prediction and digital health. However, similar to MCC, privacy and security are critical issues that governments and policymakers must address to ensure the safe use of AI and telehealth [5].

This research work focuses on developing an intelligent health monitoring system that combines IoMT sensors, machine learning algorithms, and mobile cloud computing to monitor at-risk patients in real-time during pandemics. At-risk patients refer to individuals who are more susceptible to developing severe complications or have a higher mortality rate from a particular disease. This group includes elderly individuals, people with chronic diseases, pregnant women, and children who require more attention and monitoring than healthy individuals. AI-LMS is unique in two ways: it concentrates on at-risk patients and proposes long-term monitoring required for such groups of patients. Moreover, the case study of COVID-19 highlights the importance of investigating long-term intelligent systems for at-risk patients in the healthcare sector. The proposed approach aims to improve patient outcomes and reduce the burden on healthcare systems by providing personalized monitoring and care for at-risk patients.

The remainder of this paper is written as follows. It reveals in the 'related work' section some recent studies proposing smart health monitoring systems in the context of the COVID-19 pandemic. Concerning the 'used concepts' section, it describes the essential concepts chosen in our study's strategy. The 'Proposed system' section describes the system and reveals its different components in depth. Finally, the conclusion section summarizes the findings and emphasizes the importance of our proposed study.

2 Related Work

In this section, we provide a concise overview of recent COVID-19 monitoring studies conducted from 2020 to 2023, using COVID-19 as a pertinent example for a broader understanding of pandemic monitoring.

To provide a comprehensive overview, we categorize these works into two groups: one focused solely on monitoring COVID-19 patients, and the other on both the identification and monitoring of COVID-19 patients.

2.1 Approaches for Monitoring COVID-19 Patients

Starting with the first part, we find several studies that have proposed different approaches to monitor COVID-19 cases.

Jeyaraj et al. [16] developed the Smart-Monitor system, an IoT-based automated physiological monitoring system employing AI, particularly a Deep Convolutional Neural Network (DCNN), for real-time signal prediction and visualization. This system is structured into four units: Monitoring, Processing, Visualization & Storage, and Learning, delivering an impressive real-time accuracy of 97.5% within just 65 s of computational time, enabling remote patient monitoring.

The study in [3] developed an IoT-based health monitoring system tracking vital signs (blood pressure, heart rate, oxygen, and temperature). It's especially useful in rural areas, allowing local clinics to connect with city hospitals for health status monitoring. IoT alerts providers for abnormal readings, with accuracy similar to commercial devices. It enables real-time data collection for physicians.

In [7], a COVID-19 monitoring system was introduced. It remotely assesses a patient's health using IoT sensors, Cloud, and Web layers. The system uses the Meta-Heuristics optimized Convolute Neural Network (MHCNN) to categorize health states, gathering data like temperature, heart rate, oxygen levels, and auditory signals for cough detection. It achieves 98.76% accuracy with low deviation, employing additional statistical features and advanced neural network techniques.

These studies highlight the significant impact of integrating IoT, Cloud, and AI in healthcare research, but they often overlook disease detection.

2.2 Approaches for Identifying and Monitoring COVID-19 Patients

We shall now turn our attention to the second category, which comprises works focusing on monitoring and also detecting issues.

In [14], the authors introduced a real-time framework using eight ML models to identify COVID-19 cases. They used IoT for data collection, monitoring, and treatment analysis, achieving over 90% accuracy in most models, except for Decision Stump, OneR, and ZeroR. This framework is suitable for healthcare professionals.

In [12], the authors introduced an intelligent healthcare system merging IoT and Cloud tech. It offers real-time patient monitoring, cost-effective healthcare, aids in COVID-19 screening, and provides second opinions. IoT sensors capture CT scan images, sent to the Cloud for processing using ResNet50 DL classification. The system achieved 98.6% accuracy, 97.3% sensitivity, 98.2% specificity, and a 97.87% F1-score using benchmark datasets containing 6,000 CT images.

In [4], authors presented a real-time COVID-19 monitoring framework. It comprises three layers: wearable sensors and a mobile app for patient monitoring, a fog network to handle Cloud storage and data transmission, and a ResNet-50 DL model for COVID-19 identification via X-ray scans. The model, trained on 750 images, achieved 97.95% accuracy and 98.85% specificity.

2.3 Discussion

The mentioned studies focused on general COVID-19 patient management, neglecting the specific needs of at-risk individuals like the elderly, those with chronic illnesses, pregnant women, and children. Moreover, they emphasized short-term monitoring, while our innovative approach addresses these gaps. We use AI and cloud computing for long-term monitoring of at-risk patients during pandemics (refer to Table 1).

3 Used Concepts

In this section we introduce some important concepts that will be used in our approach.

Table 1. Summary of related work

Work	Used technologies	Used for what?	Caracteristics				User	Target patient	Metrics
			Time	Real time	Storage	Used devices			
[14]	SVM, NN, NaÃfve Bayes, K-NN, Decision Table, Decision Stump, OneR, and ZeroR IoT Cloud	Detecting and monitoring	Fast	Y	Cloud	Wearable sensors	Healthcare physicians Patients Data analysis Quarantine	Suspected patient Covid-19 patient	Accuracy 90%
[12]	ResNet50 IoT Cloud	Detecting and monitoring	Fast	Y	Cloud	IoMT CT-scan sensors	Healthcare physicians	Suspected patient Covid-19 patient	Accuracy 98,6% Sensitivity 97,3% Specificity 98,2% F1-score 97,87%
[4]	ResNet50 IoT Fog Mobile application	Detecting and monitoring	Fast	Y	Cloud	IoMT Hospital tests	Healthcare staff Patient	Suspected patient Covid-19 patient	Accuracy 97,95% Specificity 98,85%
[16]	DCNN IoT Cloud	Monitoring	65s	Y	Cloud	IoMT Ni-myRio	Healthcare physcians Patient	Covid-19 patients	Accuracy 97,5%
[7]	MHCNN IoT Cloud Web layer	Monitoring	Fast	Y	Cloud	Wearable sensors	Healthcare physcians Patient	Covid-19 patients	Accuracy 98,76%
[3]	IoT Clouf	Monitoring	Fast	Y	Cloud	IoMT	Healthcare physicians	Covid-19 patients	/

3.1 AI Methods

Before introducing proposed AI methods in healthcare domain, we estimate essential to briefly reveal AI methods in general.

- *Artificial Intelligence*: AI research focuses on creating machines that simulate human thinking, including learning, reasoning, and self-correction. While the AI community works on developing intelligent computer programs, this remains a challenging endeavor. AI systems fall into categories like mimicking human thought, enhancing machine intelligence, extending human abilities with computers, and refining programming techniques for efficiency [9].
- *Machine Learning*: ML is a branch of AI enabling computers to learn and enhance their performance in specific tasks. It utilizes algorithms that learn from extensive data, uncover hidden patterns, and don't require explicit programming. ML finds applications in object detection, natural language translation, fraud detection, and speech/image recognition. It encompasses supervised, unsupervised, and reinforcement learning, with Artificial Neural Networks (ANNs) being the most versatile and widely used. ANNs, inspired by biological information processing, consist of connected units called Artificial Neurons adaptable to various ML tasks. Deep Neural Networks (DNNs), a type of ANN, enable automatic discovery of representations from raw data, known as Deep Learning (DL). However, advanced ML algorithms are often considered "black boxes", making their decision-making process less transparent unless explained otherwise [8,13].
- *Deep Learning*: DL, a subset of ML, employs Artificial Neural Networks (ANNs) with multiple layers to analyze and learn from data. These networks extract and transform features, adjusting weights and biases to minimize prediction errors. DL excels in image/speech recognition, natural language processing, and bioinformatics, often outperforming other ML methods. It's a key part of ML, emphasizing data representation learning [13].

In healthcare, AI has three key applications: living assistance, information processing, and disease diagnostics/prediction. In [18], the authors use NNs and data mining for living assistance, ML and natural language processing for information processing, and ML techniques for disease diagnostics/prediction.

In this paper, we employ AI to boost patient outcomes, enhance accuracy in diagnoses, personalize treatments, and facilitate drug discovery. AI aids decision-making, streamlines administration, lowers costs, and enhances the patient experience.

3.2 Mobile Cloud Computing

It is important to briefly define Cloud Computing and Mobile Computing before providing a definition of MCC since this latter combines them.

- *Cloud computing*: According to [11], it is a model that allows users to access and utilize a variety of computing resources over the internet, with rapid provisioning and management by the service provider.

- *Mobile Computing*: As mentioned in [21], it's a wireless technology enabling data, voice, and video transmission via wireless devices, eliminating the need for fixed physical connections. Users can access information and communicate from anywhere, anytime, without location or time constraints.

Hence, MCC combines Mobile and Cloud Computing, relocating data processing and storage to centralized Cloud platforms accessible via wireless connections. It upholds Quality of Service (QoS) via ongoing monitoring until connection termination [2].

In our work, MCC facilitates remote data access, healthcare professional collaboration, on-the-go decision-making, real-time monitoring, and telemedicine.

3.3 Internet of Medical Things

To begin discussing IoMT, it's necessary to provide a brief explanation of IoT.

- *Internet of Things*: IoT equips everyday objects, even non-electronic ones like food and clothing, with internet-based sensing, location, and control. It fosters universal connectivity, offering interconnectivity, diverse services, and immense scale per [15].
- *Internet of Medical Things*: IoMT merges medical devices with IoT, enabling healthcare professionals to monitor and manage them over the internet. This results in faster, cost-effective healthcare and is seen as the future of healthcare systems [17].

Our approach improves healthcare through remote monitoring, predictive analytics, and precise data collection for better diagnoses and personalized treatments, ultimately aiming to enhance patient outcomes, which is the core objective of our paper.

4 AI-LMS System

We recall that our main objective is to propose an intelligent system that aims to help doctors in the identification of at-risk patients, monitor their health status and intervene before their condition worsens, with the goal of leading to earlier interventions and improved patient outcomes. Let us notice that while it is currently being developed with the example of COVID patients in mind, the proposed work is designed to be adaptable to eventual future pandemics or healthcare emergencies.

In short, AI-LMS focuses to provide a flexible and proactive approach to patient care (see Fig. 1).

4.1 Departure Points

In AI-LMS, we assume the following three (3) main actors:

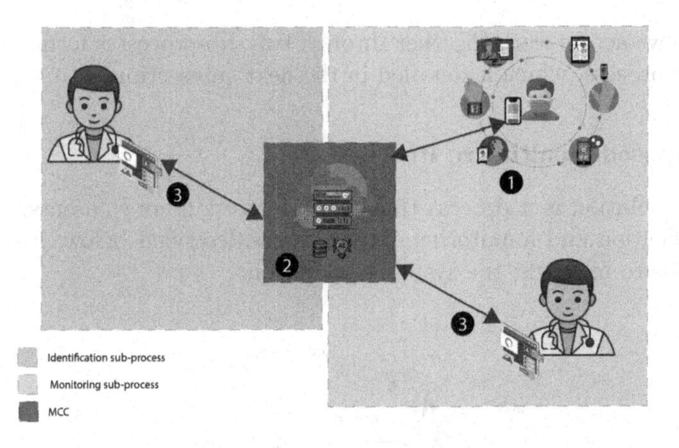

Fig. 1. Global architecture

1. **IoMT**: This first part is dedicated to helping doctors in monitoring at-risk patients by accessing their own medical information. It uses IoMT technology to achieve this objective. By this way, patients become actors in our intelligent system since they are able to receive/send alerts and notifications, thanks to IoMT. For instance, if a patient wears a smartwatch to monitor abnormal cardiac activity, then the sensors send the collected data to MCC where AI algorithms can detect any irregularities in the heart rate and rhythm. Once detected, the wearer or a healthcare provider can be alerted, leading to earlier intervention and potentially improving patient outcomes.
2. **MCC**: MCC provides storage and computation offloading to execute codes remotely for mobile devices. It serves as a centralized system with the role of providing a collection of AI models (see Subsect. 3.1) for processing and analyzing data, as well as generating notifications and alerts. Concerning Cloud infrastructure, it consists of many heterogeneous resources called servers. Let us notice that the MCC part must address data security and management concerns to ensure the privacy and confidentiality of patient data and protect against unauthorized access or data breaches.
3. **Doctors**: They are active actors and have the main role of monitoring one or more patients through efficient access to the patient's medical information. This allows for personalized care and early intervention when necessary. The system can also assist doctors in their patient care by providing notifications and alerts from the server. This helps doctors to respond proactively to any concerns. Consequently, the proposed AI-powered patient monitoring system can track a patient's vital signs, medication schedule, and medical history to provide personalized care and early intervention when necessary. It can alert doctors in real-time of any changes in the patient's condition and provide automated notifications and alerts from the server to enable proactive response to concerns.

The above actors work together through two sub-processes forming our main healthcare process, which is detailed in the next sub-section.

4.2 Proposed Healthcare Process

Our smart solution is a process that aims to assist users (patients/doctors) in the identification and monitoring sub-processes described below. Figure 2 gives more details to highlight the system's operation.

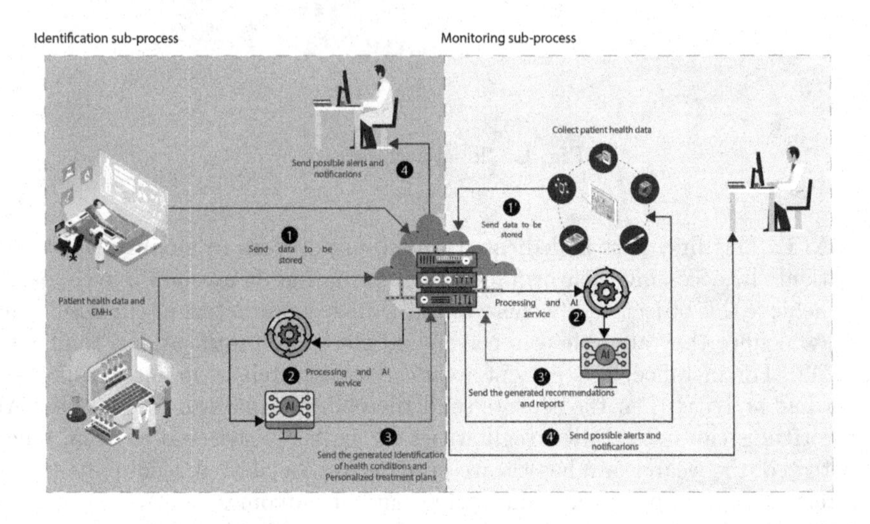

Fig. 2. Proposed AI-based Long-term Monitoring System

Identification Sub-process. In this subsection, we reveal the identification part of our smart system.

- **Input**:
  1. *Patient health data*: It refers patient's health records and may include data on vital signs, medical history, laboratory results, and medications.
  2. *Electronic Medical Records (EMRs)*: EMRs are applications used by healthcare practitioners to document and manage patient medical records within a care delivery organization [6].
- **Output**:
  1. *Identification of health state*: It refers the detection whether the patient in question is a at-risk patient or not.
  2. *Personalized treatment plan*: It refers the creation of personalized treatment plans for patients by considering their specific needs and circumstances.

- **At-risk patient identification sub-process**: The system's prime objective is to detect persons in risk (See the pink part of Fig. 2: Identification sub-process). Upon a patient's arrival, the doctor conducts clinical and biological exams. Consequently, the data of this exams is transferred and stored in MCC (1). Then, a multi layer classifier model, leveraging transfer learning techniques, analyses this data to detect whether the patient is at-risk or not (2). The chosen model uses the collected data as input and returns a classification result and the personalized treatment plan as an output(3). Once identified as at risk, the patient becomes part of the system, enabling doctors to monitor and act upon their conditions proactively (4). Let us notice that MCC component provides real-time access to patient data, facilitating efficient and personalized care. The whole identification sub-process is formalized in Algorithm 1.

Algorithm 1. At-risk patient Identification algorithm

Require: *Patient health data and EMRs*
Ensure: *Identification of health state and treatment plan*
1: Apply an ML model to analyse the collected data and identify any potential health risks or abnormalities.
2: Apply an ML model to evaluate the patient's risk level based on the analysed data.
3: **if** patient is identified as at-risk patient **then**
4: Generate alerts or notifications for the doctor.
5: **end if**
6: Suggest a personalized treatment plan.
7: Store patient data securely and confidentially.

Monitoring Sub-process. In this subsection, we reveal the monitoring part of our smart system.

- **Input**:
  1. *Patient health data*: the same data given above.
  2. *Extracted data from real-time monitoring sensors*: It refers data collected from various sources, including IoMT devices, monitoring sensors such as wearables or other connected devices, as well as Electronic Health Record (EHR is a subset of the EMR that is owned by the patient and spans episodes of care across multiple care delivery organizations within a community or region [6]).
- **Output**:
  1. *Alerts and notifications*: They include generated alerts or notifications sent to both doctors and patients in order to inform them of any abnormal event or change in the patient's health status.
  2. *Predictive Analytics and Recommendations*: They form the analysis of patient data and the provided predictions on future health outcomes or

potential health risks. Additionally, they include generated recommendations or insights for healthcare providers based on the patient's health data, such as suggested treatments or interventions.
3. *Dashboards and reports*: Referring provided dashboards or reports that summarize the patient's health status over time, including trends and patterns in their health data.

- **At-risk patient monitoring sub-process**: Once the system identifies at-risk patients, it moves on to the monitoring and support stage following (see the yellow part of the Fig. 2: monitoring sub-process). Through IoMT connected devices, vital sign data is collected regularly, and sent to the MCC (1') for storage and processing (2'). If any abnormalities are detected during data analysis (3'), then the system sends alerts to both the patient and doctor (4'). The patient is advised to contact a doctor immediately or given instructions on how to react, while the doctor receives an alert about the patient's emergency.
  Furthermore, the system generates predictive analytics and recommendations, as well as dashboards and reports to help doctors and patients monitor and manage the patient's health. The system also assists in decision-making by proposing a suitable action plan based on a ML model that uses data collected by IoMT as inputs and provides the doctor with a proposed course of action as an output. By doing this, the system allows doctors to respond proactively to any concerns, provide personalized care, and early intervention when necessary, which can significantly improve patient outcomes.

Algorithm 2. At-risk patient monitoring algorithm

Require: *Patient health data and EHRs*
Ensure: *Alerts and notifications, Predictive Analytics and Recommendations, Dashboards and reports*

```
1: while patient data available do
2:     Collect patient data from IoMT devices
3:     Pre-process data (e.g. noise removal, normalization)
4:     Extract features
5:     Predict patient risk using AI model (e.g. decision trees, SVM)
6:     if patient risk is high then
7:         Alert the doctor and the patient.
8:     else
9:         Continue monitoring patient
10:    end if
11: end while
```

Models Description. To provide a more detailed description of the AI-LMS system, we need to provide information about the models used in each sub-process (see Fig. 3).

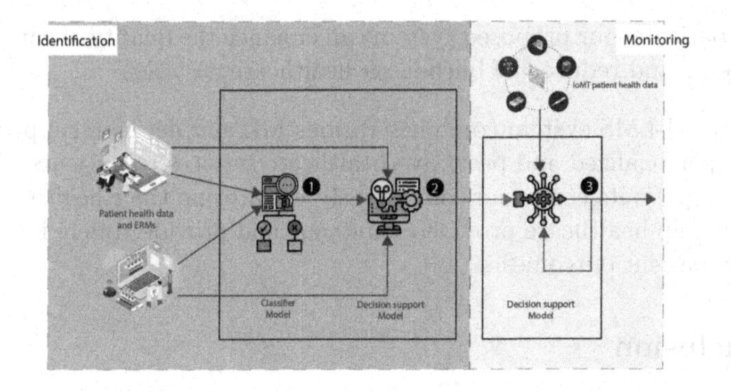

Fig. 3. Identification and monitoring models

1. *Classifier model*: It is used by the identification sub-process (1). Classifier models utilize patient health data and EMRs to classify or identify at-risk patients by analyzing various features such as vital signs, lab results, medical history, demographics, and socio-economic factors. Supervised learning algorithms such as logistic regression, decision trees, or support vector machines are trained on a labeled dataset of both at-risk and non-at-risk patients to identify a decision boundary that separates these two categories in the feature space. Moreover, the classifier model can not only identify at-risk patients but also defines their level of risk by considering factors such as age, gender, medical history, and current symptoms.
2. *Decision support model*: It is also used by the identification sub-process (2). It aims to generate alerts and treatment plans for at-risk patients and can also use ML algorithms to provide personalized recommendations based on the patient's medical history, current symptoms, and other relevant data. For example, a ML model can analyze data from thousands of patients to identify which treatments are most effective for patients with similar medical histories and symptoms. This information can then be used to generate personalized treatment plans for new patients with similar conditions. The model can also be updated over time as new patient data becomes available, allowing it to continually improve its accuracy and effectiveness.
3. *Decision support model*: It is used in the monitoring sub-process (3). Our proposed system utilizes IoMT sensors to continuously collect patient data, which is then fed into a deep learning model. This model uses advanced algorithms to identify patterns and trends in the patient's vital signs and other health metrics. If the model detects any concerning deviations from normal ranges, it sends alerts and notifications to the healthcare providers in real-time. These alerts can be used to initiate timely interventions to prevent adverse outcomes and improve patient outcomes. Furthermore, the model can provide ongoing feedback to healthcare providers, allowing for continuous refinement of treatment plans based on the patient's individual response to therapy. By leveraging the power of ML models for real-time monitoring of

at-risk patients, our proposed system can enhance the quality of care provided to patients and reduce the burden on healthcare systems.

Overall, the AI-LMS system combines various ML and decision support models to provide personalized and proactive healthcare services to patients. By identifying at-risk patients early and continuously monitoring their health status, the system can help healthcare providers intervene and provide timely interventions to improve patient outcomes.

5 Conclusion

The COVID-19 pandemic exposed weaknesses in healthcare systems worldwide, leading to a focus on intelligent healthcare systems to aid doctors in monitoring and potentially identifying at-risk patients. However, no existing work has addressed this specific concern. This paper introduces an approach combining Mobile Cloud Computing (MCC), Internet of Medical Things (IoMT), and Artificial Intelligence (AI) to enhance patient care.

The approach comprises two main stages: identification and monitoring. In the identification stage, a classifier model detects patients at risk of certain conditions, while the monitoring stage involves an AI-based system that assists doctors by analyzing IoMT data, providing alerts, recommendations, predictive analyses, and reports.

Future plans include selecting a specific scenario within this broad area, detailing AI techniques for it, and considering advanced constraints like energy autonomy, storage quality, and time constraints using formal tools and model checking techniques for verification.

References

1. Alonso, S.G., et al.: Telemedicine and e-health research solutions in literature for combatting Covid-19: a systematic review. Heal. Technol. **11**, 257–266 (2021)
2. Asrani, P.: Mobile cloud computing. Int. J. Eng. Adv. Technol. **2**(4), 606–609 (2013)
3. Bhardwaj, V., Joshi, R., Gaur, A.M.: IoT-based smart health monitoring system for Covid-19. SN Comput. Sci. **3**(2), 137 (2022)
4. El-Rashidy, N., El-Sappagh, S., Islam, S.R., El-Bakry, H.M., Abdelrazek, S.: End-to-end deep learning framework for coronavirus (Covid-19) detection and monitoring. Electronics **9**(9), 1439 (2020)
5. El-Sherif, D.M., Abouzid, M., Elzarif, M.T., Ahmed, A.A., Albakri, A., Alshehri, M.M.: Telehealth and artificial intelligence insights into healthcare during the Covid-19 pandemic. Healthcare **10**(2) (2022). https://doi.org/10.3390/healthcare10020385. https://www.mdpi.com/2227-9032/10/2/385
6. Garets, D., Davis, M.: Electronic medical records vs. electronic health records: yes, there is a difference. Policy white paper. Chicago, HIMSS Analytics, pp. 1–14 (2006)
7. Jaber, M.M.: Remotely monitoring Covid-19 patient health condition using metaheuristics convolute networks from IoT-based wearable device health data. Sensors **22**(3), 1205 (2022)

8. Janiesch, C., Zschech, P., Heinrich, K.: Machine learning and deep learning. Electron. Mark. **31**(3), 685–695 (2021)
9. Kok, J.N., Boers, E.J., Kosters, W.A., Van der Putten, P., Poel, M.: Artificial intelligence: definition, trends, techniques, and cases. Artif. Intell. **1**, 270–299 (2009)
10. Lancet, T.: Covid-19: fighting panic with information. Lancet (London, England) **395**(10224), 537 (2020)
11. Mirashe, S.P., Kalyankar, N.V.: Cloud computing (2010)
12. Nasser, N., Emad-ul Haq, Q., Imran, M., Ali, A., Razzak, I., Al-Helali, A.: A smart healthcare framework for detection and monitoring of Covid-19 using IoT and cloud computing. Neural Comput. Appl. 1–15 (2021)
13. Ongsulee, P.: Artificial intelligence, machine learning and deep learning. In: 2017 15th International Conference on ICT and Knowledge Engineering (ICT&KE), pp. 1–6. IEEE (2017)
14. Otoom, M., Otoum, N., Alzubaidi, M.A., Etoom, Y., Banihani, R.: An IoT-based framework for early identification and monitoring of Covid-19 cases. Biomed. Signal Process. Control **62**, 102149 (2020)
15. Patel, K.K., Patel, S.M., Scholar, P.: Internet of things-IoT: definition, characteristics, architecture, enabling technologies, application & future challenges. Int. J. Eng. Sci. Comput. **6**(5) (2016)
16. Rajan Jeyaraj, P., Nadar, E.R.S.: Smart-monitor: patient monitoring system for IoT-based healthcare system using deep learning. IETE J. Res. **68**(2), 1435–1442 (2022)
17. Razdan, S., Sharma, S.: Internet of medical things (IoMT): overview, emerging technologies, and case studies. IETE Tech. Rev. **39**(4), 775–788 (2022)
18. Rong, G., Mendez, A., Assi, E.B., Zhao, B., Sawan, M.: Artificial intelligence in healthcare: review and prediction case studies. Engineering **6**(3), 291–301 (2020)
19. Sakly, H., Said, M., Al-Sayed, A.A., Loussaief, C., Sakly, R., Seekins, J.: Blockchain technologies for internet of medical things (BIoMT) based healthcare systems: a new paradigm for COVID-19 pandemic. In: Sakly, H., Yeom, K., Halabi, S., Said, M., Seekins, J., Tagina, M. (eds.) Trends of Artificial Intelligence and Big Data for E-Health. IS, vol. 9, pp. 139–165. Springer, Cham (2022). https://doi.org/10.1007/978-3-031-11199-0_8
20. Sheth, H.S.K., Tyagi, A.K.: Mobile cloud computing: issues, applications and scope in COVID-19. In: Abraham, A., Gandhi, N., Hanne, T., Hong, T.-P., Nogueira Rios, T., Ding, W. (eds.) ISDA 2021. LNNS, vol. 418, pp. 587–600. Springer, Cham (2022). https://doi.org/10.1007/978-3-030-96308-8_55
21. Talukder, A., Yavagal, R.: Mobile Computing. McGraw-Hill, Inc. (2006)

Cardiovascular Anomaly Detection Using Deep Learning Techniques

Wassim Sliti[1(✉)], Seif Eddine Ben Abdelali[1], Aymen Yahyaoui[1,2], Amine Mosbah[3], and Olfa Djebbi[4]

[1] Military Academy of Fondouk Jedid Nabeul, Nabeul, Tunisia
wassimsliti1998@gmail.com
[2] Science and Technology for Defense Lab (STD), Tunis, Tunisia
[3] Laboratory of Drug Design Faculty of Pharmacy, University of Monastir, Monastir, Tunisia
[4] Emergency Department, Principal Military Hospital of Instruction of Tunis, Tunis, Tunisia

Abstract. Cardiovascular diseases (CVD) refer to a group of health conditions that affect the heart and blood vessels. This can also include arterial damage in organs such as the kidneys, the heart, the eyes, and the brain. Electrocardiograms (ECGs) are a quick, safe, and non-intrusive cardiac exploration, to check for heart rate, heart rhythm, and signs of potential heart disease. The interpretation of ECGs can be vital in determining the condition of the human body, and it is important to obtain accurate results. Deep learning techniques are being used in this work to automatically analyze ECG recordings. We regenerated certain datasets from the Physionet data bank, such as the MITBIH dataset, and others from a cardiology challenge in 2020 by performing some transformations and adaptations. We have implemented different models to detect ECG anomalies using both single and multiple-lead ECG datasets. CNN family model has the highest detection performance when trained on a single lead ECG dataset. Its performance decreased significantly when anomaly classes and lead numbers increased. In fact, accuracy passed from 0.96 to 0.60 whereas the F1 score passed from 0.98 to 0.55 when trained on a 12-lead dataset with 27 classes of anomalies. Given the regular time series pattern of the ECG signal, we propose a combination of a CNN-LSTM model for classification. This model achieved 0.668 accuracy. Combining all models into one ensemble learning model increased significantly the detection accuracy on the 12-lead ECG dataset to reach 0.82. Our combined architecture has proven to achieve state-of-the-art accuracy in ECG anomaly detection and could help health professionals better manage CVD.

Keywords: Cardiovascular Disease · ECGs · Deep Learning · Anomaly Detection

1 Introduction

Cardiovascular diseases (CVDs) are the leading cause of death worldwide, according to the World Health Organization (WHO) [16]. An estimated 17.9

M. Mosbah et al. (Eds.): MEDI 2023, LNCS 14396, pp. 286–299, 2024.
https://doi.org/10.1007/978-3-031-49333-1_21

million people die from CVD, which represents 31% of all global deaths. Heart attacks and strokes are responsible for four out of every five CVD-related deaths, with one-third of these deaths occurring before the age of 70 [16]. By measuring voltages from electrodes attached to the patient's chest, arms, and legs, an electrocardiogram (ECG) can record a patient's heart electrical signal activities for a long time [1]. ECGs are a quick, non-invasive, and safe way to check heart rate, rhythm, and activity and detect electrical heart anomalies. Cardiologists use a twelve-lead ECG to detect various cardiovascular abnormalities, and it is today's standard tool. Heart problems, on the other hand, may not always be visible on a standard 10-second recording from a 12-lead ECG performed in hospitals or clinics. As a result of the advancement of new sensing technologies, long-term ECG monitoring that tracks the patient's heart condition at all times and in any circumstance is now possible. Apple Watch, AliveCor, Omron HeartScan [18], QardioMD, and more recently, the Astroskin Smart Shirt [17] are revolutionizing cardiac diagnostics by measuring a patient's 24/7 cardiac activities and transmitting this information to cloud service to be stored and processed remotely. The massive health records are valuable sources of information and could help detect, mitigate, and predict fatal health conditions providing an appropriate real-time exploration by professionals, which is now a major limitation to the generalization of this time-consuming test. However, the rise of sophisticated algorithms to process and explore these resources in a timely fashion seems to offer a reliable alternative to the manually contacted expert assessment. Providing both robust and proactive detection of deadly heart failures is now possible and efficient using deep learning, streamlined detection of ECGs data records.

Deep learning techniques are being used in this work to automatically analyze ECG recordings. We regenerated certain datasets from the Physionet data bank, such as the MITBIH dataset, and others from a cardiology challenge in 2020 by performing some transformations and adaptations. We have implemented different models to detect ECG anomalies using both single and multiple-lead ECG datasets. CNN family model has the highest detection performance when trained on a single lead ECG dataset. Its performance decreased significantly when anomaly classes and lead numbers increased. In fact, accuracy passed from 0.96 to 0.60 whereas the F1 score passed from 0.98 to 0.55 when trained on 12 leads dataset with 27 classes of anomalies. Given the regular time series pattern of the ECG signal, we propose a combination of a CNN-LSTM model for classification. This model achieved 0.66 accuracy. Combining all models into one ensemble learning model increased significantly the detection performance on 12 leads ECG dataset to reach 0.82 accuracy. Our combined architecture has proven to achieve state-of-the-art accuracy in ECG anomaly detection and could help health professionals better manage CVD.

The remainder of this paper is structured as follows: Sect. 2 provides the background of the study, explaining the main concepts related to the cardiovascular anomaly detection research topic. In Sect. 3, previous research relevant to the study is reviewed, and the existing knowledge is linked to the new discoveries presented in this paper. Section 4 details the proposed approach, and the model generation pipeline. The used datasets and experimental results of the study are presented in Sect. 5. Finally, Sect. 6 provides a conclusion that summarizes the main points of the paper, highlights the contributions of the study, and suggests avenues for future research.

2 Background

The electrodes placed on the human chest or arms produce a variety of curves related to heart activity that may be very hard to understand. But, how would a doctor determine a cardiac condition just on such curves?

An ECG is a test that measures the electrical activity of the heart to study how it works. An electrical impulse (or "wave") travels through the heart with each heartbeat. This wave causes the heart muscle to contract and the heart to expel blood. ECGs are a quick, painless, and safe approach to looking for indicators of probable cardiac illness as well as heart rhythm and heart rate.

2.1 ECG's Components

We describe in the following the main ECG components that doctors need to interpret to make a diagnosis of heart activity. **ECG Waves:** There are four types of waves as shown in Fig. 1.

- **P wave:** The P wave is the first positive deflection on the ECG and represents atrial depolarization. In healthy individuals, there should be a P wave preceding each QRS complex. The normal duration should be less than 0.12 s.
- **Q wave:** A Q wave is any negative deflection that precedes an R wave. The Q wave represents the normal left-to-right depolarisation of the inter-ventricular septum.
- **R wave:** The R wave is the first upward deflection after the P wave. The R wave represents early ventricular depolarization.
- **T wave:** The T wave is the positive deflection after each QRS complex. It represents ventricular repolarization.

Intervals and Segments: There are two types of segments and intervals as depicted in Fig. 1.

- **PR interval:** The PR interval is the time from the onset of the P wave to the start of the QRS complex. It reflects conduction through the atrioventricular node. The normal PR interval is between 120–200 ms (0.12–0.20 s) in duration.
- **PR segment:** The PR segment is the flat, usually isoelectric segment between the end of the P wave and the start of the QRS complex.
- **QT interval:**
  The QT interval begins at the start of the QRS complex and finishes at the end of the T wave. It represents the time taken for the ventricles to depolarise and then repolarise.

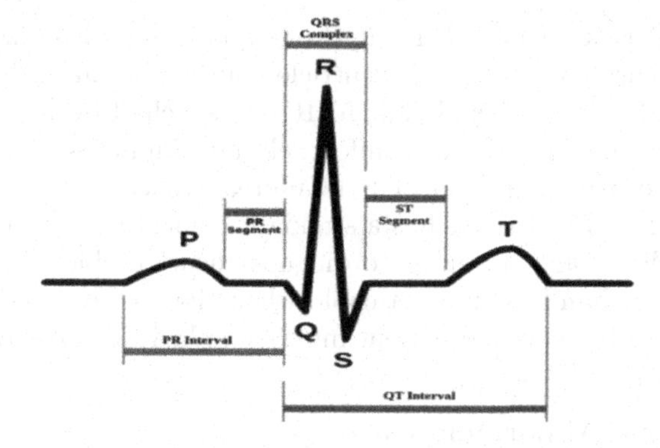

Fig. 1. The components of an ECG [24]

- **ST segment:** The ST segment is the flat, isoelectric section of the ECG between the end of the S wave and the beginning of the T wave. The ST Segment represents the interval between ventricular depolarization and repolarization. The most important cause of ST segment abnormality is elevation or depression.

2.2 Twelve Lead ECG

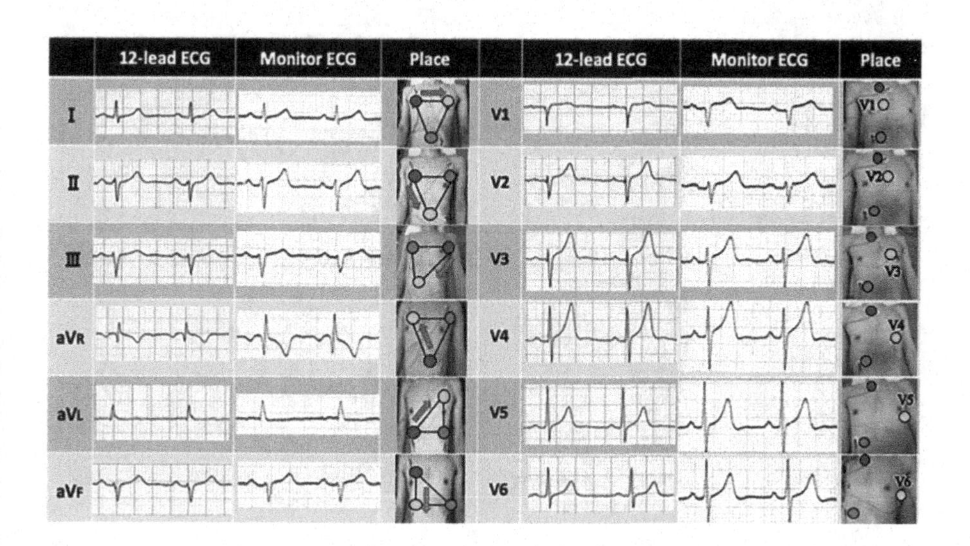

Fig. 2. Twelve lead ECG placement

Standard 12-lead ECG depicted in Fig. 2 offers frontal and horizontal views of the heart, as well as views of the left ventricle's surfaces from 12 different angles. Six limb leads (I, II, III, aVF, aVL, and aVR) and six chest leads (V1-V6) make a 12-lead ECG. For chest pain or discomfort, electrical injuries, electrolyte imbalances, medication overdoses, ventricular failure, stroke, syncope, and unstable patients, the 12-lead ECG is used as a standard clinical analysis tool. It is commonly used in clinics and hospitals to diagnose heart problems [19]. A 12-lead ECG, on the other hand, is impracticable when the patient needs to be monitored continuously because the patient must be linked to 10 electrodes.

2.3 ECG Signal Abnormalities

ECG leads can help doctors diagnose cardiovascular diseases. Arrhythmia is a heart electrical activity disorder that affects ECG signals. Doctors examine ECG signals to classify them and detect different diseases if they exist. There are various ECG abnormality classes as shown in Fig. 3 such as Atrial fibrillation (b), Left bundle branch block (c), and Atrial flutter (k).

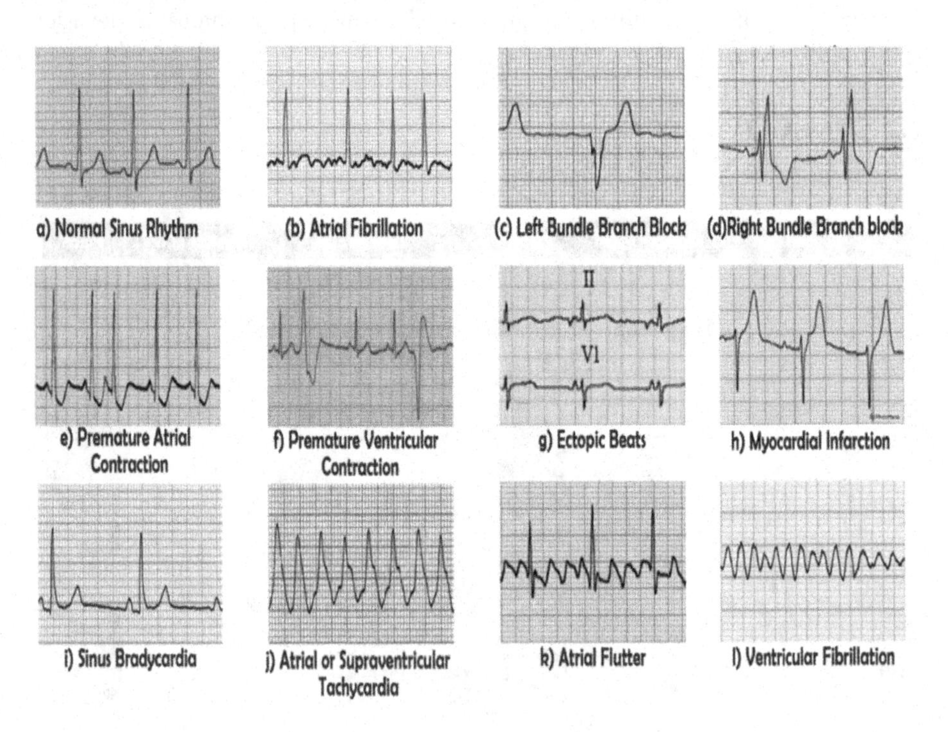

Fig. 3. Illustrating different abnormalities

3 Related Works

The 12-lead ECG is a widely available tool for screening for heart disease [3]. However, ECG interpretation requires experienced doctors who can carefully examine and identify pathological inter-beat and intra-beat patterns. This process is time-consuming and subject to inter-observer variability [4]. As a result, a precise algorithm for automated ECG pattern classification is highly desired.

Earlier studies have reported automated ECG analysis [5,6]. These methods are mainly based on frequency domain features, time-frequency analysis, and signal transformations (i.e. wavelet transform and Fourier transform). However, such techniques are incapable of capturing complex ECG signal features.

Deep learning and neural networks, particularly convolutional neural networks (CNNs), have shown promising results in a variety of fields, including computer vision and Natural language processing. CNNs have been used for ECG abnormalities detection based on ECG signals. In their work [8], authors presented a 21-layer 1D convolutional recurrent neural network to detect atrial fibrillation, which was trained on single-lead ECG data. Authors [9] proposed two deep neural networks for classifying pulse-generating rhythm and pulse-less electrical activity using short single-lead ECG segments. In [10], authors proposed an ensemble deep learning model for the automatic classification of ECG arrhythmias based on single-lead ECGs that combined the decisions of ten classifiers and outperformed a single deep classifier.

The majority of previous works have focused on the classification of one or at nine ECG abnormalities such as [11]. As a result, we aim to establish a robust model that can generalize to 27 different types of ECG abnormalities. Furthermore, most existing works only deal with single-lead ECG signals [14,15], whereas 12-lead ECGs are more commonly used in clinical practice for abnormality detection and diagnosis.

Related to that, there is a challenge called The PhysioNet/Computing in Cardiology Challenge 2020, which offers an opportunity to solve these issues by collecting data from a variety of sources with a wide variety of cardiac abnormalities. The goal is to create open-source algorithms that can detect cardiac anomalies in 12-lead ECG data. After scoring, 41 teams qualified for ranking as shown in Table 1 listing the top ten. Deep learning and convolutional neural networks were the most popular algorithmic approaches, but the results were very low.

4 Proposed Approach

We describe in this section the proposed approach as summarized in Fig. 4, and we mention details about the dataset preprocessing as well as the model creation pipeline and architectures.

Table 1. Final scores from top six official winning teams from Computing in Cardiology 2020.

Rank	Team	Score
1	Prna	0.533
2	Between a ROC and a heart place	0.520
3	HeartBeats	0.514
4	Triage	0.485
5	Sharif AI Team	0.437
6	DSAIL SNU	0.420

4.1 Data Preprocessing

Data preprocessing is a technique for converting raw data (data that has not been processed for use) into a clean data set. In other words, whenever data is collected from various sources, it is collected in raw format, which makes analysis impossible. As a result, specific steps are taken to convert the data into a small clean data set. This technique is used prior to running the Iterative Analysis. Data preprocessing is required due to the presence of unformatted data from multiple sources of datasets with different data structures. This dataset is made up of time series data with varying sampling, frequency, and value ranges, as well as many inaccurate, inconsistent, and noisy data. Therefore, to handle raw data, data preprocessing is performed.

Data Cleaning: We focused on missing data in this stage since we discovered numerous records without annotation in these different datasets, and because we don't have sufficient knowledge in this field, we couldn't manually annotate them, so we decided to discard them. We also excluded samples that are missing one or more ECG leads.

Data Integration: Since data has been collected from several sources and combined to get a reliable result, we standardized their annotations in order to combine data from various separate sources into a single massive unified dataset because each dataset has its own annotation file format.

Data Balancing: The main problem with imbalanced dataset prediction is how accurately are we actually predicting both majority and minority classes. An imbalanced dataset can lead to unbalanced classification, which is the problem of classification when the distribution of classes in the training dataset is unbalanced. To avoid this problem, we tried to resample our dataset to balance the classes, but since it is a multi-label annotated dataset, it won't be as easy to balance the classes as in the multiclass classification task. For example, the undersampling of major classes can affect minor classes. As a result, we used other datasets to oversample minor classes and undersample major classes without affecting minor classes.

The final used dataset contains 49,510 electrocardiographic samples after it has been cleaned, integrated, and balanced.

Transformation: This step is used to convert the raw data into a specified format according to the needs of the model. It's the process of modifying the format, structure, or values of data. This step is essential because transformed data may be easier for computers to use. Properly formatted and validated data improves the final results.

- **Denoising:** Denoising is usually an essential first step before analyzing waveform data. To compensate for such data corruption, an efficient denoising algorithm must be used. For that, we used the PyWavelet library and a multilevel decomposition with a discrete wavelet transformation.
- **Normalization:** This method converts numerical data into the specified range. To begin, we scaled the denoising data using RobustScaler from the Sklearn library. Typically, this is done by removing the mean and scaling to unit variance. Then, from the same library, we used the MinMaxScaler method to scale the data between −1 and 1.

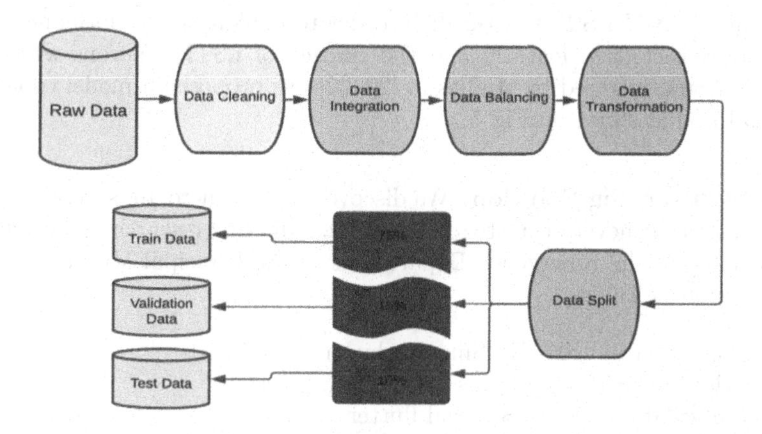

Fig. 4. Data Preparation and Preprocessing

4.2 Used Models

Throughout this project, we tested several deep learning networks and time series classification model architectures in order to achieve a better result.

We began by reviewing the papers submitted by the top-ranking teams in the physionnet competition to understand more about the models they employed before defining our own. We discovered that they all used convolutional neural networks or long short-term memory networks as models.

In order to develop times series multi-label classification models for 27 ECG anomalies, we first tested various model architectures based on convolution neural networks, as well as LSTM, Bi-LSTM, and GRU networks, even we tried their combinations (CNN with LSTM, CNN with Bi-LSTM).

This final architecture and configuration of hyperparameters was obtained after many iterations of the procedure:

- Optimizing the model parameters in the training set.
- Check the performance in the validation set.
- Manually choose new hyperparameters and architecture using insight from previous iterations.

After hours of training and testing several model architectures and neural network layer types, we settled on these three model architectures.

CNN Model: After evaluating several similar works, we discovered that their proposed models were basically made up of the 1-dimensional CNN layers and this was due to the performance of this sort of neural network in feature extraction. We tried several architectures based on the 1D CNN.

Residual Neural Network Model: We evaluated a Residual Neural Network model inspired by the model proposed in [2], and the positive results were obtained in a similar study of diagnosing 9 ECG abnormalities on 12-lead ECG.

Combined CNN-LSTM Model: In order to exploit the performance of 1D CNNs to extract deep features and the efficacy of LSTM neurons with time-series data, and inspired by studies in [20–22], we proposed a model combining CNN and LSTM shown in Fig. 5.

Ensemble Learning Solution: We discovered that there are several opposite ECG classes that never occur together after conducting research and consulting our specialist in the Emergency Department at the Principal Military Hospital of Instruction of Tunis.

- Sinus bradycardia (SB) VS Sinus tachycardia (STach)
- Sinus rhythm (SNR) VS Sinus arrhythmia (SA)
- Atrial fibrillation (AF) VS Atrial flutter (AFl)
- Aight axis deviation (RAD) VS Left axis deviation (LAD)
- Incomplete right bundle branch block (IRBBB) VS Complete right bundle branch block (RBBB)

After realizing these links between these classes, we conclude that we can export these pairs of classes into single models to simply classify them and keep the multilabel classification model for the other classes. To integrate this solution, we proposed an ensemble learning combined with 6 deep learning models. Figure 6 represents our proposed ensemble learning model.

We use the combination CNN-LSTM model for the multi-label classification model (17 other classes) since it provides us with the best result of the three models. We utilized simple CNN classifiers for the multiclass classification of each class couple. Each model from the five separate classifiers of the opposite ECG class pairs must classify the ECG data into three classes (either one of the opposite classes or none)

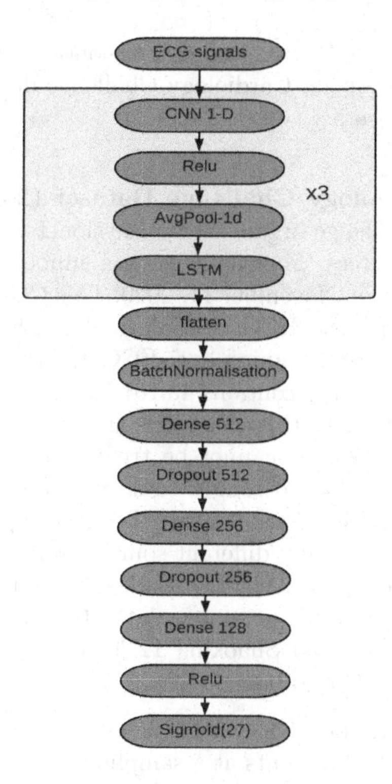

Fig. 5. CNN-LSTM network model architecture

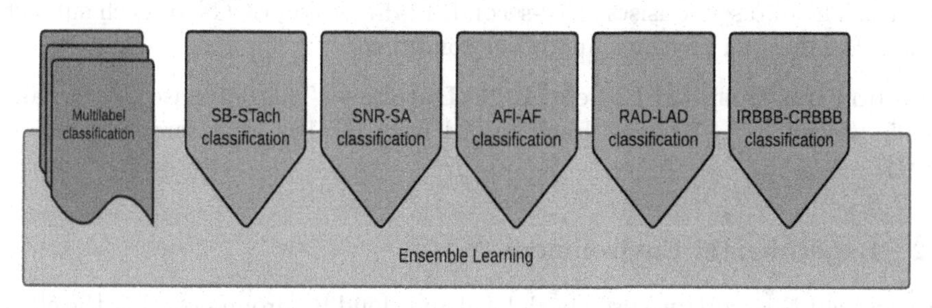

Fig. 6. Ensemble learning solution

5 Experimental Evaluation and Results

5.1 Datasets

First, we took a look at datasets used in related works, and we discovered that the majority of these datasets have limitations and can not help in classifying some ECG anomalies because most of them have only one or two ECG leads or have a very limited dataset annotations (12 anomalies at most). We initially considered using two datasets that were frequently used in related research: the MIT-BIH Arrhythmia Database and MIT-BIH Long-Term ECG Database.

These two datasets are very limited and do not contain 12 ECG leads, for that, we started with The PhysioNet Cardiology Challenge dataset (2020), and then we added two other datasets.

The PhysioNet Cardiology Challenge Dataset (2020): We came across a recent Cardiology Challenge organized by Physionet while searching through several open-source resources. The challenge was announced on April 23, 2020, and it ended at midnight on November 11, 2020. This Challenge aims to identify clinical diagnoses from 12-lead ECG recordings. The Physionet provides open-source 12-lead ECG datasets containing 27 ECG anomalies to challenge teams. The training data in this study contains 43,101 Electrocardiographic recordings from four different sources. In this challenge, the winning team got a score of 0.533, which is a low score that cannot be trusted in such a critical task. So, we considered these datasets and sought to accomplish better results in order to target detecting clinical ECG anomalies. The dataset provided was an unbalanced dataset composed of many different sources from various ECG Machines around the world, which explains why the challenge results were low. So, in order to enhance this dataset, we added two new datasets compatible with those used during the challenge: Chapman Shaoxing 12 lead ECG database and Ningbo First Hospital 12 lead ECG database.

Chapman Shaoxing 12 Lead ECG Database: This database [23] contains the 12-lead ECGs of 10,646 patients at a sampling rate of 500 Hz, with 11 common rhythms and 67 additional cardiovascular conditions labeled by professional experts. The dataset consists of 10-second, 12-dimension ECGs for each subject, as well as labels for rhythms and other conditions.

Ningbo First Hospital 12 Lead ECG Database: This database [23] contains 34905 12-lead ECG records. Each record lasts for 10 s at a sampling rate of 500 Hz.

5.2 Experimental Environment

To carry out this work, we used both local and cloud Environments. For the cloud environment, we used Google Colab Pro to train our models, which is a service that allocates large amounts of resources for machine learning purposes. This service provided us with NVIDIA Tesla T4 (16 GB VRAM) and NVIDIA Tesla P100 (12 GB VRAM) cards dedicated to accelerating ML and DL calculations, with a total allocated RAM of 36 GB.

5.3 Results

After training the three models (CNN, RNN, and CNN-LSTM) on our 12-lead dataset, we evaluate them using a test dataset. The following tables show the different results.

CNN Model Results. After training and testing this model with several dataset versions, we obtained as best result an average accuracy of 0.603.

The aggregated metrics of this model are depicted in Table 2.

Table 2. Aggregated metrics of the CNN model

	F1-score	Precision	Recall
macro AVG	0.54	0.55	0.616
weighted AVG	0.59	0.619	0.603

Residual Neural Network Model Results. For the Residual Neural Network model, we obtained an average accuracy of 0.647.

The aggregated metrics of this model are represented in Table 3.

Table 3. Aggregated metrics of the Residual Neural Network model

	F1-score	Precision	Recall
macro AVG	0.626	0.622	0.696
weighted AVG	0.635	0.651	0.649

Our Proposed CNN-LSTM Model Results. we achieved an average accuracy of 0.668. The aggregated metrics of this model are depicted in Table 4.

Table 4. Aggregated metrics of the CNN-LSTM model

	F1-score	Precision	Recall
macro AVG	0.665	0.6445	0.6920
weighted AVG	0.651	0.693	0.7105

Our Proposed Ensemble Learning Solution

After training each model separately, the five multi-class classification models summary of results are shown in Table 5.

Table 5. Results of the five dedicated classifiers

Pairs	Accuracy	Precision	Recall	F1-score
SB & STach	0.85	0.87	0.91	0.89
SNR & SA	0.90	0.92	0.93	0.92
AF & AFl	0.89	0.91	0.92	0.91
RAD & LAD	0.91	0.94	0.95	0.94
IRBBB & CRBBB	0.92	0.93	0.95	0.94

On average, the multiclass classifiers perform at 89% accuracy, 91.4% precision, 93% recall, and 92% F1-score. After exporting the 5 ECG class pairs, we still have a multi-label classification for the other 17 ECG classes using the combined CNN-LSTM model. Following the train for 200 epochs on the remaining 17 classes. Finally, we integrated all of the models into an Ensemble Learning solution and tested them across all 27 classes. Using this method on the test data, performance improved by 15% in terms of accuracy passing from 66% to 82% as shown in Table 6.

Table 6. Aggregated metrics of the Ensemble learning model

	F1-score	Precision	Recall
macro AVG	0.755	0.736	0.8336
weighted AVG	0.836	0.858	0.833

6 Conclusion

In this work, we tackled a very important problem related to cardiovascular anomaly detection. We used a public dataset combining several sources. We tested two state-of-the-art models (CNN and RNN), and we proposed a combined CNN-LSTM model that leads to better results. Finally, applying ensemble learning increased the accuracy of the results considerably. In the future, we intend to develop an early forecast of ECG Signals so that we can detect and intervene before anomalies happen.

References

1. Bonow, R.O., et al.: Braunwald's Heart Disease e-Book: A Textbook of Cardiovascular Medicine. Elsevier Health Sciences (2011)
2. Zhang, D., et al.: Interpretable deep learning for automatic diagnosis of 12-lead electrocardiogram. Iscience **24**(4), 102373 (2021)
3. Recommendations for the standardization and interpretation of the electrocardiogram: Part I: The electrocardiogram and its technology; the American college of Cardiology Foundation; and the heart rhythm society. Circulation
4. Bickerton, M., Pooler, A.: Misplaced ECG electrodes and the need for continuing training
5. Martinez, J.P., Almeida, R., Olmos, S., Rocha, A.P., Laguna, P.: A wavelet based ECG delineator. IEEE Trans. Biomed. Eng.
6. Minami, K., Nakajima, H., Toyoshima, T.: Real-time discrimination of ventricular tachyarrhythmia with Fourier-transform neural network. IEEE Trans. Biomed. Eng. **46**(2), 179–185 (1999)
7. Alexakis, C., et al.: Feature extraction and classification of electrocardiogram signals related to hypoglycaemia. Comput. Cardiol

8. Xiong, Z., Nash, M.P., Cheng, E., Fedorov, V.V., Stiles, M.K., Zhao, J.: ECG signal classification for the detection of cardiac arrhythmias using a convolutional recurrent neural network
9. Elola, A., et al.: Deep neural networks for ECG-based pulse detection during out-of-hospital cardiac arrest
10. Warrick, P.A., Homsi, M.N.: Ensembling convolutional and long short-term memory networks for electrocardiogram arrhythmia detection
11. Luo, C., Jiang, H., Li, Q., Rao, N.: Lecture Notes in Computer Science (including subseries Lecture Notes in Artificial Intelligence and Lecture Notes in Bioinformatics). LNCS, vol. 11794, pp. 55–63 (2019)
12. Huang, C., Zhao, R., Chen, W., Li, H.: Arrhythmia classification with attention-based Res-BiLSTM-Net. In: Liao, H., et al. (eds.) MLMECH/CVII-STENT -2019. LNCS, vol. 11794, pp. 3–10. Springer, Cham (2019). https://doi.org/10.1007/978-3-030-33327-0_1
13. Huang, C., Zhao, R., Chen, W., Li, H.: Arrhythmia classification with attention-based Res-BiLSTM-net
14. Andreotti, F., Carr, O., Pimentel, M.A.F., Mahdi, A., De Vos, M.: Comparing feature-based classifiers and convolutional neural networks to detect arrhythmia from short segments of ECG
15. Detection of AF and other rhythms using RR variability and ECG spectral measures
16. World health organization cardiovascular diseases cardiovascular diseases
17. Carre Technologies Inc. Hexoskin Smart Shirt
18. Omron Healthcare Asia Omron ECG Monitor HCG-801
19. Hampton, J., Hampton, J.: The ECG Made Easy e-Book. Elsevier Health Sciences (2019)
20. Md Zabirul Islam and Md Milon Islam and Amanullah Asraf
21. Md Yilin Wang and Le Sun and Dandan Peng
22. Abdulaziz M. Alayba and Vasile Palade and Matthew England and Rahat Iqbal
23. A 12-lead electrocardiogram database for arrhythmia research covering more than 10,000 patients. Sci. Data **7**, 12 (2020)
24. 12-lead electrocardiogram (EKG). Washington Hearth Rythm Center (n.d.). https://www.washingtonhra.com/ekg-monitoring/12-lead-electrocardiogram-ekg.php

Applications and Security

Real-Time Mitigation of Trust-Related Attacks in Social IoT

Mariam Masmoudi[1,4](✉), Ikram Amous[2], Corinne Amel Zayani[3], and Florence Sèdes[4]

[1] MIRACL Laboratory, FSEGS, University of Sfax, Sfax, Tunisia
mariam.masmoudi19@gmail.com
[2] MIRACL Laboratory, Enet'Com, University of Sfax, Sfax, Tunisia
ikram.amous@enetcom.usf.tn
[3] MIRACL Laboratory, FSS, University of Sfax, Sfax, Tunisia
corinne.zayani@fss.usf.tn
[4] IRIT Laboratory, Paul Sabatier University, Toulouse, France
Florence.Sedes@irit.fr

Abstract. The social Internet of Things (Social IoT) introduces novel ways to enhance IoT networks and service discovery through social contexts. However, trust-related attacks raise significant challenges with regard to the performance and reliability of these networks. In fact, malicious users exploit vulnerabilities to spread harmful services, necessitating the incorporation of a trust management mechanism in the Social IoT network. To tackle this challenge, we set forward an innovative trust management mechanism empowered by blockchain technology. Through this integration, we aim to mitigate trust-related attacks and establish a secure environment for end users. Additionally, we introduce a new consensus protocol called real-time trust-related attack mitigation Protocol using Apache Spark (ProtoSpark). This protocol which leverages Apache Spark's distributed stream processing engine to process real-time stream transactions. In the implementation of ProtoSpark, we have developed a new classifier utilizing Spark Libraries. This classifier accurately categorizes transactions as malicious or secure, enabling the protocol to make informed decisions regarding transaction validation. Our research corroborates the superiority of our classifier in terms of predicting malicious transactions, surpassing previous works and other approaches in the literature. Furthermore, our new protocol exhibits improved transaction processing times, enhancing the efficiency of the network.

Keywords: Social Internet of Things (Social IoT) · Real-Time · Trust-related attack · Blockchain · Consensus protocol · Apache Spark

1 Introduction

The Internet of Things (IoT) has revolutionized the fields of information as well as communication, and the integration of social aspects has given rise to the Social Internet of Things (Social IoT) [2]. Its core goal is to encourage people to

M. Mosbah et al. (Eds.): MEDI 2023, LNCS 14396, pp. 303–318, 2024.
https://doi.org/10.1007/978-3-031-49333-1_22

connect, strengthen social relationships, and enhance communication. Through its social functions, users can build relationships, interact, share information, and access services provided by other users' devices, all while evaluating these services, as highlighted in [2,12]. However, the Social IoT faces challenges that can threaten user security and trust, particularly with respect to trust-related attacks [18]. These attacks can manipulate service discovery and recommendation results, undermining the reliability of services and connected objects [24]. To establish a trustworthy environment, real-time mitigation of trust-related attacks is intrinsic. Despite the significance of trust management mechanisms in the Social IoT, scarce studies have focused on real-time mitigation of trust-related attacks. Existing research works often overlook specific attack types and essential security properties [3,5,7,13,15,16,27,28]. To overcome this research gap, we elaborate an innovative trust management mechanism empowered by blockchain technology. Through this integration, we aim to mitigate trust-related attacks and establish a secure environment for users. Additionally, we introduce a novel consensus protocol called real-time trust-related attack mitigation Protocol using Apache Spark (ProtoSpark), leveraging Apache Spark's capabilities in order to enable efficient and real-time processing of transactions. The core objective of ProtoSpark is to mitigate trust-related attacks in real-time by leveraging Apache Spark's machine learning library, Spark MLlib. We utilize MLlib to construct a robust classifier capable of accurately classifying transactions as malicious or secure. This classifier is trained using relevant features extracted from the transactions, enabling it to make reliable predictions. To facilitate real-time processing, ProtoSpark leverages Spark Streaming, a real-time data processing component of Apache Spark. Spark Streaming enables ProtoSpark to process streaming transactions as they arrive, allowing for prompt decision-making based on the classifier's predictions. In our previous publication [19], we identified the Proof of Trust-related Attacks (PoTA) consensus protocol, which effectively detects and mitigates of offline trust-related attacks. Expanding on the foundation laid by PoTA, we introduce the ProtoSpark protocol, leveraging the advanced capabilities of Apache Spark. ProtoSpark takes trust-related attack mitigation to the next level by enabling real-time detection and prevention of malicious transactions, thereby fostering the overall trustworthiness and reliability of the network. Our key contributions reside in proposing a novel trust management mechanism that exploits blockchain technology and Apache Spark, introducing the real-time trust-related attack mitigation consensus Protocol using Apache Spark (ProtoSpark) for real-time attack mitigation, promoting our previous work [19] through incorporating the capabilities of Apache Spark, developing a robust classifier using Apache Spark's MLlib, and integrating Spark Streaming for efficient processing of streaming transactions. The paper is structured as follows: Sect. 2 foregrounds the limitations of previous works and highlights the gaps our study aims to address. Section 3 presents our proposed approach that leverages Spark and blockchain technologies for real-time mitigation of trust-related attacks. Section 4 describes the dataset, simulation setup, and presents the results. Finally, in Sect. 5, we conclude by summarizing the key findings and suggesting potential future research directions.

2 Related Work

Trust management in the social IoT environment is a critical area that requires further research, particularly regarding trust-related attacks. To provide a concise overview, we first define these attacks: Self-Promoting Attacks (SPA) involve malicious users enhancing their trust value by providing self-recommendations to increase their chances of selection. Discriminatory Attacks (DA) consist of malicious users providing negative recommendations to undermine system trust. Bad-Mouthing Attacks (BMA) tarnish the trust value of benevolent users through negative recommendations. Ballot-Stuffing Attacks (BSA) artificially inflate malicious users' trust value. Opportunistic Service Attacks (OSA) begin with high-quality service to gain trust, then shift to BSA or BMA. On-Off Attacks (OOA) involve random BMA, BSA, or SPA to avoid consistent identification [2,7,9,18]. These trust-related challenges necessitate robust strategies. Existing research suggests two primary approaches: detection-based [2,7,11,13–16,18,27], and prevention-based [3,19]. Detection-based approaches focus on identifying attacks after they occur, providing valuable information for incident response. From this perspective, multiple trust management mechanisms have been developed. In the work [27], the authors introduced a trust management mechanism that considers direct and indirect experiences as features, utilizing a weighted sum method for aggregation. Similarly, [7] built up a novel trust management model incorporating service feedback, intimacy, sociability, and transaction importance, which is effective in detecting OSA and OOA attacks. Moreover, [28] identified two trust management methods for detecting BMA attacks, though these methods depend on a centralized node for collecting and calculating trust values, which poses a potential point of failure. Therefore, it is recommended to explore decentralized approaches that eliminate the need for a centralized entity in trust management mechanisms. Likewise, in our previous works [2,18], we focused on attack detection. Initially, we adopted a machine learning method to classify users' interactions into attack and non-attack categories. Next, we enhanced our model using deep learning techniques. Additionally, we categorized attacks into five different types (BMA, BSA, SPA, DA, and non-attack) and invested seven features, including reputation, honesty, quality of provider, user similarity, direct experience, rating frequency, and rating trend. However, while detection-based approaches offer deeper and better insights into attack patterns and facilitate incident response, they cannot stop attacks in real-time and may incur downtime and data loss. To address this issue, preventive measures at the transaction level seem to be fundamental to mitigate the occurrence of attacks. Prevention-based approaches offer proactive security measures [6], preventing attacks before they happen. These approaches monitor user behavior and block malicious activities at the source, minimizing damage and system compromise. For example, a trust management mechanism elaborated by [3] emphasizes non-real-time attack prevention through the incorporation of direct and indirect observations. These features are aggregated using the Kalman filter technique, enabling the prevention of OOA attacks. To ensure comprehensive security, it is crucial that prevention measures encompass all types of trust-related attacks, including BMA, BSA, SPA, DA, OSA, and OOA. Additionally, it is important to recognize that both attack detection and prevention play significant roles in maintaining security. Each approach has its own strengths and weaknesses. Therefore, an effective security strategy needs to combine both approaches to mitigate attacks in real-time, aiming to achieve maximum protection while minimizing risk. In this respect, in our previous work [19], we set forward the Proof of Trust-related Attacks (PoTA) consensus protocol as an offline solution for preventing trust-related attacks. However, the challenge of achieving real-time transaction processing using traditional machine learning techniques persists. To overcome this challenge, there is a need for a new stream processing engine that can efficiently handle stream transactions, in combination with blockchain technology, for real-time processing of blockchain transactions. Numerous studies [8,23] compared different stream processing engines, such as Apache Spark, Flink, Storm, Hive, and others, revealing that there is no single best option among these frameworks. Each framework has its own strengths and weaknesses, and the choice depends on specific needs and requirements. Recent efforts have been afforded to integrate the distributed processing capabilities of Apache Spark with the Bitcoin blockchain [25], resulting in enhanced data security and integrity. Through combining blockchain's secure transaction handling with Spark's real-time processing and analysis capabilities, networks can benefit from both security and real-time processing. Apache Spark's distributed processing capability makes it highly appropriate for real-time attack detection, enabling swift identification and response to potential threats. Several studies [4,17,20–22,26], explored the use of Spark, particularly its MLlib (Machine Learning library) and Spark Streaming, for real-time detection of attacks and intrusions. These approaches, leveraging Spark's machine learning algorithms and stream processing capabilities, yielded promising results in effectively detecting various forms of attacks. In the light of the significance of Apache Spark for real-time attack identification, we propose incorporating this framework into our consensus protocol as an alternative to traditional learning techniques reported in the previous work [19].

3 Trust Management Mechanism for Real-Time Mitigation of Trust-Related Attacks Based on Blockchain and Apache Spark

Although existing trust management mechanisms are sophisticated and intricate, they prove to be unable to provide real-time mitigation of all types of trust-related attacks, leaving the system vulnerable to malicious activities and compromising its reliability. To address this challenge, we propose a proactive trust management mechanism that analyzes network transactions in real-time, detects malicious transactions, and identifies the type of attack being carried out. By promptly interrupting and canceling malicious transactions, our mechanism effectively prevents their propagation within the network. Through this proactive approach, we significantly promote the security and reliability of the social IoT network, ensuring a trustworthy environment for users. Our innovative trust management relies on the integration of blockchain technology with a new consensus protocol called ProtoSpark, which leverages Apache Spark. This integration aims to enhance system security and reliability through mitigating trust-related attacks in real-time. ProtoSpark analyzes transactions to identify malicious behavior, ensuring the integrity of the blockchain. Legitimate transactions are validated and integrated to the blockchain, while malicious transactions are rejected and added to a blacklist for preventive measures. This approach fosters transparency and consistency in recording transactions, reinforcing the trustworthiness of the system.

Figure 1 illustrates the integrated architecture of our trust management mechanism, which combines Spark and blockchain technologies. This architecture consists of four main phases: composition, aggregation, propagation, and update. The composition phase involves selecting relevant features and creating the transaction dataset. In the aggregation phase, trust features are aggregated to create a classification model in order to enact a real-time attack detection. The propagation and update phases incorporate the real-time attack mitigation step to ensure network security. Detailed explanations will be provided in the following sections.

3.1 Composition Phase

The first phase of our trust management mechanism corresponds to the composition phase, which involves the selection of relevant features and the creation of a transaction dataset. This phase plays a key role in predicting the nature of transactions and distinguishing between malicious and secure transactions within the social IoT network. The selected features provide a detailed description of user behaviors, enabling the identification and mitigation of trust-related attacks. In our previous research [19], we identified seven key features that contribute to the accurate prediction of malicious transactions and the mitigation of various trust-related attacks. We will provide a brief overview of each feature below. It is noteworthy that these features were extensively described in our prior published papers in [2,18,19]:

- **Trust value**: represents the overall trust value of the user u_i in the network. It is denoted as $\mathbf{TrV}(u_i)$ and is calculated by dividing the number of positive interactions of u_i by the total number of performed interactions.
- **Vote**: The vote or rate $\mathbf{rt}(u_i, u_j)$ stands for the rating assigned by user u_i to the service s_k offered by user u_j.
- **Vote similarity or veracity**: $\mathbf{VSim}(u_i)$ refers to the similarity between the vote given by user u_i to the service s_k and the other votes provided by other users in the network.
- **Users Similarity**: $\mathbf{Sim}(u_i, u_j)$ represents the similarity between user u_i and user u_j, indicating how similar or alike they are to each other.

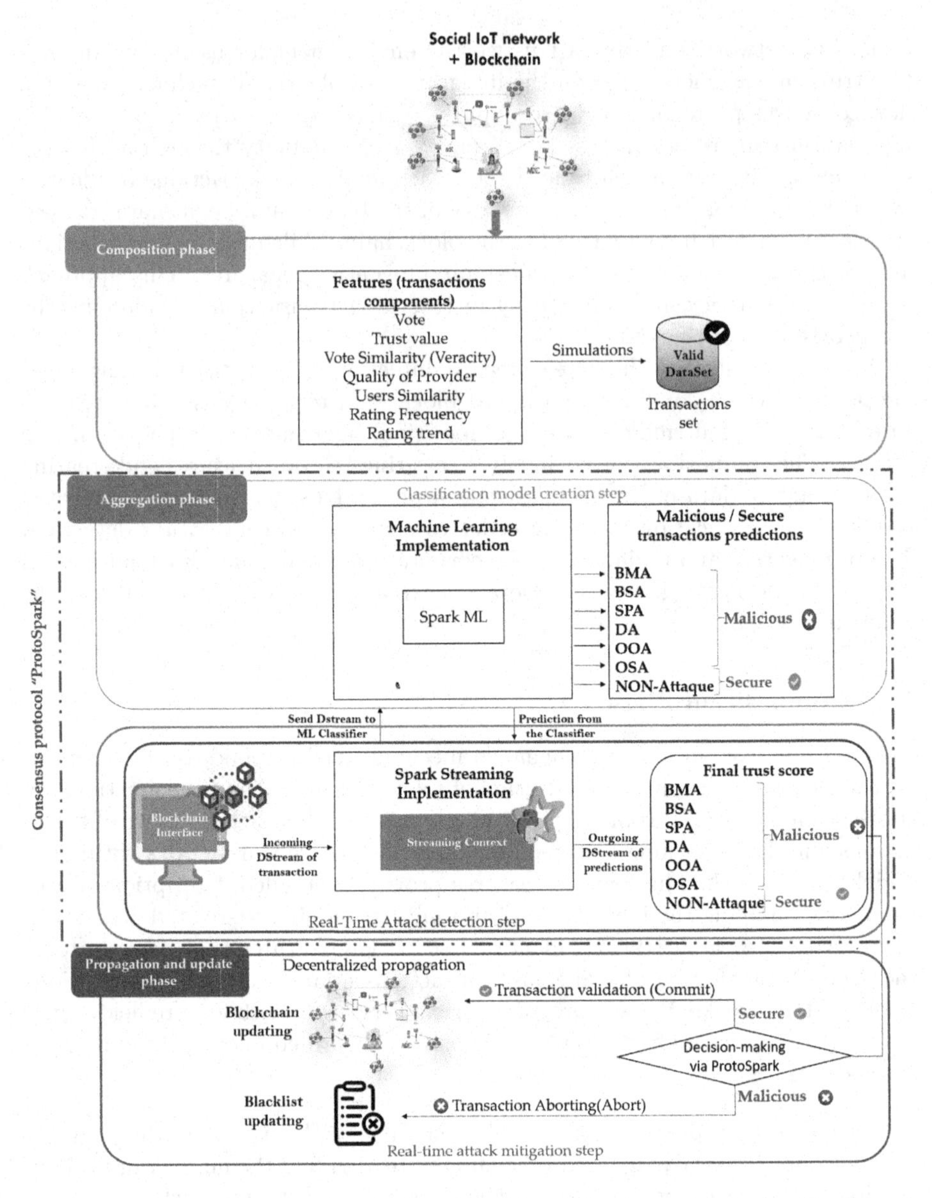

Fig. 1. Architecture for real-time trust-related attack mitigation using Spark and Blockchain

- **Quality of Provider**: refers to the quality of service provided by user u_i, referred to as $\mathbf{QoP}(u_i)$. It indicates the level of service quality, whether good or poor.

- **Rating-Frequency**: corresponds to the frequency with which user u_i provides votes to user u_j and is denoted $\mathbf{RateF}(u_i, u_j)$.
- **Rating-Trend**: $\mathbf{RateT}(u_i)$ quantifies the proportion of positive votes given by user u_i out of the total number of votes they have provided.

After selecting the features (or transaction elements), they will be invested to analyze the user actions within the social IoT network. This analysis helps determine whether a transaction is malicious or secure, indicating the type of attack conducted by the malicious user. The next stage involves gathering and storing all transactions in a dataset resting on these selected elements. The dataset is then partitioned into three subsets to facilitate further analysis and mitigation of trust-related attacks, allowing for a more focused and accurate examination. The first subset is the training set, which is used to generate a classification model, or classifier, that can accurately classify transactions. This model will be utilized to predict the class of new transactions in real-time. The second subset is the test set, which serves to assess the performance of the classifier. Finally, the third subset is the streaming set, which enables the generation of real-time transactions and the prediction of their class using the pre-trained classifier. The source of the collected and stored transactions in our dataset will be extensively described in the experimental section. Furthermore, each row in the dataset is structured as follows: Vote (u_i, s_k), Trust value of user (u_i), Trust value of user (u_j), Vote similarity or veracity of user (u_i), Quality of provider of user (u_i), Quality of provider of user (u_j), Vote tendency of user (u_i), Vote tendency of user (u_j), Frequency of votes from user u_i to user u_j (u_i, u_j), Similarity between the two users (u_i, u_j), and the class indicating whether it is secure or corresponds to one of the trust-related attack types (BMA, BSA, SPA, DA, OSA, or OOA).

3.2 Aggregation Phase

The trust aggregation phase, as part of our trust management mechanism, involves combining selected trust features to calculate a trust score for each transaction, indicating the predicted attack type. This phase encompasses two essential steps: the classification model creation step as well as the real-time attack detection step. The classification model creation step uses machine learning techniques, leveraging Apache Spark, to analyze transactions and create transaction classifier. This classifier is then used by the real-time attack detection step in order to predict the class of each new transaction in real-time, determining whether it is secure or an attack (malicious). Grounded on this classification, our ProtoSpark protocol makes informed decisions to either validate or abort the transaction.

3.2.1 Classification Model Creation Step

As we mentioned real-time attack detection is essential for mitigating various types of attacks in real-time, including BMA, BSA, SPA, DA, OSA, and OOA, through identifying malicious users. To achieve this target, we conducted a thorough analysis of user-generated transactions to promptly detect any malicious activity and interrupt it using our protocol. Therefore, ProtoSpark is designed as a classification system that categorizes transactions into seven distinct classes, including:

1. spa-attack class, when a user performs an SPA-type attack.
2. the bma-attack class, when a user performs a BMA-type attack.
3. da-attack class, when a user performs a DA-type attack.
4. bsa-attack class, when a user performs a BSA-type attack.
5. the ooa-attack class, when a user performs an OOA-type attack.
6. the osa-attack class, when a user performs an OSA-type attack.
7. the secure class, when a user has not performed any of the above attacks.

The ProtoSpark protocol leverages machine learning techniques, specifically supervised learning algorithms implemented in Apache Spark's MLlib, serving to analyze transactions and create transaction classifiers. In our research, our central focus is upon pairs of users (u_i , u_j) and their past transactions analysis. We calculated feature values that describe the transactions between (u_i , u_j). These feature values serve as input for the machine learning model during the training phase. To identify the most effective and reliable model for accurate transaction classification, we trained multiple algorithms, including support vector machines (SVMs), Naive Bayes, decision tables, and linear regression (LR). After conducting a comparative evaluation, we selected the linear regression (LR) model owing to its high performance and promising results. This model will be used to classify new transactions and predict their nature (secure or malicious) within the ProtoSpark protocol.

3.2.2 Real-Time Attack Detection Step

In this section, we will address into the functioning of the Spark Streaming-based module, emphasizing its crucial role in proactively detecting trust-related attacks.

The main objective of this second step is to predict the labels of real-time transactions, ensuring a continuous prediction process for streaming transactional data. Through incorporating real-time attack detection into our approach, we build up a robust and responsive system that actively protects the network against potential attacks. To start with, new transactions are generated from the stream set, consisting of pre-prepared transactions obtained through the blockchain application interface, as depicted in Fig. 1. These transactions are subsequently directed to a real-time transactional data stream for immediate analysis. Leveraging the capabilities of the Spark Streaming library, we efficiently process the transactional data streams in parallel, utilizing the distributed processing capabilities of Spark. In order to assign labels to new stream of transactions, we apply the previously created classifier from the first step. To ensure reliable results, a data pre-processing step is undertaken before real-time processing. This involves creating a schema that specifies the names and types of features used in transactions, to interpret data correctly. This schema guarantees the alignment between the data read from the blockchain-generated transaction source and the data types required for the real-time attack detection algorithm.

Once data pre-processing is complete, the real-time attack detection algorithm progresses to a crucial stage. First, a Spark Streaming context is created, enabling the handling of continuous transaction streams for real-time processing and instant decision-making. Next, the previously created classifier is loaded, as it plays a vital role in predicting transaction labels and identifying potential trust-related attacks. Afterwards, through the"readStream" function, new stream of transactions will be read from the blockchain application, ensuring accurate interpretation based on the defined schema. Upon reading the transaction stream, the transformation function is applied. This function invests the classifier to make instant predictions on each transaction, allowing proactive identification of potentially malicious transactions before they trigger harm. Additionally, the ProtoSpark protocol takes immediate and appropriate actions to counter and mitigate detected attacks in the subsequent phase of the approach.

3.3 Propagation and Update Phases

The propagation and update phase is intrinsic for maintaining the security and reliability of transactions in the social IoT network. This phase rests on securely handling the trust scores generated during real-time transaction analysis. Through applying effective methods, we ensure that appropriate actions are taken to mitigate trust-related attacks in real-time and secure that trust scores and transactions are correctly propagated and updated in our network.

3.3.1 Real-Time Attack Mitigation Step

After running real-time analysis of transaction flows, our ProtoSpark consensus protocol takes immediate action to counter, prevent and mitigate detected attacks. Relying on the analysis results, two actions can be taken: validating secure transactions or aborting malicious ones. For secure transactions, our protocol creates a new blockchain block by grouping verified transactions with other reliable transactions. This new block is next permanently added to the existing blockchain, ensuring the integrity and security of the blockchain. The validated transactions are finalized and included in the blockchain, guaranteeing the inclusion of only legitimate and reliable transactions. On the other side, when a transaction is identified as malicious, it is not validated. Instead, it is added to a specific blacklist that keeps track of suspicious and malicious transactions within the network. This action isolates insecure transactions and prevents them from compromising the overall integrity of the blockchain. Through canceling and excluding these malicious transactions from the validation process, our protocol effectively mitigates attacks and maintains the reliability of the network.

3.3.2 Transactions and Trust Score Propagating in the Network

In this step, we consider the method for propagating the trust scores obtained in the previous phase. We have the option to choose between a centralized or decentralized method based on the network requirements and ongoing transactions.

However, it is significant to highlight the limitations of the centralized method, which involves a central node responsible for calculating and storing trust parameters. While being simple to implement, this approach lacks scalability, especially in networks with billions of devices and high transaction volumes. Therefore, the decentralized method is preferred referring to its scalability and ability to handle the limitations of the centralized approach. This decentralized method involves nodes autonomously propagating trust information to other nodes they encounter or interact with, without relying on a central entity. This is where the use of blockchain technology becomes advantageous. Indeed, utilizing blockchain provides several important advantages. Its transparency as well as immutability ensure the permanent storage of trust scores, maintaining the integrity of information. The traceability feature of the blockchain enables a complete history of transactions and trust score updates, facilitating verification and detection of fraudulent behavior. The reliability and resilience of the blockchain ensure the system's proper functioning even in the face of node failures or attacks. The enhanced security provided by the blockchain protects transactions and trust scores against attacks and falsification. The interoperability of the blockchain enables the exchange of trust data between different applications and systems, promoting collaboration and widespread utilization of trust scores. These advantages emphasize the significance of incorporating blockchain technology in the propagation and updating phase of our trust management mechanism.

3.3.3 Real-Time Update Method

Real-time updating of trust in the social IoT network is crucial to ensure system reliability and security. Our trust management mechanism offers a real-time update approach, enabling continuous updating of trust scores after each transaction. This ensures that trust scores accurately reflect the dynamic changes in the network. There are several advantages for using real-time updates. Firstly, they provide up-to-date trust information in real-time, allowing users to make informed decisions and engage in secure interactions with trusted users. Secondly, real-time updates promote system reliability by avoiding delays and inconsistencies associated with time- or event-based methods. Eventually, users benefit from an improved experience with immediate visibility of trust in the network, facilitating the selection of trusted users and participation in secure interactions.

4 Simulation and Performances Evaluation

In this section, we will provide an overview of the dataset and elaborate the implementation details and pertinent results of our trained classification model and the ProtoSpark consensus protocol.

4.1 Dataset

Due to the lack of real-world data in many research areas, simulations are often invested to assess model performance. As far as our research work is concerned, we were fortunate to have access to the "Sigcomm[1]" database, which provided us with real-world data for simulations. This database contains valuable information about 76 user profiles, 711 interests, 531 social relationships, 32000 interactions, 300 devices and 364 services. By leveraging this dataset, we were able to generate realistic transactions and interactions within the social IoT network. This enabled us to evaluate the performance of our approach in real-time trust-related attack detection and mitigation.

Simulations were performed using the dataset to generate various types of transactions, including secure and malicious ones. For example, in the case of SPA attacks, a malicious user with a smartphone or a smart tablet manipulated its own trust value by giving high ratings to its own service, aiming to deceive future service requesters. The simulation outcomes were recorded in a CSV file, encompassing a total of 3,590 transactions. Each transaction was labeled as secure or malicious, with specific attack types. Indeed, there were 150 transactions labeled as BMA, 150 as BSA, 145 as SPA, 110 as DA, 310 as OSA, 165 as OOA, and 2,560 transactions classified as secure.

4.2 Evaluation Metrics

To assess the effectiveness, goodness of fit and reliability of the proposed features and the classifier in terms of predicting malicious transactions, we will employ standard classification metrics, such as **F1-score**, **Recall**, **Precision**, and **Feature Importance Score (FIS)**[2]. However, to evaluate our consensus protocol, we will primarily use two key metrics: **throughput (processing rate)** and **prediction rates of malicious transactions**. The throughput metric measures the number of transactions processed within a specified time frame. It provides deeper insight into the efficiency of our protocol in handling a large volume of transactions, indicating its scalability and performance. On the other side, the prediction rates of malicious transactions focus on the accuracy of our protocol in identifying and predicting malicious transactions (attacks). This metric evaluates how effectively ProtoSpark can detect and classify transactions as either secure or malicious, providing a measure of its effectiveness in mitigating trust-related attacks.

4.3 Results and Comparative Analysis

We carried out experiments to evaluate the performance of our approach, which includes proposed features, classifier, and ProtoSpark protocol. The subsequent subsections provide a detailed analysis of the results, highlighting the effectiveness and capabilities of our approach in mitigating trust-related attacks.

---

[1] https://crawdad.org/thlab/sigcomm2009/20120715/.

[2] https://machinelearningmastery.com/calculate-feature-importance-with-python/.

4.3.1 Experimental Evaluation of the Proposed Features

Relying on the analysis of Fig. 2, we can draw the outcomes of the proposed features and their respective Feature Importance Scores (FIS). Notably, the vote feature obtained the highest score, corroborating its significant role in detecting malicious transactions. This observation aligns with our previous assertion indicating the importance of votes as a key feature in real-time attack detection [19]. The VSim, TrV, RateF, Sim, QoP, and RateT features displayed similar scores, suggesting that they possess comparable discriminative capabilities in detecting malicious transactions.

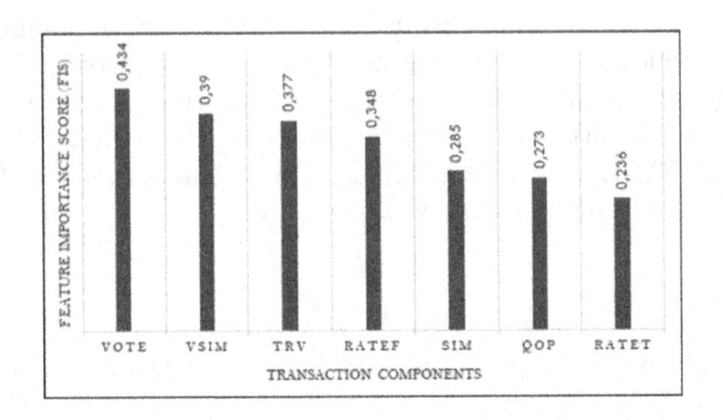

Fig. 2. Results of the proposed features

4.3.2 Experimental Evaluation of the Spark MLlib-Based Classification Model

In this section, we evaluate the performance of our classifier in classifying transactions as malicious or secure using the Spark MLlib machine learning library. We report the results for each attack type individually and compare the precision, recall, and F1-score of our classifier to those of other studies in literature.

4.3.2.1 Predictive Performance of Attack Type Label

The experiments conducted to evaluate our Spark MLlib-based classifier for predicting individual attack types (BMA, BSA, SPA, DA, OOA, and OSA) confirmed its enhanced performance. The F1-score results summarized in Fig. 3 indicated significant improvements compared to those recorded in our previous work [19]. These findings are indicative of the effectiveness and feasibility of the design and implementation refinements introduced to the classifier, leveraging the capabilities of Apache Spark.

4.3.2.2 Predictive Performance of All Attack Type Labels

This study aimed to assess the performance of our new classifier in terms of accurately classifying and identifying different types of attacks in transactions. Comparative analyses were undertaken, including comparisons to our previous works [18,19] and other works in literature [1,2], using the same dataset (Sigcomm). Furthermore, it is worth noting that [1] have already proven the effectiveness of their approach compared to various other works [5,10,16]. Therefore, there is no need to include these specific works in the comparative analysis figure. The results, summarized in Fig. 4, reveal that our new classifier achieves higher F1-scores compared to previous works and other approaches reported in [1,2]. The utilization of Apache Spark's MLlib contributes to the improved prediction accuracy and performance of our classifier, surpassing standard machine learning techniques used in other approaches.

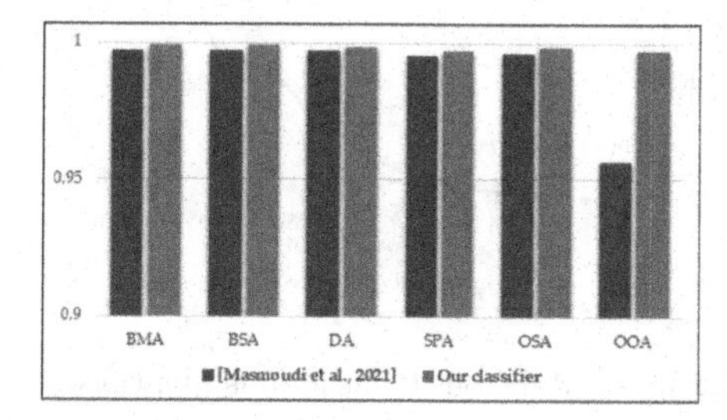

Fig. 3. Predictive performance of attack type label

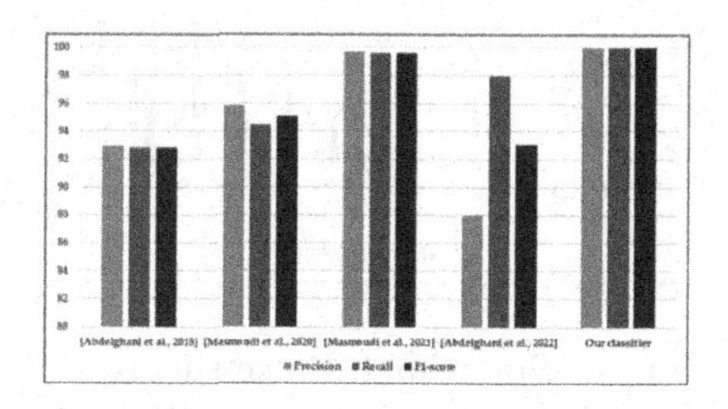

Fig. 4. Predictive Performance of All Attack Type Labels and comparison with related works

4.3.3 Experimental Evaluation of ProtoSpark Consensus Protocol

In order to evaluate the performance of our proposed ProtoSpark protocol, we will compare its processing rate and its accuracy in predicting malicious transactions to our previous protocol reported in [19]. It is worth noting that our previous protocol has already been compared to the widely used Proof of Work (PoW) protocol and demonstrated superior capabilities. Therefore, this comparison will provide additional insights into the advancements achieved by our new ProtoSpark protocol.

4.3.3.1 Throughput Metric: Processing Rate

The comparison displayed in Fig. 5 validates the higher efficiency of our ProtoSpark protocol in processing a larger number of transactions within a shorter timeframe compared to our previous protocol [19]. Despite the increasing workload, ProtoSpark consistently exhibits its capability to handle higher transaction volumes efficiently. This outcome underscores the effectiveness of Spark Streaming as a powerful tool for managing larger transaction loads in a time-efficient manner.

4.3.3.2 Malicious Transactions Identification

Our experimental results were indicative of the significant improvements achieved with our ProtoSpark protocol in terms of accurately predicting and mitigating malicious transactions compared to our previous protocol [19]. Through leveraging Spark MLlib, we developed a highly effective classifier that enhances the overall accuracy of identifying different types of attacks. This novel approach contributes to the promotion of transaction security and establishes a new benchmark in comprehensive attack mitigation. The use of Apache Spark enables real-time processing, ensuring prompt detection and mitigation of trust-related attacks (Fig. 6).

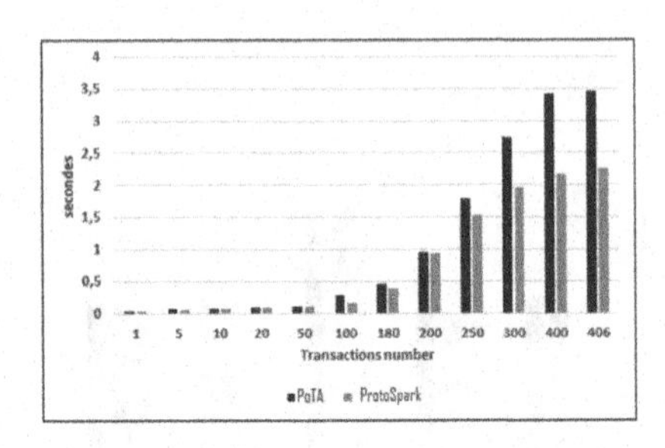

Fig. 5. Throughput: Processing rate

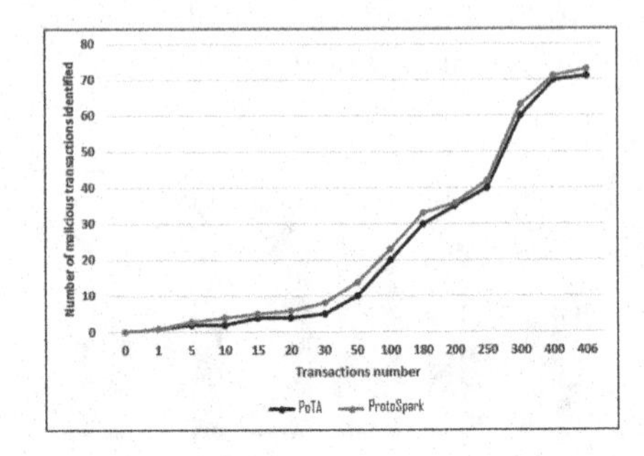

Fig. 6. Malicious transactions identification

5 Conclusion

To sum up, we would assert that the current study introduced a novel trust management mechanism that utilizes blockchain technology and Apache Spark as to mitigate trust-related attacks in real-time and enhance the security of user interactions. The integration of blockchain technology was complemented by the implementation of a new consensus protocol called real-time trust-related attack mitigation Protocol using Apache Spark (ProtoSpark). Leveraging the capabilities of Apache Spark and its MLlib and Spark Streaming libraries, we successfully elaborated ProtoSpark as a robust and powerful consensus protocol that incorporates a promising classifier capable of real-time prediction and prevention of malicious transactions. The assessment of our classifier implemented using Spark MLlib confirmed its superior performance compared to previous literature works [1,2,18,19]. This improvement highlights the accuracy and efficiency achieved in predicting malicious transactions. The ProtoSpark protocol also exhibited significant enhancements in processing rate and efficiency, enabling it to handle a larger volume of transactions in a shorter time frame. The improved prediction accuracy further reinforces its ability to identify and abort malicious transactions. These advancements establish ProtoSpark as a more effective and workable solution compared to existing approaches. From this perspective, this work pioneers the real-time mitigation of all types of trust-related attacks, demonstrating its novelty and potential impact. As a final note, the newly developed classifier proved to be worthwhile, reliable and promising. In this respect, it can be extended in several ways. This involves exploring the potential of Apache Spark's Deep

Learning library to enhance the classifier further. Another outstanding future research direction corresponds to real-world implementation and assessment of the classifier and ProtoSpark protocol in order to evaluate their performance in practical scenarios. Additionally, extending the research to address other types of attacks in social IoT environments using similar techniques can be regarded as an equally pertinent area of interest.

References

1. Abdelghani, W., Amous, I., Zayani, C.A., Sèdes, F., Roman-Jimenez, G.: Dynamic and scalable multi-level trust management model for social internet of things. J. Supercomput. **78**(6), 8137–8193 (2022)
2. Abdelghani, W., Zayani, C.A., Amous, I., Sèdes, F.: Trust evaluation model for attack detection in social internet of things. In: Zemmari, A., Mosbah, M., Cuppens-Boulahia, N., Cuppens, F. (eds.) CRiSIS 2018. LNCS, vol. 11391, pp. 48–64. Springer, Cham (2019). https://doi.org/10.1007/978-3-030-12143-3_5
3. Abderrahim, O.B., Elhdhili, M.H., Saidane, L.: TMCoI-SIOT: a trust management system based on communities of interest for the social internet of things. In: 2017 13th International Wireless Communications and Mobile Computing Conference (IWCMC), pp. 747–752. IEEE (2017)
4. Awan, M.J., et al.: Real-time DDOs attack detection system using big data approach. Sustainability **13**(19), 10743 (2021)
5. Chen, Z., Ling, R., Huang, C.M., Zhu, X.: A scheme of access service recommendation for the social internet of things. Int. J. Commun Syst **29**(4), 694–706 (2016)
6. Dagorn, N.: Détection et prévention d'intrusion: présentation et limites (2006)
7. Ekbatanifard, G., Yousefi, O.: A novel trust management model in the social internet of things. J. Adv. Comput. Eng. Technol. **5**(2), 57–70 (2019)
8. Gorasiya, D.V.: Comparison of open-source data stream processing engines: spark streaming, flink and storm (2019)
9. Guo, J., Chen, R., Tsai, J.J.: A survey of trust computation models for service management in internet of things systems. Comput. Commun. **97**, 1–14 (2017)
10. Jayasinghe, U., Lee, G.M., Um, T.W., Shi, Q.: Machine learning based trust computational model for IoT services. IEEE Transa. Sustain. Comput. **4**(1), 39–52 (2018)
11. Jmal, R., Masmoudi, M., Amous, I., Zayani, C.A., Sèdes, F.: Apache spark based deep learning for social transaction analysis. In: 19th International Conference on Web Information Systems and Technologies (2023)
12. Kalaï, A., Zayani, C.A., Amous, I., Abdelghani, W., Sèdes, F.: Social collaborative service recommendation approach based on user's trust and domain-specific expertise. Futur. Gener. Comput. Syst. **80**, 355–367 (2018)
13. Kowshalya, A.M., Valarmathi, M.: Trust management for reliable decision making among social objects in the social internet of things. IET Netw. **6**(4), 75–80 (2017)
14. Kowshalya, A.M., Valarmathi, M.: Trust management in the social internet of things. Wireless Pers. Commun. **96**, 2681–2691 (2017)
15. Senthil Kumar, J., Sivasankar, G., Selva Nidhyananthan, S.: An artificial intelligence approach for enhancing trust between social IoT devices in a network. In: Hassanien, A.E., Bhatnagar, R., Khalifa, N.E.M., Taha, M.H.N. (eds.) Toward Social Internet of Things (SIoT): Enabling Technologies, Architectures and Applications. SCI, vol. 846, pp. 183–196. Springer, Cham (2020). https://doi.org/10.1007/978-3-030-24513-9_11

16. Lee, G., Truong, N.: A reputation and knowledge based trust service platform for trustworthy social internet of things. In: Innovations in Clouds, Internet and Networks (ICIN) (2016)
17. Marir, N., Wang, H., Feng, G., Li, B., Jia, M.: Distributed abnormal behavior detection approach based on deep belief network and ensemble SVM using spark. IEEE Access **6**, 59657–59671 (2018)
18. Masmoudi, M., Abdelghani, W., Amous, I., Sèdes, F.: Deep learning for trust-related attacks detection in social internet of things. In: Chao, K.-M., Jiang, L., Hussain, O.K., Ma, S.-P., Fei, X. (eds.) ICEBE 2019. LNDECT, vol. 41, pp. 389–404. Springer, Cham (2020). https://doi.org/10.1007/978-3-030-34986-8_28
19. Masmoudi, M., Zayani, C.A., Amous, I., Sèdes, F.: A new blockchain-based trust management model. In: 25th International Conference on Knowledge-Based and Intelligent Information & Engineering Systems, pp. 389–404. Elsevier (2021)
20. Patil, N.V., Krishna, C.R., Kumar, K.: SS-DDoS: Spark-based DDoS attacks classification approach. In: Security and Resilience of Cyber Physical Systems, pp. 81–90. Chapman and Hall/CRC (2022)
21. Patil, N.V., Krishna, C.R., Kumar, K.: SSK-DDoS: distributed stream processing framework based classification system for DDoS attacks. Cluster Comput. **25**, 1–18 (2022)
22. Patil, N.V., Rama Krishna, C., Kumar, K.: S-DDoS: apache spark based real-time DDoS detection system. J. Intell. Fuzzy Syst. **38**(5), 6527–6535 (2020)
23. Perera, S., Perera, A., Hakimzadeh, K.: Reproducible experiments for comparing apache flink and apache spark on public clouds. arXiv preprint arXiv:1610.04493 (2016)
24. Roopa, M., Pattar, S., Buyya, R., Venugopal, K.R., Iyengar, S., Patnaik, L.: Social internet of things (SIoT): Foundations, thrust areas, systematic review and future directions. Comput. Commun. **139**, 32–57 (2019)
25. Rubin, J.: Btcspark: Scalable analysis of the bitcoin blockchain using spark. Dec **16**, 1–14 (2015)
26. Saravanan, S., et al.: Performance evaluation of classification algorithms in the design of apache spark based intrusion detection system. In: 2020 5th International Conference on Communication and Electronics Systems (ICCES), pp. 443–447. IEEE (2020)
27. Talbi, S., Bouabdallah, A.: Interest-based trust management scheme for social internet of things. J. Ambient. Intell. Humaniz. Comput. **11**, 1129–1140 (2020)
28. Xia, H., Xiao, F., Zhang, S.s., Hu, C.Q., Cheng, X.Z.: Trustworthiness inference framework in the social internet of things: a context-aware approach. In: IEEE INFOCOM 2019-IEEE Conference on Computer Communications, pp. 838–846. IEEE (2019)

Hybrid Data-Driven and Knowledge-Based Predictive Maintenance Framework in the Context of Industry 4.0

Fidma Mohamed Abdelillah[1(✉)], Hamour Nora[1], Ouchani Samir[2], and Benslimane Sidi-Mohammed[3]

[1] CESI Lineact, Lyon, France
{mafidma,nhamour}@cesi.fr
[2] CESI Lineact, Aix-en-Provence, France
souchani@cesi.fr
[3] LabRI-SBA Lab, Ecole Superieure en Informatique, Sidi Bel Abbes, Algeria
s.benslimane@esi-sba.dz

Abstract. The emergence of Industry 4.0 has heralded notable progress in manufacturing processes, utilizing sophisticated sensing and data analytics technologies to maximize efficiency. A vital component within this model is predictive maintenance, which is instrumental in ensuring the dependability and readiness of production systems. Nonetheless, the heterogeneous characteristics of industrial data present obstacles in realizing effective maintenance decision-making and achieving interoperability among diverse manufacturing systems. This paper addresses these obstacles by introducing a hybrid approach that harnesses the power of ontologies, machine learning techniques, and data mining to identify and predict potential anomalies in manufacturing processes. Our work concentrates on designing an intelligent system with standardized knowledge representation and predictive capacities. By bridging the semantic divide and enhancing interoperability, ontologies enable the amalgamation of various manufacturing systems, thereby optimizing maintenance decision-making in real-time. As demonstrated in the experimental results, this approach not only ensures system reliability but also fosters a seamless, integrated, and efficient production landscape.

Keywords: Industry 4.0 · Industrial Cyber-Physical System · Predictive Maintenance · Chronicle Mining · Ontology · SWRL Rules

1 Introduction

Smart factories in the Industry 4.0 era, have emerged as transformative manufacturing environments that harness cutting-edge technologies to revolutionize production processes. These smart factories leverage sophisticated sensing technologies and data analytics to gain real-time insights into their operations. They can optimize manufacturing processes by analyzing vast amounts of data, leading to higher production efficiency and reliability [5]. Artificial Intelligence (AI)

M. Mosbah et al. (Eds.): MEDI 2023, LNCS 14396, pp. 319–337, 2024.
https://doi.org/10.1007/978-3-031-49333-1_23

techniques, such as machine learning and data mining, play a pivotal role in this context. These advanced AI methods enable the factories to identify patterns, trends, and anomalies within the data, helping them detect potential issues and predict maintenance needs in advance [7].

The complexity of industrial data introduces significant challenges in achieving seamless interoperability across manufacturing systems. The diverse nature of this data leads to the emergence of complex knowledge structures, creating what is known as a semantic gap issue. This gap impedes effective communication and sharing of information between different components and systems within the manufacturing environment. As a result, the full potential of data-driven decision-making and automation remains untapped [14]. Moreover, industrial systems, particularly Cyber-Physical Systems (CPS), operate in knowledge-intensive domains that demand uniform and standardized knowledge representation [18]. A cohesive and consistent approach to knowledge representation becomes indispensable to enable real-time reasoning and automated decision-making. Overcoming these hurdles is crucial to harnessing the power of Industry 4.0 technologies and fully embracing the potential of predictive maintenance in intelligent manufacturing systems.

Amidst the diverse data sources and industrial requirements, effective predictive maintenance in Industry 4.0 faces numerous challenges. The semantic gap issue and lack of uniform knowledge representation present obstacles to smooth interoperability and automated decision-making, ultimately limiting the overall effectiveness of predictive maintenance systems [18]. Additionally, maintenance processes may encounter difficulties in detecting and addressing issues proactively. Hence, By focusing on bridging the semantic gap, ensuring uniform knowledge representation, and enabling real-time reasoning, we aim to develop a robust and efficient predictive maintenance framework that empowers industries to proactively tackle maintenance issues and optimize their manufacturing operations.

In this paper, we introduce a Hybrid data-driven and knowledge-based Predictive Maintenance for industry Systems (HPMS). HPMS employs a combination of statistical AI technologies such as machine learning and data mining, along with symbolic AI technologies. It utilizes logic rules generated from chronicle patterns and domain ontologies for ontology reasoning, as well as SQWRL queries for retrieving temporal information on failure points. This integrated approach aims to enable the automatic detection of machinery anomalies and accurate prediction of future events, leading to enhanced efficiency and effectiveness of predictive maintenance in various industrial settings. By harnessing HPMS, industries can achieve automated anomaly detection and precise event prediction, optimizing their maintenance practices and overall operational efficiency. The primary contributions of this paper are as follows.

1. Developing and implementing a Hybrid Predictive Maintenance System (HPMS) that merges both statistical and symbolic AI approaches, specifically leveraging machine learning and chronicle mining methodologies.

2. Employing Semantic Web Rule Language (SWRL) rules, which are derived from chronicle patterns and domain ontologies, to enhance its ontology reasoning capabilities.
3. Integrating Semantic Query-Enhanced Web Rule Language (SQWRL) queries to extract temporal information regarding failure points, thereby augmenting its real-time reasoning capabilities.
4. Upon detection and prediction of failures, clustering the failure points by taking into account the minimum time to failure. This facilitates effective and efficient prioritization of the said points.

The paper begins with a comprehensive literature review 2, exploring existing approaches in predictive maintenance and Industry 4.0, with a particular focus on data-driven and hybrid approaches. The proposed 'HPMS' framework, which integrates ontologies and data-driven techniques, is thoroughly explained in Sect. 3. The experimental setup is then outlined in Sect. 4, detailing the dataset and chosen machine learning algorithms for evaluation. Section 5 concludes with a summary of contributions and potential future research directions to improve 'HPMS' implementation.

2 Related Work

In the Industry 4.0 era, significant research efforts were dedicated to automating and improving smart manufacturing processes. AI-based methods have shown promising results, especially in the realm of predictive maintenance for Industry 4.0 tasks. By surveying the existing AI-based smart manufacturing and predictive maintenance approaches, this section categorized them into four distinct categories: (i) data-driven; (ii) physical model-based; (iii) knowledge-based; and (iv) hybrid model-based.

2.1 Data-Driven Approaches

In recent times, data-driven approaches have emerged as a significant solution for smart manufacturing and predictive maintenance. IWSNs, CPS, and Internet of Things (IoT) are employed together to collect and intelligently process big industrial data, aiding in decision-making. The exponential growth of data volume, coupled with the rapid advancement in data acquisition technologies, has heightened attention towards data-driven methods for the predictive maintenance of industrial equipment [23]. Sezer et al. [19] proposed a cost-effective CPS architecture for monitoring machining conditions, utilizing cloud-stored data and a Recursive Partitioning and Regression Tree model to predict part rejection while integrating time series analysis and machine learning. Zhang [24] proposed an online data-driven framework for bearing RUL[1] prediction using deep CNN.

[1] Remaining Useful Life: The length of time a machine is likely to operate before it requires repair or replacement.

The method employed the Hilbert-Huang transform for preprocessing, a nonlinear degradation indicator for learning, and a support vector regression model for RUL prediction. Dual-task deep LSTM networks were introduced by Miao et al. [13] for aeroengine degradation assessment and RUL prediction simultaneously, yielding more reliable results tailored to each aeroengine's health state. Also, De Luca et al. [6] presented a DL-based approach for PdM tasks. They utilized a highly efficient architecture with a multi-head attention (MHA) mechanism, achieving superior results in terms of RUL estimation while maintaining a compact model size. Experimental results on the NASA dataset demonstrated the approach's effectiveness and efficiency.

2.2 Physical Model-Based Approaches

Physical models are models that utilize physical laws, often from first principles, to quantitatively characterize the behavior of a failure mode. Such models typically employ mathematical representations of the physical behavior of a machine's degradation process to calculate the RUL of the machinery. The mathematical representation captures how the monitored system responds to stress on both the macroscopic and microscopic levels [20]. In Heyns et al. [8], Gaussian mixture models (GMMs) are used to detect faults in vibration signals, especially indicating potential gear damage. The method computes the negative log-likelihood (NLL) of signal segments, measuring their deviation from a healthy gearbox's reference density distribution. By synchronously averaging the NLL discrepancy signal, a clear representation emerges, offering insights into the gear damage's nature and severity. Also, Tiwari et al. [21] introduced Gaussian Process Regression (GPR) for tracking bearing features that incorporate uncertainty in predictions and is used to evaluate and predict. Three GPR models with different covariance functions are explored for feature tracking and RUL assessment. In another study by Wu et al. [22], the primary focus centers on predicting RUL of lithium-ion batteries, enhancing the dependability and safety of battery-powered systems. This research employs an empirical degradation model alongside the particle filter (PF) algorithm to facilitate real-time parameter updates.

2.3 Knowledge-Based Approaches

A system based on knowledge uses a knowledge base to store a computational model's symbols in the form of domain statements and performs reasoning by manipulating these symbols. These systems determine appropriate decisions by measuring the similarity between a new observation and a databank of previously described situations. Knowledge-based approaches are categorized into **knowledge graphs and ontologies**, **rule-based systems**, and **fuzzy systems**.

In a rule-based system, knowledge is represented using "IF-THEN" rules, comprising a knowledge base housing rules, a facts base storing inputs, and an inference engine applying rules to the facts base to derive new insights [1]. The term "knowledge graph" is often synonymous with **"ontology"**, which is explicitly defined as "a precise description or identification of conceptualization for a domain of interest" [1]. Ontologies necessitate formal logic to distinctly define concepts and relationships [1]. Fuzzy-knowledge-based models employ fuzzy logic, akin to rule-based systems, using IF-THEN rules. Fuzzy logic manages partial truth values between true and false, quantifying the degree of truth or falsity and resonating with human perceptions.

Konys [11] offered structured guidance for knowledge management-based sustainability assessment, advocating ontology as a means of conceptualizing knowledge and enhancing its accessibility and reusability. Mehdi et al. [12] introduced "sigRL", a semantic rule language, facilitating efficient diagnostic programming with a high-level, data-independent vocabulary, diagnostic ontologies, and queries. They presented the diagnostic system "SemDia", showcasing sigRL's role in effective and efficient diagnostics.

2.4 Hybrid Model-Based Approaches

Hybrid model-based approaches involve combining the strengths of different methodologies, such as physical models, data-driven techniques, and knowledge-based systems. These approaches seek to leverage the benefits of each individual method to enhance predictive accuracy and flexibility. A hybrid model-based predictive maintenance task can be classified into two main approaches: The series hybrid model and the parallel hybrid model [26].

Zhou et al. [25] devised a pioneering feature-engineering-based machine learning approach that operates on real production data. This technique delves into sequences of welding instances obtained from manufacturing lines, amalgamating engineering insights and data science methodologies. By employing sophisticated feature engineering tactics bolstered by domain knowledge, this approach captures intricate dependencies across welding sequences, filling a gap in the existing literature. Qiushi [17] introduced an innovative fusion of evidential theory tools and semantic technologies to bolster predictive maintenance. They harnessed the Evidential C-means (ECM) algorithm, among other tools, to assess failure criticality by considering time constraints and maintenance costs. This method simultaneously leveraged domain ontologies and rule-based extensions to formalize domain knowledge, enabling precise forecasts of future failures' timing and criticality.

Klein [10] incorporated expert knowledge of class or failure mode-dependent attributes into a specialized neural network framework. Accompanied by an attribute-wise encoding approach based on 2D convolutions, this strategy facilitated the sharing of knowledge through the utilization of filters among analogous data streams. Furthermore, Cao et al. [4] presented an ontology-driven strategy to enhance predictive maintenance, combining fuzzy clustering techniques with semantic technologies. This approach analyzed historical machine data to understand the significance of failures. Utilizing semantic technologies for analyzing fuzzy clustering outputs, it achieved precise predictions of failure timing and criticality. The outcome was a domain-specific ontology enriched with predictive maintenance knowledge, supported by a set of SWRL predictive rules for informed assessments of machinery failure timing and impact.

2.5 Learned Lessons

This literature review provides a comprehensive analysis of existing approaches for predictive maintenance in Industry 4.0, identifying the strengths and weaknesses of different methods and offering insights into their suitability and effectiveness in various industrial contexts. The study emphasizes that there is no one-size-fits-all solution for predictive maintenance in Industry 4.0. The primary objective of this state-of-the-art review was to address the gap in the existing literature on predictive maintenance approaches.

With respect to the studied state-of-the-art, we present a novel data-driven Knowledge-based system for Predictive Maintenance in Industry 4.0 (HPMS). HPMS moves beyond the limitations of the aforementioned approaches by combining the power of data-driven methods with a uniform representation of knowledge. This integration enhances predictive maintenance capabilities by leveraging statistical AI technologies like machine learning and chronicle mining alongside symbolic AI techniques.

3 HPMS Framework

Our developed HPMS framework strategically amalgamates statistical and symbolic AI techniques to devise a hybrid predictive maintenance mechanism. It incorporates machine learning, chronicle mining, domain ontologies, and logic rules, effectively capitalizing on both *data-driven* and *knowledge-driven* approaches. HPMS employs Semantic Web Rule Language (SWRL) rules, which are engendered from chronicle patterns and ontology reasoning, to autonomously detect anomalies in machinery and prognosticate future events. Figure 1 delineates the HPMS framework, showcasing its five key processing steps devised to augment the predictive maintenance capabilities of an industrial system.

3.1 Failure Chronicle Mining

In industrial environments, maintenance data is typically represented as timestamped sequences. To identify frequently occurring patterns in such data, Sequential Pattern Mining (SPM) is used as an instrumental technique. Initially explored for analyzing customer purchase behavior, SPM targets the discovery of sequential patterns with support exceeding a predetermined threshold.

However, traditional SPM algorithms fall short in capturing time intervals between elements and items. In the realm of predictive maintenance, the emphasis is on temporal patterns, specifically chronicles. Chronicles are unique sequential patterns that represent event sequences with associated time intervals (temporal constraints). In the predictive maintenance domain, failure chronicles are created to capture the temporal patterns preceding machinery failures. These play a pivotal role in pre-empting machine anomalies and amalgamating frequent chronicle mining with semantic approaches to enhance predictive maintenance tasks. Definition 1 describes formally a chronicle.

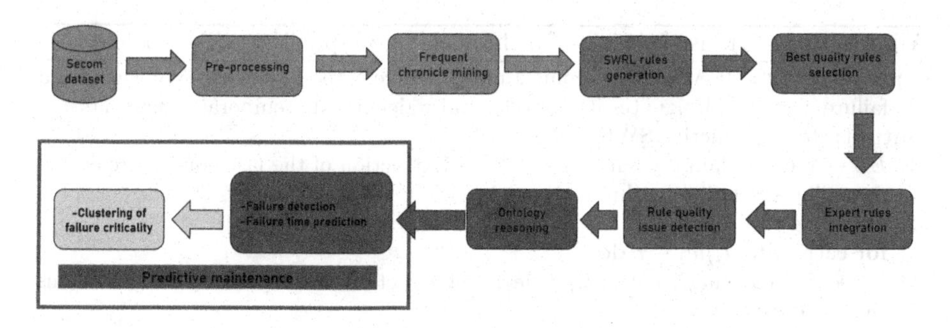

Fig. 1. An overview of the proposed HPMSframework.

Definition 1 (Chronicle). *A chronicle is a pair $C = (E, T)$ such that:*

1. $E = \{e_1, \ldots, e_n\}$ *is a set of events, where* $\forall e_i \in E$ *we have* $e_i \prec_E e_{i+1}$ *such that* $\prec_E$ *is pre-order relation that defines the sequence orders of the events in* E.
2. $T = \{t_{ij} : 1 \leq i < j \leq |E|\}$ *is a set of temporal constraints on* E *such that for all pairs* (i, j) *satisfying* $i < j$, t_{ij} *is denoted by* $e_i[t_{-ij}, t_{+ij}]e_j$. *A pair of events* (e_1, t_1) *and* (e_2, t_2) *are said to satisfy the temporal constraint* $e_1[t_-, t_+]e_2$ *if* $t_2 - t_1 \in [t_-, t_+]$.

3.2 Predictive Rule Generation

In this part, we focus on SWRL rule generation, a crucial step in HPMS for rule-based reasoning in predictive maintenance. Following rule generation, we rigorously prune to remove low-quality rules, retaining only relevant and accurate ones, thus improving the system effectiveness.

3.2.1 SWRL Rules Generation

During this stage, a set of predictive SWRL rules are created based on the extracted frequent failure chronicles. These rules constitute logical IF-THEN statements facilitating rule-based reasoning. Algorithm 1 automatically transforms the extracted chronicles into a set of SWRL predictive rules by extracting distinct event types from a failure chronicle, determining their order and temporal constraints, and subsequently formulating rules as implications between the antecedent and consequent sections. Algorithm 1 extracts the last non-failure events before the failure in the chronicle (lines: 1–2) and processes each time interval to obtain atoms representing the interval, preceding event (lines: 3–4), and subsequent event (line: 5). These atoms form the antecedent part of the SWRL rule (lines: 6–7). It then extracts the time constraint between the last event and the failure event, creating another atom representing this constraint (lines: 9–10). These atoms form the consequent part of the SWRL rule (lines: 11–12). By creating an implication between the antecedent and consequent (lines: 14–15), the algorithm generates a predictive SWRL rule (R) for reasoning and prediction in the context of the failure chronicle.

Algorithm 1. Transformation of a chronicle into a predictive SWRL rule.

Input: $\mathcal{F} = (\Sigma, \tau)$: A failure chronicle model wherein the last event type signifies a failure event, Σ designates its episode, and τ denotes its temporal constraints.
Output: R: A predictive SWRL rule
1: $EL \leftarrow$ LastNonfailureEvent($\mathcal{F}$) ▷ Extraction of the last non-failure events preceding the failure within a chronicle.
2: $R \leftarrow \emptyset$, $A \leftarrow \emptyset$, $C \leftarrow \emptyset$
3: **for** each $e_i[t_{ij}, t_{ij}]e_j \in \tau$ **do**
4: $pe \leftarrow$ PrecedingEvent($e_i[t_{ij}, t_{ij}]e_j$) ▷ Extraction of the preceding event of this time interval.
5: $se \leftarrow$ SubsequentEvent($e_i[t_{ij}, t_{ij}]e_j$) ▷ Extraction of the subsequent event of this interval.
6: $AtomA \leftarrow [t_{ij}, t_{ij}] \wedge pe \wedge se$
7: $A \leftarrow AtomA \wedge A$
8: **end for**
9: **for** each $el \in EL$ **do**
10: $ftc \leftarrow$ FailureTimeConstraint($el, \mathcal{F}$) ▷ Extraction of the time constraint between the last event prior to the failure and the failure event.
11: $AtomC \leftarrow el \wedge ftc$
12: $C \leftarrow AtomC \wedge C$
13: **end for**
14: $R \leftarrow (A \rightarrow C)$ ▷ Generation of rule R as an implication between the antecedent and consequent.
15: **Return** R

3.2.2 Best Quality Rules Selection

Upon completing the rule generation process, real-world data often contains imprecision, and rule-based classification and prediction can suffer from overfitting issues, leading to rules of low quality [3]. we conduct a subsequent rule pruning step to eliminate these rules. To achieve this, we adopt a multi-objective optimization strategy that considers two crucial measures: rule accuracy and rule coverage.

By analyzing the correlation between each decision rule (R) and the target class (F) through a contingency table, which tabulates data categories and their frequencies defined in Table 1.

The accuracy is defined as Accuracy(R) $= \frac{n_{\mathrm{af}}}{n_a}$ and the rule coverage is defined by Coverage(R) $= \frac{n_{\mathrm{af}}}{n_f}$ where n_{af} denotes examples where both the antecedent and consequent of the rule are true, while $n_{\mathrm{a\bar{f}}}$ represents cases where the antecedent is true, but the consequent is not. Similarly, $n_{\mathrm{\bar{a}f}}$ signifies scenarios where the antecedent is false, but the consequent is true, and $n_{\mathrm{\bar{a}\bar{f}}}$ denotes instances where neither the antecedent nor the consequent holds. These derived accuracy and coverage metrics serve as evaluative indicators of the overall rule quality.

Table 1. Contingency table for evaluation.

	F is true	F is not true	
A is true	n_{af}	$n_{a\bar{f}}$	n_a
A is not true	$n_{\bar{a}f}$	$n_{a\bar{f}}$	$n_{\bar{a}}$
	n_f	$n_{\bar{f}}$	N

In our multi-objective optimization process, we employ the fast non-dominated sorting algorithm, also known as the Pareto sorting algorithm [15]. This algorithm efficiently selects high-quality rules by prioritizing accuracy and coverage as key objectives, andeffectively capturing the subset with the best balance of these measures.

3.3 Expert Rule Integration

The process of expert rule integration involves incorporating domain experts' knowledge by integrating expert rules to supplement the existing chronicle rule base. This integration aims to address the inherent incompleteness of the chronicle rule base. During this process, rule quality issues such as *redundancy*, *conflict*, and *subsumption* are examined to ensure effective reasoning performance when combining diverse rules. To automatically detect rule quality measures, Algorithm 2 was proposed that includes functions for extracting atom sets from expert rules and subsequent steps for identifying issues with the chronicle rule base. The algorithm starts by initializing an updated rule base (C') with the existing chronicle rule base (C). It then iterates through each rule (R) in C to identify potential issues with an expert rule (Re) (1–2). If ChroRedundancy(Re, R) is true, indicating redundancy, the algorithm prints a ChroRedundancy message and retains the original rule base (C') (3–4). If ChroSubsumes(Re, R) is true, indicating subsumption, it removes R from C' and integrates Re (6–8). Similarly, if ChroConflict(Re, R) is true, indicating a conflict, it removes R from C' and integrates Re. After evaluating all rules, the algorithm returns the updated rule base (C') with the integrated expert rule. Thus, HPMS integrates an expert rule into a chronicle rule base while ensuring that the resulting rule base remains consistent and effective in capturing knowledge for predictive maintenance tasks.

3.4 Failure Detection and Prediction

HPMS incorporates the failure detection and prediction process which leverages ontology reasoning for detecting anomalies and potential failures in the manufacturing environment. To enable this, HPMS utilizes the Manufacturing Predictive Maintenance Ontology (MPMO) as the domain ontology which encapsulates domain-specific concepts and relationships, providing a formal representation of key elements in the designed manufacturing environment, combined with generated SWRL rules to perform reasoning tasks.The SWRL rules act as a set of logical implications that define relationships and dependencies between different elements in the manufacturing process. These rules are created based on domain-specific knowledge and expertise. They describe how events, conditions, and patterns observed in the data can lead to potential failures or anomalies.

In the domain of intelligent systems, ontologies play a vital role in encapsulating domain knowledge. Within HPMS, the Manufacturing Predictive Maintenance Ontology (MPMO) [2] has been developed to define concepts and relationships within chronicles. The definitions of key concepts and relationships in

Algorithm 2. Enhancing Detecting Issues in Expert Rule Integration within the Chronicle Rule Base.

Input: A chronicle rule base $\mathcal{C}$ which contains a set of failure chronicles, and an expert rule R_e which is in the form of a failure chronicle.
Output: $\mathcal{C}'$: The integrated rule base.

1: $\mathcal{C}' \leftarrow \mathcal{C}$.
2: **for** each $R \in \mathcal{C}$ **do**
3:     **if** ChroRedundancy(R_e, R) **then** ▷ Rule redundancy issue detected.
4:         ▷ A ChroRedundancy message is printed, no change in the rule base.
5:     **end if**
6:     **if** ChroSubsumes(R_e, R) **then** ▷ Rule subsumption issue detected.
7:         Remove(R, $\mathcal{C}'$) ▷ Remove the subsumed chronicle rule from the rule base.
8:         Integrate(R_e, $\mathcal{C}'$) ▷ Integrate the expert rule into the rule base.
9:     **end if**
10:     **if** ChroConflict(R_e, R) **then** ▷ Rule conflict issue detected.
11:         Remove(R, $\mathcal{C}'$) ▷ Remove the conflict chronicle rule from the rule base.
12:         Integrate(R_e, $\mathcal{C}'$) ▷ Integrate the expert rule into the rule base.
13:     **end if**
14: **end for**
15: **return** $\mathcal{C}'$

the MPMO ontology have been formalized based on fundamental notions discussed in Sect. 3.1. Figure 2 illustrates the main modules in the MPMO ontology. To enhance reusability, we have employed the ontology modularization method during development, resulting in three small, reusable modules: the *Condition Monitoring* Module, the *Manufacturing* Module, and the *Context* Module.

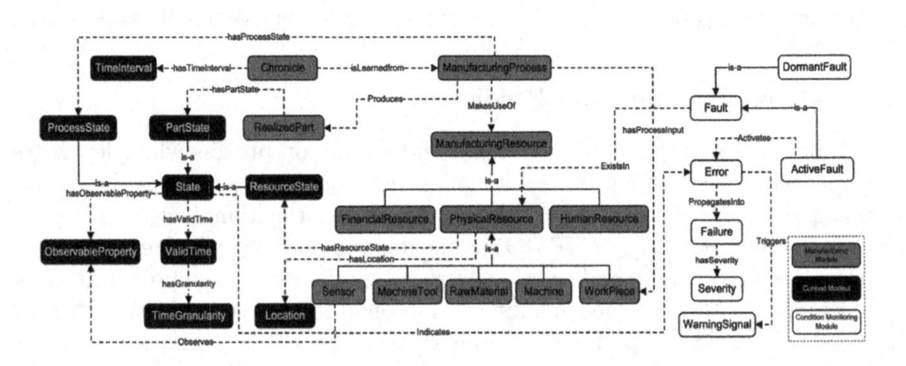

Fig. 2. Main modules in MPMO ontology [3].

When executing the reasoning process, HPMS employs a reasoner capable of interpreting and applying the SWRL rules to the MPMO ontology. The reasoner performs semantic-based inference, matching the facts and rules in the ontology to derive new knowledge to semantic-based inference [9]. By applying ontology reasoning, HPMS is capable of effectively detecting machinery failures.

Furthermore, HPMS extends beyond simple detection by utilizing SWRLAPI[2] to retrieve temporal information about failure points thanks to the chronicle format, SQWRl queries are generated and executed, enabling HPMS to predict the timing of these failures. This predictive capability facilitates predictive maintenance planning and intervention to mitigate the impact of potential failures. By leveraging the knowledge and inference capabilities provided by MPMO and the SWRL rules, HPMS achieves accurate and timely identification of machinery failures. The predictive aspect of HPMS adds an extra layer of foresight, enabling maintenance teams to take proactive measures to prevent or minimize the impact of future failures.

3.5 Failure Criticality Prediction

The final step in our predictive maintenance approach is the *failure criticality prediction*, a crucial process for prioritizing maintenance actions. After detecting the failure points and predicting their temporal information, our goal is to cluster the predicted failure points based on their minimum time to failure. The latter serves as a critical indicator of the urgency and severity of each potential failure. Through this clustering process, we can efficiently allocate resources and schedule maintenance activities. To achieve this, we apply the K-means clustering algorithm, which groups the failure points into three levels based on their minimum time to failure. By analyzing similarities and dissimilarities between the failure points, the algorithm assigns them to different clusters, allowing the identification of patterns and relationships among the failures. Once the failure points are clustered, we assign criticality levels to each cluster based on the urgency and potential impact of the failures. Clusters with shorter minimum time to failure and higher potential consequences are deemed more critical, necessitating immediate attention and intervention. Following clustering, the criticality of each failure group is determined based on the assigned clusters and their characteristics, enabling effective prioritization of maintenance efforts.

The failure criticality prediction process provides valuable insights into the maintenance prioritization strategy. Furthermore, the clustering results can be visualized through graphical representations to facilitate decision-making and enhance situational awareness. Thus, maintenance teams can easily identify high-risk clusters and allocate resources accordingly, ensuring optimal utilization of resources and minimizing the impact of failures on production and safety. Consequently, HPMS leverages a powerful combination of ontology reasoning, SWRL rules, and data-driven approaches to effectively detect potential failures in the manufacturing environment. The process involves the extraction of frequent failure chronicles, the transformation of these chronicles into SWRL predictive rules, and the application of rule pruning using a multi-objective optimization approach. This results in a refined set of high-quality rules for failure prediction. Furthermore, the failure criticality prediction step ensures the prioritization

[2] SWRLAPI is a standalone SWRL API-based application.

of maintenance actions, enabling efficient resource allocation and mitigation of potential failures' impact.

4 Implementation and Experimental Results

In this section, we showcase experimental results from HPMS. We outline the settings, detail the dataset characteristics, and provide illustrative figures depicting HPMS steps and outcomes.

4.1 Data Preparation

In our settings, we have used UCI SECOM[3] dataset comprises 1567 data records with 590 attributes, representing test points, each consisting of 589 numerical attributes and a timestamp indicating the recording time. The dataset contains a binary label for pass/fail status, with 104 instances of production failures out of 1567 records. The data is stored in a raw text file, with timestamps allowing temporal analysis of events. The SECOM dataset was chosen due to its temporal aspect, facilitating the study of predictive maintenance systems and anomaly detection over time. Being derived from real-world industrial scenarios, the dataset presents complex and diverse patterns that reflect challenges encountered in actual industrial environments, enhancing the value of the experiments. To ensure the accuracy and efficiency of our failure prediction task, we perform essential preprocessing steps on the UCI SECOM dataset.

- **Feature selection**: Not all the data in the dataset is relevant for this specific task, as some may contain noise or be redundant. To address this challenge in high-dimensional data, we employ a feature selection technique to identify and retain only the 10 relevant subsets of features.
- **Discretization**: After the feature selection step, the dataset undergoes data preprocessing to transform continuous variables into a discrete representation. This is achieved through data discretization. Discretization involves dividing continuous variables into a finite set of intervals, where each interval is associated with a specific range of data values. In our case, we discretize the dataset by creating 20 bins for each variable, representing discrete numeric intervals. Integer values from 1 to 20 are used to represent these intervals (Listing 1.1). This discretization step enables us to handle the data effectively and prepare it for subsequent analysis.

```
2 4 5 7 9 11 13 15 17 19,1 3 5 7 9 11 13 15 17 20,2 4 5 7
    9 11 14 15 18 19,
```

Listing 1.1. the nominal attributes obtained from the discretization step

[3] The UCI SECOM (Semiconductor Manufacturing) dataset consists of manufacturing operation data and the semiconductor quality data.

- **Sequentialization**: Following the data discretization, the sequentialization process organizes the data in the form of (event-timestamp) pairs. Within each data sequence, the last event represents a failure, while the events preceding it are considered ‘normal’. The first given sequence (Listing1.2) starts with events 2, 4, 5, 7, 10, 11, 13, 15, 17, 19, each followed by a timestamp denoted by “< timestamp > ”. These events are separated by “−1” from the next event in the sequence. After the last event “19” and its corresponding timestamp, we encounter the pattern “−1 −1 −1 −1”, which indicates the end of this failure sequence. By extracting frequent failure sequences, we can identify patterns leading up to failures and gain insights into failure prediction.

```
1 2 4 5 7 10 11 13 15 17 19 <1199911860> -1 2 4 5 7 10 11 13
    15 17 19 <1199916300> -1 2 4 5 7 10 11 13 15 17 19
    <1199919900> -1 2 4 5 7 10 11 13 15 17 19 <1199920740>
    -1 2 4 5 7 10 11 13 15 17 19 <1199921040> -1 2 4 5 7 10
    11 13 15 17 19 <1199921280>  -1 2 4 5 7 9 11 13 15 17
    19 <1199944560> -1 -1 -1 -1
```

Listing 1.2. an extracted failure sequence.

With the completion of these preprocessing steps, we have prepared the UCI SECOM dataset for chronicle mining and the development of HPMS.

4.2 Chronicle Mining

After preprocessing the UCI SECOM dataset, we employ the **Clasp-CPM** algorithm[4] to obtain a set of **frequent failure chronicles** to describe failure events and their temporal constraints. This involves analyzing historical data sequences from manufacturing processes to identify sequential patterns **(SP)** of failures and their temporal relationships. By mining these chronicles, we extract common event sequences that precede failures. In a failure chronicle, the last node represents a failure event. These frequent failure chronicles will be transformed into SWRL rules for failure prediction. Figure 3 illustrates a graph-based representation of a chronicle. The integers associated with each node are nominal attributes from the discretization step. Those integers (e.g., 4 5 7 11 13 19 in the upper episode) collectively form the description for an event. An event with an integer value of 0 indicates a failure, which is typically the last event within a chronicle, while edges represent time intervals between events.

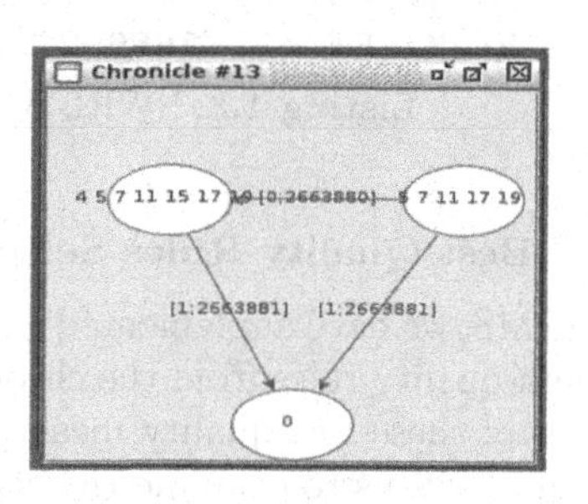

Fig. 3. The graphical representation of a failure chronicle.

[4] ClaSP is used for discovering closed sequential patterns in sequence databases. It can be run using SPMF – an open-source software and data mining library – https://www.philippe-fournier-viger.com/spmf/.

4.3 SWRL Rules Generation

To generate SWRL predictive rules for ontology reasoning, we utilize the set of frequent failure chronicles obtained from the previous step that will be transformed into rules using Algorithm 1. By analyzing failure chronicle data, the algorithm identifies unique events contributing to failure occurrences and represents them as distinct event types. This enables us to understand the sequence and temporal constraints among these events. As a result of this process, a set of SWRL rules is formulated, which act as logical implications between the antecedent (body), describing normal events and their temporal constraints, and the consequent (head), containing information about the temporal constraints between normal events and the failure event parts. In this rule (Listing 1.3), the antecedent (body) of the SWRL rule includes various conditions represented by the statements **ManufacturingProcess(?s)**, **hasEvent(?s, ?e1)**, and **hasItem(?e1,2)**, etc. These conditions specify the normal events and their corresponding items, which are associated with the ManufacturingProcess represented by **?s**. The consequent (head) of the rule, represented by **hasMinF(?e1, 1)** and **hasMaxF(?e1, 2656921)**, provides information about the temporal constraints between normal events and the failure event, indicated by **?e1**. The **hasMinF** and **hasMaxF** relations are used to define the minimum and maximum failure values associated with the specific event ?e1.

```
1 ManufacturingProcess(?s) ^ hasEvent(?s, ?e1)  ^ hasItem(?e1
    ,2) ^ hasItem(?e1,4) ^ hasItem(?e1,5) ^ hasItem(?e1,7) ^
     hasItem(?e1,10) ^ hasItem(?e1,19) -> hasMinF(?e1, 1) ^
    hasMaxF(?e1, 2656921)
```

Listing 1.3. SWRL rule generated from a failure chronicle

4.4 Best Quality Rules Selection

In HPMS, we have implemented a rule pruning module to carefully select a subset of high-quality rules from the chronicle rule base. First, we calculate the average values of these two quality measures across various levels of chronicle support[5] which allows us to examine the correlation between them. Then, we employ the fast non-dominated sorting algorithm for multi-objective optimization, which ranks rules based on Pareto dominance. By considering the rules within the first Pareto Front, we have selected a subset of three high-quality rules that strike a balance between Accuracy and Coverage measures. These selected rules are retained for rule-based reasoning in our chronicle rule base, as shown in Fig. 4a. To evaluate the performance of each rule, we define a fitness function as the product of two crucial measures: **Accuracy** and **Coverage** (Fig. 4b). These measures play a vital role in assessing the quality of SWRL rules. By pruning out less robust rules, we ensure that HPMS makes more accurate and efficient predictions for better decision-making in industrial environments.

[5] An occurrence of a chronicle $\mathcal{C}$ in a sequence $\mathcal{S}$ is a subsequence of $\mathcal{S}$ that satisfies all temporal constraints in $\mathcal{C}$. The support of a chronicle $\mathcal{C}$ in a set $\mathcal{D}$ of sequences is the number of sequences in $\mathcal{D}$ in which $\mathcal{C}$ occurs.

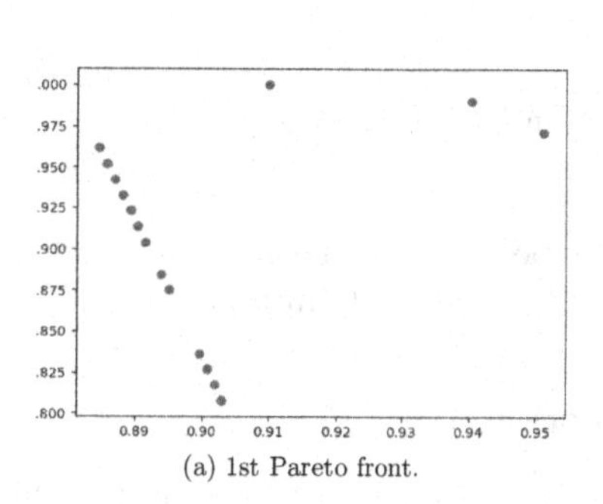

(a) 1st Pareto front.

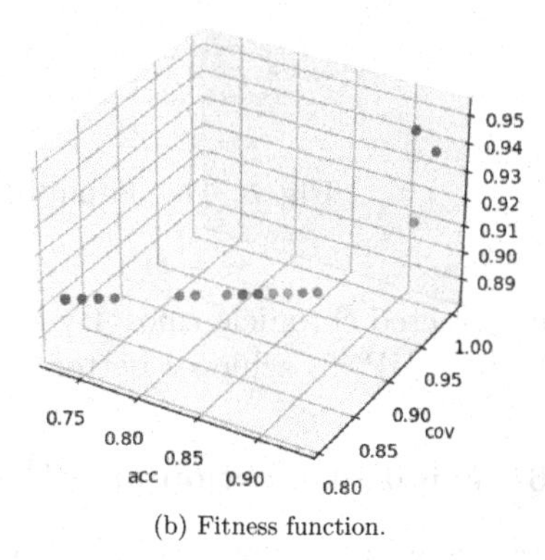

(b) Fitness function.

Fig. 4. Fast non-dominated algorithm output visualization.

4.5 Expert Rules Insertion

After obtaining a set of high-quality rules, the next phase involves integrating expert rules to complement the chronicle rule base. The purpose of this step is to enhance the overall fitness of the rule base and improve the performance of failure prediction.

The expert rules, stored separately from the chronicle rule base, can be imported into HPMS by assigning it to the expert profile. HPMS also detects potential quality issues with the rules, which are then displayed at the lower part of the interface. Expert rules have a similar format to the chronicle rules but may differ in rule atoms within the antecedent or consequent parts (Listing 1.4).

```
1 Expert rule 1:
2 ManufacturingProcess(?s) ^ hasEvent(?s, ?e1)  ^ hasItem(?e1
    ,2) ^ hasItem(?e1,4) ^ hasItem(?e1,5) ^ hasItem(?e1,7) ^
     hasItem(?e1,11) ^ hasItem(?e1,17) ^ hasItem(?e1,14569)
    -> hasMinF(?e1, 1) ^ hasMaxF(?e1, 2648721)
```

Listing 1.4. An expert rule with diffrent atoms

After obtaining a set of best-quality rules, users can proceed to the expert rule integration phase by either opening the expert rule file and pushing the stored expert rules into the system or by directly entering the rules. However, if any quality issues arise, HPMS automatically performs appropriate actions based on the decision-making process outlined in Algorithm 2. For instance, if a conflict is detected between the integrated expert rule and the chronicle rule base, HPMS automatically removes the conflicting chronicle rule and adds the expert rule to the rule base as shown in Fig. 5. Since the fitness values of expert rules are assigned as 1, they are considered to have higher quality than chronicle rules. Therefore, in case of any issues, HPMS always **prioritizes the expert rule over the chronicle rule**.

This expert rule integration phase allows users to enhance the rule base's overall fitness by incorporating domain-specific knowledge and complementing

```
==>Oops,Input expert rule has conflict with the chronicle rule with index 4,
chronicle rule 4 has been removed from the rule base,
the expert rule has been added to the rule base.
```

Fig. 5. The output of the HPMS in case of conflict.

the extracted chronicle rules. By integrating expert rules alongside the chronicle rule base, HPMS achieves better performance in terms of failure prediction.

4.6 Failure Detection and Time Prediction

After integrating expert rules, the next phase involves failure detection. For this purpose, we utilize the Drools rule engine [16] to perform ontology reasoning on the individuals present in the MPMO ontology. MPMO serves as the domain ontology, and the Drools rule engine efficiently processes the rules and populated individuals, allowing for effective failure detection. Once the ontology reasoning is completed, SQWRL queries are generated to retrieve the prediction results by processing the antecedent of SWRL rules as pattern specifications for queries (Listing 1.5). The SQWRL queries aim to retrieve the minimum time and maximum time related to failures, resulting in a set of prediction results.

We obtained 27 rows of prediction results as shown in Fig. 6a. SWRLTAB table contains 4 columns indicating the time span between a specific event and a future failure. After the initial round of prediction, we further enhance the rule base by integrating additional expert rules. The expanded rule base, comprising 13 rules, is used for the second round of prediction. Utilizing this updated rule base, we conduct ontology reasoning and obtain 41 rows , each indicating the temporal information of a specific failure from the SQWRL query. This iterative process of rule base enhancement and ontology reasoning leads to progressive updates, enabling HPMS to detect and predict more potential failures and their respective time of occurrence. Thus, empowering HPMS to provide more comprehensive failure detection and prediction capabilities in manufacturing processes.

```
1 hasEvent(?mp, ?e) ^ hasMaxF(?e, ?xf) ^ sf:
    ManufacturingProcess(?mp) ^ hasMinF(?e, ?nf) -> sqwrl:
    select(?mp, ?e, ?nf, ?xf) ^ sqwrl:columnNames("
    ManufacturingProcess", "Event", "MinTimeToFail", "
    MaxTimeToFail")
```

Listing 1.5. The generated SQWRL querie

4.7 Failure Criticality

In HPMS, we incorporated the K-means clustering algorithm using the Weka library to perform failure criticality analysis based on the minimum time to failures. By setting the value of k to 3, representing three clusters (high, medium, and low), we aimed to categorize the failure points into different levels of criticality. After applying the K-means algorithm to the obtained 27 rows of prediction

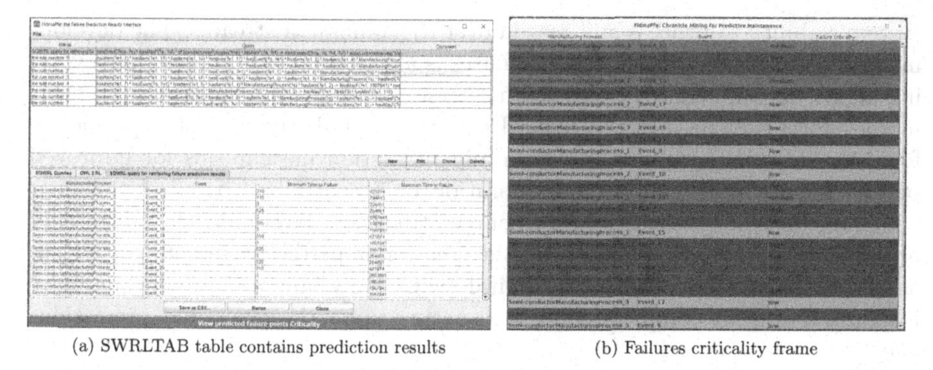

(a) SWRLTAB table contains prediction results

(b) Failures criticality frame

Fig. 6. Failure detection and Time prediction

results, we successfully clustered the data into three distinct groups representing different levels of failure criticality. To present these predictions in a clear and accessible manner, we have organized them in a table format with three columns: "Manufacturing Process," "Events," and "Criticality." Each row in the table corresponds to a specific predicted failure point, and the level of criticality for each point is indicated by the color assigned to the row. To represent the varying levels of criticality, we adopted a color-coding scheme, using light gray to represent low criticality, medium gray to represent medium criticality, and dark gray to represent high criticality as shown in Fig. 6b. By using the K-means clustering technique and visualizing the results in a table with color-coded criticality levels, we were able to effectively assess the level of danger associated with each failure point. This analysis provides valuable insights into the criticality of failures and helps prioritize necessary actions or interventions based on their potential impact.

As seen, we believe that the experimental results of our failure criticality process have demonstrated the effectiveness and usefulness of our predictive maintenance framework, HPMS.

5 Conclusion

In this work, we introduced a **hybrid predictive maintenance system** – HPMS– employing ontologies and machine learning within Industry 4.0, to navigate issues arising from industrial data heterogeneity and semantic gaps. HPMS utilizes ontologies for standardized knowledge representation and machine learning for failure prediction, significantly enhancing production efficiency. The efficacy of this approach was affirmed through the implementation of a tool and rigorous experimental scenarios. Encouraging outcomes from HPMS underscore the potential for predictive maintenance advancements and future research avenues.

The future work will center on several key areas. First, the investigation of stream reasoning techniques will address real-time data handling, enhancing the system's ability to process and reason over dynamic data streams. The development of adaptive ontologies and rule bases is also crucial to accommodate evolving knowledge and context, ensuring the system's agility and accuracy. Furthermore, efforts will be directed toward enhancing predictive models, specifically focusing on refining failure detection and prediction mechanisms for improved performance. Additionally, the exploration of decentralized-based deployment will allow for efficient scaling and resource utilization, leveraging the advantages of the blockchain platforms. Finally, the creation of predictive maintenance visualization tools coupled with the digital twins paradigm will provide intuitive insights, aiding in comprehending and acting upon the generated predictions effectively.

References

1. Ayadi, A.: Semantic approaches for the meta-optimization of complex biomolecular networks. Ph.D. thesis, Université de Strasbourg; Institut supérieur de gestion (Tunis) (2018)
2. Cao, Q., et al.: Smart condition monitoring for industry 4.0 manufacturing processes: an ontology-based approach. Cybern. Syst. **50**, 1–15 (2019)
3. Cao, Q., et al.: Combining evidential clustering and ontology reasoning for failure prediction in predictive maintenance. In: ICAART (2), pp. 618–625 (2020)
4. Cao, Q., Samet, A., Zanni-Merk, C., de Bertrand de Beuvron, F., Reich, C.: An ontology-based approach for failure classification in predictive maintenance using fuzzy c-means and SWRL rules. Procedia Comput. Sci. **159**, 630–639 (2019)
5. Dalenogare, L.S., Benitez, G.B., Ayala, N.F., Frank, A.G.: The expected contribution of industry 4.0 technologies for industrial performance. Int. J. Prod. Econ. **204**, 383–394 (2018). ISSN 0925–5273
6. De Luca, R., Ferraro, A., Galli, A., Gallo, M., Moscato, V., Sperli, G.: A deep attention based approach for predictive maintenance applications in IoT scenarios. J. Manuf. Technol. Manag. **34**(4), 535–556 (2023)
7. Frank, A.G.: Industry 4.0 technologies: implementation patterns in manufacturing companies. Int. J. Prod. Econ. **210**, 15–26 (2019)
8. Heyns, T., et al.: Combining synchronous averaging with a Gaussian mixture model novelty detection scheme for vibration-based condition monitoring of a gearbox. Mech. Syst. Sig. Process. **32**, 200–215 (2012)
9. Khamparia, A., Pandey, B.: Comprehensive analysis of semantic web reasoners and tools: a survey. Educ. Inf. Technol. **22**, 3121–3145 (2017)
10. Klein, P., Weingarz, N., Bergmann, R.: Enhancing siamese neural networks through expert knowledge for predictive maintenance. In: Gama, J., et al. (eds.) ITEM/IoT Streams -2020. CCIS, vol. 1325, pp. 77–92. Springer, Cham (2020). https://doi.org/10.1007/978-3-030-66770-2_6
11. Konys, A.: An ontology-based knowledge modelling for a sustainability assessment domain. Sustainability **10**(2), 300 (2018)
12. Mehdi, G., et al.: SemDia: semantic rule-based equipment diagnostics tool. In: Proceedings of the 2017 ACM on Conference on Information and Knowledge Management, pp. 2507–2510 (2017)
13. Miao, H., et al.: Joint learning of degradation assessment and RUL prediction for aeroengines via dual-task deep LSTM networks. IEEE Trans. Ind. Inform. **15**(9), 5023–5032 (2019)

14. Nilsson, J., Sandin, F.: Semantic interoperability in industry 4.0: survey of recent developments and outlook. In: 2018 IEEE 16th International Conference on Industrial Informatics (INDIN), pp. 127–132. IEEE (2018)
15. Ortega Lopez, G., Filatovas, E., Garzon, E.M., Casado, L.G.: Non-dominated sorting procedure for pareto dominance ranking on multicore CPU and/or GPU. J. Global Optim. **69**, 1–21 (2017)
16. Proctor, M.: Drools: a rule engine for complex event processing. In: Schürr, A., Varró, D., Varró, G. (eds.) AGTIVE 2011. LNCS, vol. 7233, pp. 2–2. Springer, Heidelberg (2012). https://doi.org/10.1007/978-3-642-34176-2_2
17. Cao, Q., et al.: Combining chronicle mining and semantics for predictive maintenance in manufacturing processes. Semant. Web **11**(6), 927–948 (2020)
18. Romeo, L., Loncarski, J., Paolanti, M., Bocchini, G., Mancini, A., Frontoni, E.: Machine learning-based design support system for the prediction of heterogeneous machine parameters in industry 4.0. Expert Syst. Appl. **140**, 112869 (2019)
19. Sezer, E., Romero, D., Guedea, F., Macchi, M., Emmanouilidis, C.: An industry 4.0-enabled low cost predictive maintenance approach for smes. In: 2018 IEEE International Conference on Engineering, Technology and Innovation (ICE/ITMC), pp. 1–8. IEEE (2018)
20. Sikorska, J., et al.: Prognostic modelling options for remaining useful life estimation by industry. Mech. Syst. Sig. Process. **25**(5), 1803–1836 (2011). ISSN 0888–3270
21. Tiwari, R., Sharma, N., Kaushik, I., Tiwari, A., Bhushan, B.: Evolution of IoT & data analytics using deep learning. In: 2019 International Conference on Computing, Communication, and Intelligent Systems (ICCCIS), pp. 418–423. IEEE (2019)
22. Wu, Y., Li, W., Wang, Y., Zhang, K.: Remaining useful life prediction of lithium-ion batteries using neural network and bat-based particle filter. IEEE Access **7**, 54843–54854 (2019)
23. Zhang, W., Yang, D., Wang, H.: Data-driven methods for predictive maintenance of industrial equipment: a survey. IEEE Syst. J. **13**(3), 2213–2227 (2019). https://doi.org/10.1109/JSYST.2019.2905565
24. Zhang, B.: Bearing performance degradation assessment using long short-term memory recurrent network. Comput. Ind. **106**, 14–29 (2019)
25. Zhou, B., et al.: Machine learning with domain knowledge for predictive quality monitoring in resistance spot welding. J. Intell. Manuf. **33**(4), 1139–1163 (2022)
26. Zonta, T., da Costa, C.A., Zeiser, F.A., de Oliveira Ramos, G., Kunst, R., da Rosa Righi, R.: A predictive maintenance model for optimizing production schedule using deep neural networks. J. Manuf. Syst. **62**, 450–462 (2022)

Enhancing Semantic Image Synthesis: A GAN-Based Approach with Multi-Feature Adaptive Denormalization Layer

Karim Magdy[1(✉)], Ghada Khoriba[1,2], and Hala Abbas[1,3]

[1] Computer Science Department Faculty of Computers and Artificial Intelligence, Helwan University, Helwan, Egypt
{karimmagdy_csp,ghada_khoriba}@fci.helwan.edu.eg

[2] School of Information Technology and Computer Science (ITCS), Nile University, Giza, Egypt
ghadaKhoriba@nu.edu.eg

[3] Faculty of Computer Studies, Arab Open University, Cairo, Egypt
hala.abbas@aou.edu.eg

Abstract. Semantic image synthesis, a pivotal task in image-to-image translation, has been widely addressed using generative adversarial network (GAN) models. However, existing GAN-based approaches often suffer from inadequate incorporation of structural and spatial information, resulting in unsatisfactory quality of the synthesized images and a pronounced disparity between photo-realistic and generated images. In this paper, we propose a novel GAN-based methodology to address these limitations, enabling the generation of high-resolution images from semantic label maps while bridging the quality gap and preserving detailed information in the generated outputs. The proposed approach leverages a two-step process, starting with a local binary pattern convolutional generator that produces a local binary pattern feature map. Subsequently, a global convolutional generator is fed with the segmentation map and the feature map through a learned modulation scheme facilitated by a multi-feature adaptive denormalization layer (MFADE) during the training process to generate photo-realistic images. Extensive experiments using Cityscapes, ADE20K, and COCO-stuff datasets validate the performance of our proposed method and showcase its accuracy and robustness in addressing semantic image synthesis tasks, thereby paving the way for its potential applications in enhancing urban sensing and data analytics in Smart Cities. The source code is available at https://github.com/karimmagdy/ULBPGAN.

Keywords: Semantic Image Synthesis · Generative Adversarial Networks (GANs) · Local Binary Pattern (LBP)

1 Introduction

Recently, computer vision and image processing fields have been used to solve many life challenges. Semantic image synthesis, considered one of these chal-

M. Mosbah et al. (Eds.): MEDI 2023, LNCS 14396, pp. 338–351, 2024.
https://doi.org/10.1007/978-3-031-49333-1_24

Fig. 1. Semantic image synthesis results of our framework on challenging datasets.

lenges, generates images from user-specified semantic layouts or changes one visual representation of an object or scene into another form. Also, semantic image synthesis generates photo-realistic images using only a semantic segmentation map, which means this form has a variety of applications. Semantic image synthesis tasks can be solved by one of the generative model techniques. Generative models are among the essential machine learning techniques that aim to generate an algorithm to analyze and detect the probability of the data distribution. Many generative models have been introduced in recent years [1–6] that differ from each other in how the probability distribution is computed. In other words, they differ in whether the density function is an explicit density or an implicit density. Also, the generative models use additional information or assumptions such as variational [1], Markov chains [5,6], or other assumptions. Unlike previous methods, generative adversarial networks (GANs) [3] do not rely on assumptions about input information.

The GAN model has made significant progress regarding the resolution and quality of generated images, which makes it one of the most crucial approaches in recent years. This is because GAN algorithms depend on direct samples from the underlying density function without relying on any assumptions, such as in variational, Markov chains, or other methods related to generative models. GAN can also learn complex and highly dimensional data and generate realistic samples from latent space. Despite the great success achieved so far, applying GAN to semantic image synthesis tasks still poses enormous challenges. As proposed in our previous survey paper [7], we discuss the challenges in semantic image synthesis and focus on GAN and its variants. Also, we include the benefits of Local Binary Pattern (LBP) and Normalization layers in detail (Fig. 1).

With the introduction of the Pix2Pix [8] model in 2017, the semantic image synthesis field has grown significantly, especially when using GAN. The Pix2Pix model employs a conditional GAN consisting of a generator with U-Net [9] architecture and a convolutional PatchGAN classifier discriminator. This method is proposed for different applications of image-to-image translation, such as converting from semantic labels to street scenes, semantic labels to facades, and day to night. However, their approach has failed for high-resolution image generation tasks. Pix2PixHD [10] resolved this issue, which can synthesize high-resolution photo-realistic images without hand-crafted losses using a perceptual loss. This loss extracts deep features from real and fake images with a pre-trained classification network, and the distance between these features is minimized to make fake images indistinguishable from real data.

To stabilize the training, Pix2PixHD used a pre-trained VGG network on imageNet [11]. Also, using the VGG network significantly changes the quality of the generated images compared to the previous methods. Despite all these benefits of the VGG loss, it comes with additional computational requirements as it adds an extra network to the model. Moreover, in Pix2PixHD, the quality of the results has improved without using pre-trained networks. However, the layout information cannot be well preserved in the generator. In Pix2Pix and Pix2PixHD, there is an encoder part to feed the segmentation map of the generator part; however, in SPADE [12], the feeding of the segmentation map is encoded in the learned modulation, so the segmentation map is not applied in the first layer of the SPADE generator.

Consequently, in SPADE, the encoder part from the generator, which was commonly used in the previous architectures, has been discarded. After SPADE's great success, several researchers followed SPADE and took it as a baseline for their models, for example, CC-FPSE [13]. CC-FPSE is composed of a generator with conditional convolutions that employs a weight prediction network and uses a feature-pyramid semantic-embedding discriminator. OASIS [14] aims to achieve high-fidelity images by relying on adversarial supervision. OASIS implements a segmentation-based discriminator that can provide spatial and semantic-aware supervision. CLADE [15] proposes a new normalization layer similar to SPADE that can reduce the parameter and computation overhead.

SCGAN [16] presents spatially variant and appearance-correlated operations and introduces a dynamic weighted network to strengthen semantic relevance. INADE [17] uses instance adaptive stochastic sampling to solve the diversity problem of generated images. RESAIL [18] introduces a feature normalization layer that uses both a semantic mask and retrieval-based guidance. It also uses distorted ground-truth images during the training. SDM [19] uses denoising diffusion probabilistic models instead of GANs and introduces classifier-free guidance. It also extracts the semantic information in the form of noise by designing a new network that handles a noisy input and semantic masking.

Normalization layers have recently been investigated in many tasks that aim to solve real-world problems, so many modern deep networks use normalization layers or introduce a new normalization to produce fine details for samples such as adaptive instance normalization (AdaIn) [20], layer normalization [21], instance normalization [22], batch normalization [23], and local response normalization [24].

Normalization layers are also widely used in semantic image synthesis methods such as SPADE [12] that propose a spatially adaptive normalization layer. SEAN [25] takes SPADE as a starting point and improves it by adding per-region style encoding. SPADE has inspired TSIT and StyleGAN [20], introducing spatially adaptive and adaptive instance normalization layers. Moreover, most of the proposed semantic image synthesis techniques use both conventional batch

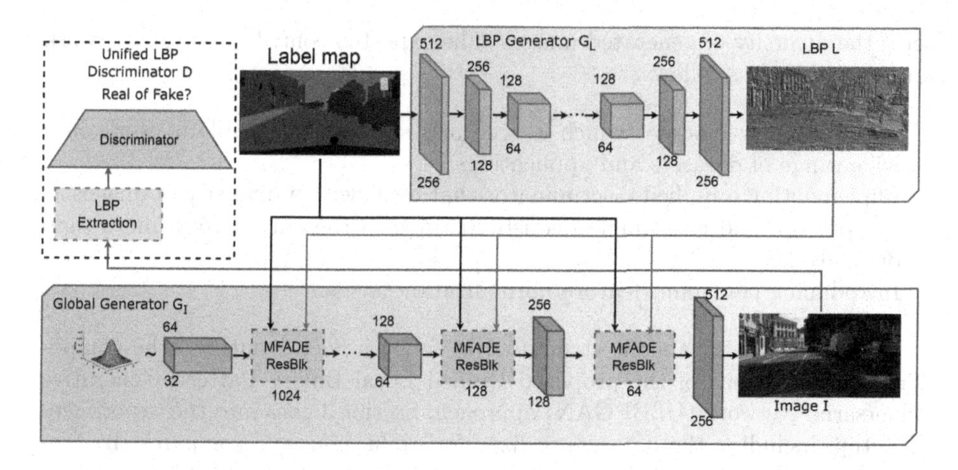

Fig. 2. Full illustration of our modules.

normalization layers and instance normalization layers to improve the efficiency of the generated images. However, there is a significant gap between the quality of photo-realistic and synthesized images. This gap is caused by training stability problems such as image generation diversity and the lack of semantic information.

Local binary pattern (LBP) [26] is a texture feature extraction technique that is easy and simple to implement, and it has been used for various applications such as face recognition [27], image classification [28], image reconstruction [29], and texture classification [30]. Recently, much research has discussed the power of LBP, its simplicity, and how it provides accurate results. Dhingra et al. [31] discuss the most prominent texture feature extraction techniques, including LBP. When applying convolution neural networks, Wei et al. [32] use LBP as a feature extraction technique.

In this paper, we aim to tackle these challenges by proposing a new approach that can produce high-resolution images from semantic label maps and reduce the effect of the issues found in the semantic image synthesis methods, such as the lack of both semantic information and detailed features of the generated image, and the diversity of generated images. Also, our approach produces outperforming results according to the evaluating protocol employed. For stability issues, an LBP convolutional generator is proposed to directly generate the LBP feature maps from semantic label maps. These LBP feature maps represent detailed semantic information, and at the same time, they are used to extract spatial features. For balancing the semantic information from overloading and modulating the activations in the generator layers, we propose a multi-feature adaptive denormalization layer (MFADE), which is achieved through learning transformation by adding weights to both semantic label maps and generated LBP features convolution layers. Finally, we add a unified LBP discriminator to solve the diversity of generated images. The expected contributions of our work are summarized as follows:

- Designing new modules which is accurate, stable, and flexible to work on a wide range of datasets and applications.
- Implementing a unified discriminator that deals with a diversity of images as one pattern and can simultaneously distinguish the output from more than one pattern.
- Introducing new multi-feature normalization layers.

The rest of this paper is structured as follows: Sect. 2 outlines the components and functions of the proposed Unified Local Binary Pattern Generative Adversarial Network (ULBPGAN) approach. Section 3 goes into the experimental setup, including the datasets utilized for evaluation, and compares the proposed method to state-of-the-art techniques. Also, it focuses on additional comments and insights derived from the experimental results. Section 4 concludes the paper by suggesting future research directions to improve the proposed approach.

2 The Proposed Unified Local Binary Pattern Generative Adversarial Network (ULBPGAN)

A novel framework for semantic image synthesis using GAN and LBP texture extraction techniques has been proposed to generate photo-realistic images. This framework, ULBPGAN, comprises 1) an LBP generator (G_L). 2) a global generator G_I with MFADE layers. 3) A unified LBP discriminator D.

With the LBP extraction technique, our framework generates fine-grained details with spatial and structural awareness for each object inside the image, capable of generating diverse results. The LBP generator is proposed to produce LBP feature maps L. Multi-Feature Adaptive DE-normalization (MFADE) layers are designed by adding weights to both semantic label maps and generated LBP features convolutional layers to modulate global generator activations' biases and scales. We also propose a unified LBP discriminator D that can transform the diversity of generated images into a unified LBP feature map. As clarified in Fig. 2, a Gaussian noise map is given as a latent space input for the global generator. The feature representation is extracted from the Label map and the LBP generator's generated LBP feature map G_L. Both will be input to multi-feature adaptive denormalization layers in the global generator. We exploit the standard multi-scale patch-based discriminators, combining with a unified feature LBP extraction.

2.1 Local Binary Pattern Generator

LBP is an operator used to describe a rectangular block's spatial features and texture. Additionally, LBP is an effective technique that involves thresholding the neighborhood of each pixel using the window mean, window medium, or the actual value of the pixel as thresholds, then extracting an LBP code that represents the spatial feature for that pixel

As discussed in the previous sections, there is a huge gap between the realistic image and the synthesized image quality due to the lack of structured information and spatial features. An LBP generator G_l has been proposed to address these limitations. The encoder part of the LBP generator feeds the semantic label map; then, the LBP generator outputs the estimated LBP feature map L. Meanwhile, we used the semantic label map and generated the LBP feature map L to guide the image process in the global generator, as shown in Fig. 3. The loss function for the local binary pattern generator G_l is defined as:

$$\mathcal{L}_{G_l} = \mathbb{E}_L[log(\tilde{D}(L))] - \mathbb{E}_{L_g}[log(1 - \tilde{D}(L_g))] \tag{1}$$

where L_g is the local binary pattern features extracted from the ground truth image, and,

$$\tilde{D}(L_g) = sigmoid(G_l(L_g)) - \mathbb{E}_L[G_l(L)] \tag{2}$$

$$\tilde{D}(L) = sigmoid(G_l(L)) - \mathbb{E}_{L-g}[G_l(L_g)] \tag{3}$$

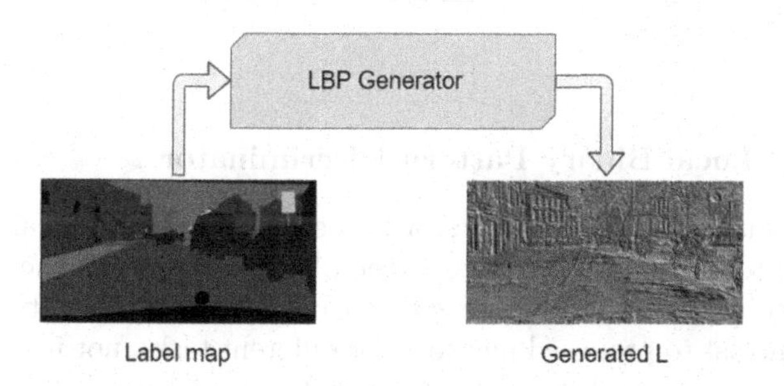

Fig. 3. Local Binary Pattern Generator

2.2 Multi-Feature Adaptive Denormalization (MFADE)

The global generator is a deep network designed to output a photo-realistic image. To prevent semantic information from overloading and to learn from structure information from the semantic label map M and the generated LBP feature map L, we propose MFADE, a new normalization technique.

As in previous normalization techniques [12,25], MFADE modulates the biases and scales of global generator activations. MFADE is inspired by both SPADE [12] and SEAN [25]. In contrast to SPADE and SEAN, we use two inputs: the semantic label map M and the generated LBP feature map L from the LBP generator as shown in Fig. 4. Let h denote the input of MFADE block in the global generator network G_i for a batch of N samples. Let h denote the input of the MFADE block in the global generator network G_i for a batch of N samples. Let H, W, and C be the activation map's height, width, and number of channels, respectively. The modulated activation (n ϵ N, c ϵ C, y ϵ H, x ϵ W) is given by:

$$\gamma_{c,y,x}(L,M)\,\frac{h_{n,c,y,x}}{\sigma_c} + \beta c, y, x(L,M) \tag{4}$$

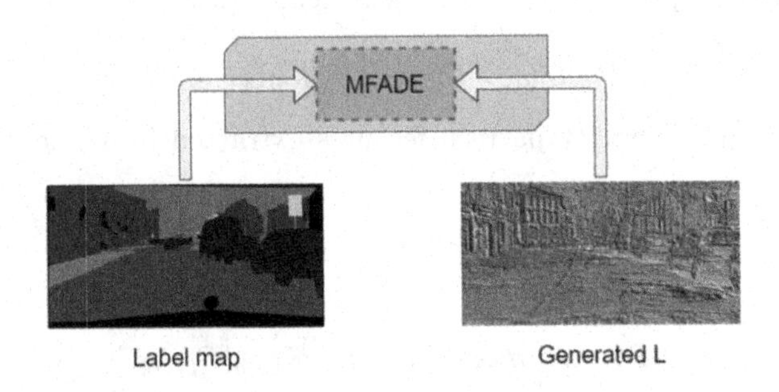

Fig. 4. Multi-Feature Adaptive Normalization

2.3 Unified Local Binary Pattern Discriminator

We develop a unified LBP discriminator to distinguish between real/fake images and LBP features from both generators (the LBP generator and the global generator). A unified LBP discriminator is capable of solving the diversity of generated images. In contrast to proposed methods for different tasks, not just for semantic image synthesis tasks, we do not distinguish between real and fake photo-realistic images, but instead, we use the LBP extraction module to extract the LBP features and then discriminate real/fake LBP features with our generated LBP feature map L, as shown in Fig. 5. The loss function for the local binary pattern generator G_l is defined as:

$$\mathcal{L}_{D_l} = \mathbb{E}_{L_g}[log(\tilde{D}(L_g))] - \mathbb{E}_L[log(1 - \tilde{D}(L))] \tag{5}$$

Table 1. Comparison between our method and other work using FID score, accuracy, and mIOU on multiple public datasets. A lower FID score, higher mIOU, and higher accuracy indicate better performance

Method	ADE20K				Cityscapes			
	mIOU↑	accu↑	FID↓	epoch↓	mIOU↑	accu↑	FID↓	epoch↓
Pix2pixHD	23.7	35.5	91.4	50	49.5	48.9	80.6	50
SPADE	44.1	42.9	38.0	50	50.1	49.7	58.7	50
CC-FPSE	45.0	45.1	37.5	200	52.2	51.5	50.5	200
CLADE	44.1	42.5	56.6	200	50.4	49.6	56.6	200
SDM	43.4	40.1	**33.2**	200	48.5	48.0	**47.7**	200
Ours	**45.1**	**45.3**	35.0	**43**	**53.2**	**52.4**	55.6	**50**

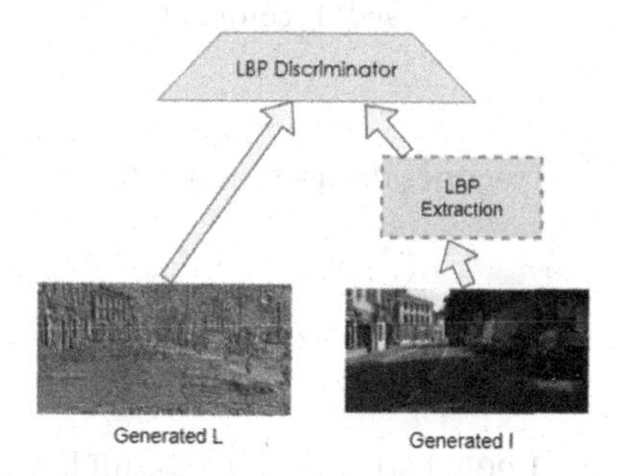

Fig. 5. Unified Local Binary Pattern Discriminator

LBP extraction is considered the most important part of our model as it enables spatial and structural feature extraction from the generated image. We provided PyTorch implementation for the LBP extraction. We greatly reduce the running time algorithm by four times using a PyTorch tensor GPUs instead of the direct LBP algorithm.

3 Results

Implementation details. We adopt ADAM [33] optimizer with a 0.0001 learning rate for all generators and 0.0004 learning rate for all discriminators. Synchronized BatchNorm was used and we also use the ADAM solver with $\beta_1 = 0$ and $\beta_2 = 0.999$. Detailed proposed MFADE model hyperparameters are shown in Table 2.

In addition to the normalization layers, we applied spectral normalization [34] for all networks: the global generator, LBP generator, original discriminator, and finally, the unified LBP discriminator.

We conducted all our experiments and obtained all the results from 16 NVIDIA Tesla $K80$ GPUs. We train 50 epochs for both ADE20k and Cityscapes datasets. The resolution for training and generated images is set to 256 for height

Table 2. MFADE Model Hyperparameters

Hyperparameter	Value/Range
Backbone Network	Pre-trained ResNet
Loss Function	GAN Loss + L1 Reconstruction Loss
Optimizer	Adam
Initial Learning Rate	0.0002
Batch Size	Depends on hardware resources
Image Size	256×256 or 512×256 pixels
Number of Layers	Varies
Number of Channels	Varies
Dropout Rate	Varies
Spectral Normalization	Used in conjunction with MFADE

and 512 for width for the Cityscapes dataset, and 256 for height and 256 for width for other datasets.

3.1 Datasets

We use Cityscapes [35], ADE20K [36] and COCO-stuff [37] datasets in our experiments.

- Cityscapes dataset is widely used in semantic image synthesis. The number of images in the Cityscapes dataset for training is 3000 images. It is 500 images for validation, and the number of semantic classes is 35.
- ADE20K dataset contains 22,210 images (20,210 training and 2,000 validation images), covering 150 semantic classes of challenging scenes.
- COCO-stuff dataset includes the highest diversity of indoor and outdoor scenes. It also contains more than 23,000 images which is a huge number compared to other datasets used in semantic image synthesis tasks because it is considered the most challenging dataset. The number of images in the COCO-stuff dataset for training is 18,000 images and for validation is 5000 images, and the number of semantic classes is 182.

3.2 Performance Metrics

The same evaluation protocol from previous work was used for evaluating our method. We used pixel accuracy (accu), mean Intersection over-Union (mIoU), and Fr'echet Inception Distance (FID). In recent years, the FID score has been the most frequently adopted metric for evaluating the quality of GAN models. This score is focused on the quality and diversity of generated images compared to real images. To evaluate the semantic segmentation for the generated image, we used both mIoU and pixel accuracy metrics. Besides evaluating semantic segmentation, mIoU also measures the quality of the generated images.

Fig. 6. Cityscapes dataset visual comparison

3.3 Baselines

We took Pix2pixHD [10], and SPADE [12], as baselines. We conduct our experiments to validate the efficiency of our work compared to most leading work for semantic image synthesis. CC-FPSE [13] is considered a leading semantic image synthesis model, but we could not afford the GPU consumption so it is not applicable to compare our result with it.

3.4 Discussion

Quantitative Comparisons. Using the datasets mentioned above (ADE20K, Cityscapes, and COCO-stuff) and the previous performance metrics (FID, mIOU, and pixel accuracy), we made a comparison between the proposed method and baseline methods (pix2pixHD and SPADE).

We retrieve the photo-realistic image generated from the global generator, and by using the LBP extraction technique we extracted the spatial and structural features from this image and discriminated these features using the LBP discriminator to gain a detailed object in the image. We achieve a sample quality that was superior to the previous GAN-based method.

As shown in Table 1, a comparison has been made between our method and other work using FID score, accuracy, mIOU, and the number of trained epochs on ADE20K(Indoor and outdoor) and Cityscapes datasets. The better performance is indicated by a lower FID score, higher mIOUm, number of trained epochs, and higher accuracy. The results generated from our work outperform Pix2PixHD, SPADE, CC-FPSE, CLADE, and SDM. For ADE20K, our method obtained the best results for all evaluation metrics. For Cityscapes, we archived

Fig. 7. ADE20K dataset (Indoor and outdoor) visual comparison. Our method produces better-detailed information for the generated image.

the best mIoU and pixel accuracy evaluation. We noted that SDM archived a lower FID score in the Cityscapes and ADE20K datasets. That is because SDM has an image encoder that processes the real image and transforms it into Gaussian noise.

Quantitative Results. Visual comparisons between our method and SPADE are shown in Figs. 6 and 7. Our proposed method shows the best results and significantly reduces the gap between photo-realistic and generated images. Also, there is more detailed information in the generated image, especially in the ADE20K dataset. In addition to the better quality of the generated images, our proposed method can generate a diversity of images with more stable training. Also, in the visual comparisons, we can see that our results have more detailed information for each object in the generated image. This is because the generated LBP image from the LBP generator provides detailed features alongside the spatial features.

4 Conclusion

Image generation diversity and the lack of semantic information problems for semantic image synthesis tasks using generative adversarial networks were investigated in this paper. First, a local binary pattern feature map was constructed based on the local binary convolutional generator. Then, normalization layers based on the segmentation map and the feature map were created to feed the global generator with special and feature information, which ultimately generated the output image. In addition, the diversity of generated images was handled by introducing the unified local binary pattern discriminator. Also, establishing new models using the same criteria employed in the proposed model for any generative adversarial network application-based tasks, especially image-to-image translation tasks instead of semantic image synthesis, is a valuable topic for future work. The main challenges we faced were the computational power and the runtime for the single training process, which took an average of 16 days according to all datasets used to get our results. Also, we faced problems regarding memory capacity, which led us to downsize the batch size to 1 for the Cityscapes dataset.

Acknowledgments. We acknowledge BA-HPC, The Bibliotheca Alexandrina High-Performance Computing, for their facility that offers us merit-based access to a High-Performance Computing cluster to conduct our experiments.

Funding Information. "This research received no external funding".

Conflicts of Interest. Declare conflicts of interest or state "The authors declare no conflict of interest."

References

1. Kingma, D.P., Welling, M.: Auto-encoding variational bayes, arXiv preprint arXiv:1312.6114 (2013)
2. Kingma, D.P., Mohamed, S., Rezende, D.J., Welling, M.: Semi-supervised learning with deep generative models. In: Advances in Neural Information Processing Systems, pp. 3581–3589 (2014)
3. Goodfellow, I., et al.: Generative adversarial nets. In: Advances in Neural Information Processing Systems, pp. 2672–2680 (2014)
4. Van den Oord, A., Kalchbrenner, N., Espeholt, L., Vinyals, O., Graves, A., et al.: Conditional image generation with PixelCNN decoders. In: Advances in Neural Information Processing Systems, pp. 4790–4798 (2016)
5. Salakhutdinov, R., Hinton, G.: Deep Boltzmann machines. In: Artificial Intelligence and Statistics, pp. 448–455 (2009)
6. Bengio, Y., Laufer, E., Alain, G., Yosinski, J.: Deep generative stochastic networks trainable by backprop. In: International Conference on Machine Learning, pp. 226–234 (2014)
7. Karim Magdy, G.K., Abbas, H.: Semantic image synthesis manipulation for stability problem using generative adversarial networks A survey. Comput. Inf. Bull. (2021)
8. Isola, P., Zhu, J.-Y., Zhou, T., Efros, A.A.: Image-to-image translation with conditional adversarial networks. In: Proceedings of the IEEE/CVF Conference on Computer Vision and Pattern Recognition (CVPR), pp. 1125–1134 (2017)
9. Ronneberger, O., Fischer, P., Brox, T.: U-Net: convolutional networks for biomedical image segmentation. In: Navab, N., Hornegger, J., Wells, W.M., Frangi, A.F. (eds.) MICCAI 2015. LNCS, vol. 9351, pp. 234–241. Springer, Cham (2015). https://doi.org/10.1007/978-3-319-24574-4_28
10. Wang, T.-C., Liu, M.-Y., Zhu,J.-Y., Tao, A., Kautz, J., Catanzaro, B.: High-resolution image synthesis and semantic manipulation with conditional GANs. In: Proceedings of the IEEE/CVF Conference on Computer Vision and Pattern Recognition (CVPR), pp. 8798–8807 (2018)

11. Deng, J., Dong, W., Socher, R., Li, L.-J., Li, K., Fei-Fei, L.: ImageNet: a large-scale hierarchical image database. In: IEEE Conference on Computer Vision and Pattern Recognition, vol. 2009, pp. 248–255. IEEE (2009)
12. Park, T., Liu, M.-Y., Wang, T.-C., Zhu, J.-Y.: Semantic image synthesis with spatially-adaptive normalization. In: Proceedings of the IEEE/CVF Conference on Computer Vision and Pattern Recognition (CVPR), pp. 2337–2346 (2019)
13. Liu, X., Yin, G., Shao, J., Wang, X., et al.: Learning to predict layout-to-image conditional convolutions for semantic image synthesis. In: Advances in Neural Information Processing Systems, pp. 570–580 (2019)
14. Sushko, V., Schönfeld, E., Zhang, D., Gall, J., Schiele, B., Khoreva, A.: You only need adversarial supervision for semantic image synthesis. arXiv preprint arXiv:2012.04781 (2020)
15. Tan, Z., et al.: Efficient semantic image synthesis via class-adaptive normalization. IEEE Trans. Pattern Anal. Mach. Intell. **44**, 4852–4866 (2021)
16. Wang, Y., Qi, L., Chen, Y.-C., Zhang, X., Jia, J.: Image synthesis via semantic composition. In: Proceedings of the IEEE/CVF Conference on Computer Vision and Pattern Recognition (CVPR), pp. 13 749–13 758 (2021)
17. Tan, Z., et al.: Diverse semantic image synthesis via probability distribution modeling. In: Proceedings of the IEEE/CVF Conference on Computer Vision and Pattern Recognition (CVPR), pp. 7962–7971 (2021)
18. Shi, Y., Liu, X., Wei, Y., Wu, Z., Zuo, W.: Retrieval-based spatially adaptive normalization for semantic image synthesis. In: Proceedings of the IEEE/CVF Conference on Computer Vision and Pattern Recognition (CVPR), pp. 11 224–11 233 (2022)
19. Wang, W., et al.: Semantic image synthesis via diffusion models. arXiv preprint arXiv:2207.00050 (2022)
20. Karras, T., Laine, S., Aila, T.: A style-based generator architecture for generative adversarial networks. In: Proceedings of the IEEE Conference on Computer Vision and Pattern Recognition, pp. 4401–4410 (2019)
21. Ba, J.L., Kiros, J.R., Hinton, G.E.: Layer normalization. arXiv preprint arXiv:1607.06450 (2016)
22. Ulyanov, D., Vedaldi, A., Lempitsky, A.: Instance normalization: the missing ingredient for fast stylization. arXiv preprint arXiv:1607.08022 (2016)
23. Ioffe, S., Szegedy, C.: Batch normalization: accelerating deep network training by reducing internal covariate shift. arXiv preprint arXiv:1502.03167 (2015)
24. Krizhevsky, A., Sutskever, I. Hinton, G.E.: Imagenet classification with deep convolutional neural networks. In: Advances in Neural Information Processing Systems, pp. 1097–1105 (2012)
25. Zhu, P., Abdal, R., Qin, Y., Wonka, P.: Sean: image synthesis with semantic region-adaptive normalization. In: Proceedings of the IEEE/CVF Conference on Computer Vision and Pattern Recognition (CVPR), pp. 5104–5113 (2020)
26. Ojala, T., Pietikäinen, M., Harwood, D.: A comparative study of texture measures with classification based on featured distributions. Pattern Recogn. **29**(1), 51–59 (1996)
27. Ahonen, T., Hadid, A., Pietikäinen, M.: Face recognition with local binary patterns. In: Pajdla, T., Matas, J. (eds.) ECCV 2004. LNCS, vol. 3021, pp. 469–481. Springer, Heidelberg (2004). https://doi.org/10.1007/978-3-540-24670-1_36
28. Jia, S., Deng, B., Zhu, J., Jia, X., Li, Q.: Local binary pattern-based hyperspectral image classification with superpixel guidance. IEEE Trans. Geosci. Remote Sens. **56**(2), 749–759 (2017)

29. Wu, H., Zhou, J.: Privacy leakage of sift features via deep generative model based image reconstruction. arXiv preprint arXiv:2009.01030 (2020)
30. Ojala, T., Pietikainen, M., Maenpaa, T.: Multiresolution gray-scale and rotation invariant texture classification with local binary patterns. IEEE Trans. Pattern Anal. Mach. Intell. **24**(7), 971–987 (2002)
31. Dhingra, S., Bansal, P.: Experimental analogy of different texture feature extraction techniques in image retrieval systems. Multimedia Tools Appl **79**(37), 27391–27406 (2020)
32. Wei, X., Yu, X., Liu, B., Zhi, L.: Convolutional neural networks and local binary patterns for hyperspectral image classification. Eur. J. Remote Sens. **52**(1), 448–462 (2019)
33. Kingma, D.P., Ba, J.: Adam: a method for stochastic optimization. arXiv preprint arXiv:1412.6980 (2014)
34. Miyato, T., Kataoka, T., Koyama, M., Yoshida, Y.: Spectral normalization for generative adversarial networks. arXiv preprint arXiv:1802.05957 (2018)
35. Cordts, M., et al.: The cityscapes dataset for semantic urban scene understanding. In: Proceedings of the IEEE Conference on Computer Vision and Pattern Recognition, pp. 3213–3223 (2016)
36. Zhou, B., Zhao, H., Puig, X., Fidler, S., Barriuso, A., Torralba, A.: Scene parsing through ade20k dataset. In: Proceedings of the IEEE Conference on Computer Vision and Pattern Recognition, pp. 633–641 (2017)
37. Caesar, H., Uijlings, J., Ferrari, V.: Coco-stuff: Thing and stuff classes in context. In: Proceedings of the IEEE Conference on Computer Vision and Pattern Recognition, pp. 1209–1218 (2018)

Towards an Effective Attribute-Based Access Control Model for Neo4j

Adil Achraf Bereksi Reguig[1(✉)], Houari Mahfoud[1], and Abdessamad Imine[2]

[1] Abou-Bekr Belkaid University & LRIT Laboratory, Tlemcen, Algeria
{adilachraf.bereksireguig,houari.mahfoud}@univ-tlemcen.dz
[2] Université de Lorraine & Loria-Cnrs-Inria, Nancy, France
abdessamad.imine@loria.fr

Abstract. The graph data model is increasingly used in practice due to its flexibility in modeling complex real-life data. However, some security features (e.g., access control) are not receiving sufficient attention from researchers since that graph databases are still in their infancy. Existing access control models do not rise to the finest granularity level of data, use expensive methods for data filtering or explicitly enforce access control rules in application code which may lead to several data security breaches. Based on the most popular graph database system Neo4j and its query language Cypher, this paper provides an Attribute-based Access Control (ABAC) support to Neo4j which makes the model more fine-grained and allows to specify more expressive access policies. Next, we provide a rewriting algorithm that transforms an arbitrary Cypher query into a safe one that enforces the underlying access control policy by returning only authorized data. Contrary to most existing solutions that use non-practical query languages, the proposed solution can be integrated easily within the Neo4j database system.

Keywords: Graph Database · Cypher · Neo4j · Access Control · ABAC · Query Rewriting

1 Introduction

Several works in the literature showed that the graph data model provides more flexibility, efficiency and scalability when it comes to deal with large and highly interconnected data. Graph database systems (e.g. Neo4j) are being used and studied by both researchers and practitioners for many applications such as fraud detection, knowledge discovery, business analytics, recommendation systems, data integration and cleaning. Traditional access control models proposed for relational data cannot be adapted for graph data. Indeed, fulfilling the main requirements of access control (i.e. fine-grained, context management and efficiency for preventing non-authorized data access) is still a challenge in graph data type context due to it schema-less characteristic and the lack of standards neither for data model nor for query language, which makes the design

M. Mosbah et al. (Eds.): MEDI 2023, LNCS 14396, pp. 352–366, 2024.
https://doi.org/10.1007/978-3-031-49333-1_25

of a graph access control model more intriguing. We consider the most popular graph database system, Neo4j[1], and its query language Cypher [8]. Neo4j provides a Role-Based Access Control (RBAC) model which is not suitable enough for applications that require fine-grained control over data access and permissions for different resources in the graph. One scenario that could highlight a limitation of this model is when dealing with dynamic authorization rules. For example, in an Electronic Health Record system, if the role *Doctor* allows reading patient *Health Record* (*HR*), so each doctor will be granted to access any patient HR even though this latter is not being treated by him. This limitation raises a need of an access control model that considers attributes of the context.

Therefore, we propose an extension of the Neo4j RBAC model by allowing specification of more expressive access rules that can be applied to the finest granularity level of data. Precisely, our extension incorporates attributes to access authorization of Neo4j which yields to an ABAC model. Our model takes all advantages of the Neo4j model, overcomes its limits, and allows access policies to be specified based on information of the user and the data being accessed. Based on a large class of Cypher queries, we propose a rewriting algorithm, we enforce our access policies by rewriting any Cypher query into a safe one that returns only accessible data.

Contributions and Road-map. The main contributions of this paper are as follows: *1)* We give a thorough study of the Neo4j access control model and we show its limits (Sect. 2); *2)* We propose an ABAC model for graph databases that aims to overcome limits of the Neo4j model (Sect. 3); *3)* By considering a practical fragment of the Cypher query language, we provide a rewriting algorithm that translates queries from this fragment into safe ones (Sect. 4); *4)* We conduct an experimental study to show effectiveness and efficiency of our approach (Sect. 5). Further details can be found at the online version[2].

Related Work. Table 1 compares the most important access control models for graph databases. These models differ in:

Data type. Several access control models have been proposed for graph modeled data including especially works designed for RDF triplestores [2] and those for *Property Graph* (*PG*) [3,9,11,12,14]. Unlike models for RDF graphs which focus on semantic relationships [15], those for PGs require to handle fine-grained permissions to nodes, edges and properties level.

Access Control Model (ACM). Several graph database systems (e.g. Neo4j, TigerGraph) provide a RBAC model [10,12] which assigns permissions to users based only on their roles. This model lacks granularity when it comes to specify permissions for a specific user or level of the data. To overcome this limit, several works have combined the RBAC with other ACMs: e.g. [2] with *Mandatory Access Control* (*MAC*); [15] with *Mandatory* and *Discretionary Access Control* (*DAC*); [9] with *Attribute-Based Access Control* (*ABAC*). Some studies [11,14]

[1] https://neo4j.com/.
[2] https://github.com/adil-ber/AC-for-Graph-DB.

Table 1. Comparative table of related works.

Graph	Work	ACM	(–)	Granularity	Reinforcement
RDF	J. Chabin et al. [2]	RBAC+MAC		node,rel	Query Rewriting
PG	S. Rizvi et al. [14]	ABAC+ReBAC		node,rel	Query Rewriting
	A. Mohamed et al. [11]	ABAC+ReBAC	✓	node,rel	Brute-force
	C. Morgado et al. [12]	RBAC		node	Brute-force
	M. Valzelli et al. [15]	DAC+MAC +RBAC	✓	node	Views
	S.Clark et al. [3]	ReBAC	✓	node,rel	Not mentioned
	D.Hofer et al. [9]	RBAC+ABAC	✓	node,rel	Query Rewriting
	Neo4j ACM [1]	RBAC	✓	node,rel,attr	At Query time
	Our work	RBAC+ABAC	✓	node,rel,attr	Query Rewriting

focused on *Attribute* and *Relation-Based Access Control* (*ReBAC*) that naturally express the connections and associations between a subject and a resource, enabling intuitive access decisions based on their inherent relationships [3].

Fine-grained level. Nodes, relationships and attributes are entities that compose any PG. However, existing ACMs focus only on nodes and relationships [3,9, 11,14], with some designed solely for nodes [12,15]. Some works [12,14] allow definition of only positive permissions. The Neo4j system allows definition of unconditional positive (+) and negative (–) access permissions to all entities of the PG.

Reinforcement mechanism. The commonly used mechanisms are: *i*) *Brute-force* which involves either executing a query as-is or rejecting it if it contains any unauthorized element; *ii*) *Two-steps* by fetching the data generated by the query, then filtering the outcomes before they are returned to the user; *iii*) *Views* allows the administrator to build personalized views that represents accessible data only; and *iv*) *Query rewriting* where queries are modified to include access control filters prior to their execution. The rewriting principle has been recently considered for graph databases: [14] transforms user's query into another query that returns authorized data only, while [9] introduces access control filters to the methods of the Neo4j Object Graph Mapper framework.

Our work differs from existing ones in: *i)* we allow controlling access to each entity of the PG; *ii)* we propose expressive access control policies with negation, conditions and several kinds of access privileges; *iii)* we use the query rewriting principle that has demonstrated high efficiency for relational data [4] as well as graph data [14]; and *iv)* we propose a practical solution by considering a large class of Cypher queries and by showing how to integrate it within Neo4j, the most popular graph database system.

2 Background

We next discuss the Cypher query language as well as the Neo4j RBAC model.

2.1 Cypher Queries

For the sake of clarity and along the same lines as [14], we consider a specific subset of the Cypher query language that includes the most commonly used features[3]. The syntax of a Cypher query Q is given as follows:

$$
\begin{array}{l}
Q \texttt{ ::= } \mathcal{M}\ \mathcal{W}\ \mathcal{R} \\
\mathcal{M} \texttt{ ::= Match } \alpha \mid \mathcal{M}\ \mathcal{M} \\
\mathcal{W} \texttt{ ::= } \epsilon \mid \texttt{Where } \beta \\
\mathcal{R} \texttt{ ::= Return } \varphi \\
\alpha \texttt{ ::= } \alpha' \mid \alpha' \texttt{,} \alpha \\
\alpha' \texttt{ ::= } \texttt{(}m\texttt{)} \mid \texttt{(}m\texttt{)-[}m\texttt{]->}\alpha' \mid \texttt{(}m\texttt{)<-[}m\texttt{]-}\alpha' \\
m \texttt{ ::= v} \mid \texttt{v\{}\delta\texttt{\}} \mid \texttt{:l} \mid \texttt{:l\{}\delta\texttt{\}} \mid \texttt{v:l} \mid \texttt{v:l\{}\delta\texttt{\}} \\
\delta \texttt{ ::= a:val} \mid \delta\texttt{,}\delta \\
\beta \texttt{ ::= } \beta \texttt{ AND } \beta \mid \beta \texttt{ OR } \beta \mid \texttt{NOT } \beta \mid \texttt{Exists\{ Match } \alpha\ \mathcal{W}' \texttt{ \}} \mid \texttt{(v.a op val)} \\
\mathcal{W}' \texttt{ ::= } \epsilon \mid \texttt{Where } \beta' \\
\beta' \texttt{ ::= } \beta' \texttt{ AND } \beta' \mid \beta' \texttt{ OR } \beta' \mid \texttt{NOT } \beta' \mid \texttt{(v.a op val)} \\
\varphi \texttt{ ::= v} \mid \texttt{v.a} \mid \varphi\texttt{,}\varphi
\end{array}
$$

A Cypher query consists of three types of statements: *Match* ($\mathcal{M}$), *Where* ($\mathcal{W}$) and *Return* ($\mathcal{R}$) statements. In essence, the *Match* statement defines the patterns to search for in the data graph. The *Where* statement adds conditions to these patterns. Upon finding matches for these patterns in the data graph, the *Return* statement specifies the data components to be presented to the user. Each Cypher query contains one or more *Match* statements, each declared using the *Match* clause. A single *Match* statement (Match α) is formed by one or more path patterns (α'). A path pattern is a sequence of nodes and/or relationships. A node (m) (or a relationship $[m]$) is defined by a label l, an optional variable v, and an optional list δ of attribute conditions. Notably, all nodes and relationships are labeled. If a variable v appears in a *Match* statement without an associated label, it is assumed that the label of v has been declared in a previous *Match* statement. Optional *Where* statements can follow *Match* statements to filter the patterns. The *Where* statement applies filtering using simple conditions (e.g., v.a op val) that operate on attributes within the *Match* statements. These conditions employ operators (=, <>, <,<>,<=,>=) or built-in functions (in, contains, starts with, ends with). Complex conditions, referred to as path conditions, are path patterns intersecting with the *Match* statement's patterns to filter nodes/relationships. These complex conditions are expressed using the Exists function, which validates the existence of specified path patterns. We adopt a simplified syntax of the Exists function for clarity. Finally, the *Return* statement is employed to present nodes, relationships, and/or attributes that are linked to the *Match* statements.

Example 1. A property graph is given in Fig. 1 to represent a health information system. Nodes represent entities such as *Doctor*, *Health Records* (HR) and *Event*, while edges represent relationships between these entities. We consider the following Cypher queries:

---

[3] It is worth noting that our findings can be extended to cover other features of the language as well.

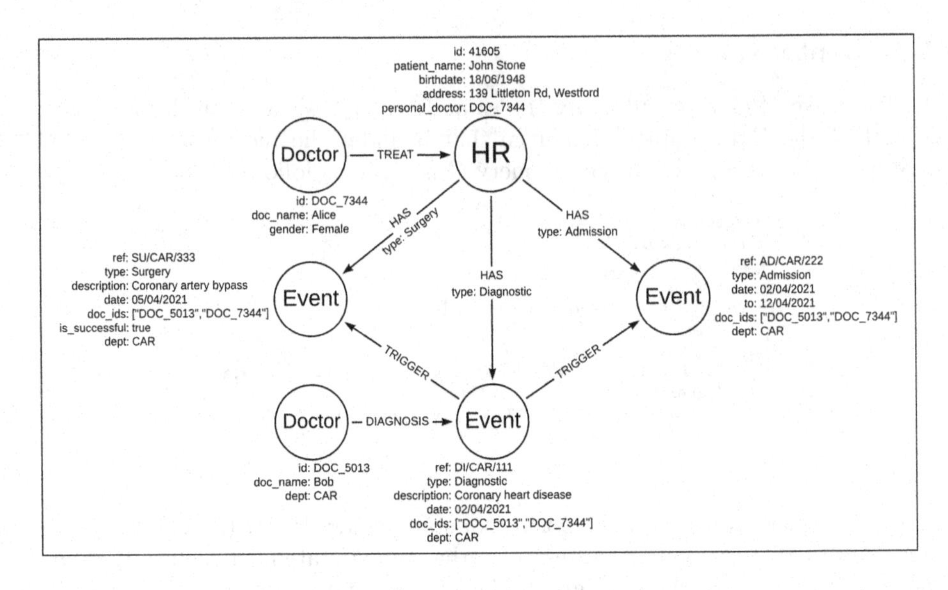

Fig. 1. Property Graph sample.

```
Q1 : Match (d:Doctor) Where d.gender='female' AND Exists{
        Match (e:Event{Description:'Coronary heart disease'})
        Where d.ID IN e.doc_ids
    } return d.doc_name

Q2 : Match (hr:HR{patient_name:'John Stone'})-[r:HAS]->(e:Event),
        (e)<-[r2:DIAGNOSIS]-(d:Doctor)
        return d
```

$Q1$ returns names of woman doctors which are involved in events related to coronary heart disease. $Q2$ provides all information of doctors who were involved in diagnosing the patient named *John Stone*. □

2.2 The Neo4j Access Control Model

We next discuss the Neo4j Role-Based Access Control model.

Syntax and Semantics. ACPs created within Neo4j are *closed*, which means that all entities of the underlying data graph (i.e. nodes, relationships and attributes) are by default unauthorized. For each user role, the administrator can grant/deny privileges to some entities. Three kinds of read privileges are supported:

1. Traverse: this privilege allows the user query to traverse nodes and/or relationships of the data graph without examining their attributes.
2. Read: this privilege allows the user query to access to attributes of some nodes and relationships of the graph.
3. Match: having this privilege on some node (resp. relationship) is semantically the same as having both Traverse and Read privileges on this node (resp. relationship). That is, this privilege allows the user query to traverse an entity as well as to read its attributes.

```
ACP ::= ACR | ACR, ACP
ACR ::= T | R | M
T := (Grant|Deny) Traverse on Graph G (Nodes|Relationships) E to R
E := * | E'
E' := l | l , E'
R := (Grant|Deny) Read{A} on Graph G (Nodes|Relationships) E to R
A := * | A'
A' := a | a , A'
R := (Grant|Deny) Match{A} on Graph G (Nodes|Relationships) E to R
```

Fig. 2. Grammar of the Neo4j Access Control Model.

Neo4j allows definition of positive and negative permissions, specified respectively with the clauses Grant and Deny. Given the above, an *Access Control Policy* (*ACP*) can be written, as a set of *Access Control Rules* (*ACRs*), following the grammar of Fig. 2. Despite the privilege type, each *ACR* specifies the graph G, the user role R and the set E of entities (i.e. nodes and relationships) for them the privilege will be applied. The entities are specified either by *, i.e. all nodes/relationships are concerned; or a finite set of nodes/relationships labels. In addition, an *ACR* with Read or Match privilege specifies in addition a set A of attributes (or * for all attributes) that can(not) be read.

Notice that a denied Match privilege is a little bit confusing. Its semantics depends on whether a concrete attributes set is specified or it is just *. A denied Match with a concrete attributes set specifies that those attributes cannot be read while the corresponding entities can be traversed. However, if a denied Match is specified with * as attributes set, then both traversing the entities and reading their attributes are denied.

Example 2. Consider the following *ACRs* specified for *Doctor* role:

```
GRANT MATCH{patient_name,age} ON GRAPH * NODES HR TO Doctor
DENY MATCH{address} ON GRAPH * NODES HR to Doctor
DENY MATCH{*} ON GRAPH * NODES HR to Doctor
```

The first one allows traversing all nodes labeled HR as well as reading their attributes patient_name and age. This ACR can be written differently as follows:

```
GRANT TRAVERSE ON GRAPH * NODES HR TO Doctor
GRANT READ{patient_name,age} ON GRAPH * NODES HR TO Doctor
```

The second ACR denies access to attribute address of nodes labeled HR. The third ACR denies traversing nodes labeled HR as well as reading their attributes. □

As any access control model, conflicts may arise when opposite permissions are given for the same entity. In such a case, deny privileges take precedence over grant privileges. More details about this resolution are given in Sect. 3.2.

Reinforcement Mechanism. Neo4j reinforces its access control policies dynamically at query time. Precisely, when the user asks the execution of a

Cypher query, the Neo4j's query planner establishes the most efficient way to execute this query by using index, caches as well as access control privileges of the system. That is, entities requested by the query are found efficiently and are also filtered to keep only authorized ones. This process differs from *Brute-force* approaches in the sense that the query is not rejected if it attempts to find some unauthorized entities. Instead, the system purifies the output, delivering solely the authorized entities that were originally requested by the query.

Limits. As any RBAC model, an ACP defined within Neo4j for some role will be applied for all users who are assigned that role, while assigning a personalized policy for each user may lead to role explosion and complicated administration of those policies [6]. Moreover, many real-life scenarios often require to consider not only user information, but also specific contextual factors and properties of the resources being accessed. Such fine-grained access control is necessary to accurately reflect the intricacies of different situations and ensure appropriate security measures. Furthermore, access control policies may be relatively big in practice [16] and must be written within a simplified syntax so that the administrator could express exactly what she/he needs. However, as mentioned before, a denied Match privilege may lead to confusion, and its semantics will be more difficult to understand when considering update rights.

3 Our Access Control Model

We propose an ABAC model for Cypher queries that keeps the expressive power of the Neo4j's model and enhances it with conditional permissions.

3.1 Syntax and Semantics

We consider the grammar of the Neo4j's access control model (Fig. 2) and we make two fundamental changes. Firstly, we add the following rules:

```
T := (Grant|Deny) Traverse on Graph G (Nodes|Relationships) E to R WHERE C
C := C AND C | C OR C | NOT C | (C) | @a op val | $a op val
```

Where `a` is an attribute, `val` is a value and `op` is an operator or a built-in function as described for the Cypher syntax. Intuitively, our additional rules allow defining granted/denied traverse permissions based on some conditions. Notice that `@a` refers to attribute of the data graph while `$a` refers to an attribute of the user/context (e.g. user degree, department number). Moreover, `val` is a value belonging either to the data graph or the user/context. Such conditions allow taking into account properties of the data graph, the user and/or the context where expressing permissions. It is worth noting that we do not assign conditions to read privileges in order to avoid circular permissions[4] which may lead to ambiguous, inconsistent and vulnerable policies. That is why we assign conditions only to Traverse privileges.

[4] An attribute A is accessible depending to the value of an attribute B and vice versa.

Table 2. Conflict Resolution

/////		Grant Traverse		Deny Traverse		Deny Read
		uncond.	**cond. c2**	**uncond.**	**cond. c2**	
Grant Traverse	**uncond.**	Yes	Yes	No	not (c2)	/////
	cond. c1	Yes	c1 or c2	No	c1 and not(c2)	
Deny Traverse	**uncond.**	No	No	No	No	
	cond. c1	not(c2)	c2 and not(c1)	No	not(c1 or c2)	
Grant Read						No

Secondly, we omit the use of the Match privilege as it may lead to confusion and obfuscated semantics as explained in Sect. 2. This will not hinder practicability of our approach since the semantics of Match can be expressed using Traverse and Read privileges.

Example 3. We consider the data graph of Example 1 and we define two user roles (*Doctor* and *Administrator*) with different permissions. A *doctor* has access to (a) all his information; (b) events in which he is involved and (c) health record of all patients that he treats. An *Administrator* is authorized to access (a) all health records without their attributes; (b) all events with their associated attributes except 'doc_ids' attribute; and (c) relationships 'HAS' having a property type equals to 'Surgery'. These permissions are expressed in the same order via the following ACRs:

```
GRANT TRAVERSE ON GRAPH * NODES Doctor TO Doctor WHERE @ID=$doctorID
GRANT TRAVERSE ON GRAPH * NODES Event TO Doctor WHERE $doctorID in @doc_ids
GRANT TRAVERSE ON GRAPH * NODES HR TO Doctor WHERE @personal_doc=$doctorID
GRANT TRAVERSE ON GRAPH * NODES HR TO Administrator
GRANT TRAVERSE ON GRAPH * NODES Event TO Administrator
GRANT READ{*} ON GRAPH * NODES Event TO Administrator
DENY READ{doc_ids} ON GRAPH * NODES Event TO Administrator
GRANT TRAVERSE ON GRAPH * RELATIONSHIPS HAS TO administrator WHERE @type='Surgery'
```

Where the use of @ (resp. $) refers to attribute of the data graph (resp. system). □

3.2 Conflict Resolution

As our approach is an extension of the model proposed by Neo4j, we adopt the same conflict resolution mechanism as Neo4j in which negated access take precedence over granted access. Details about our conflict resolution rules are given in Table 2 where *Yes* (resp. *No*) specifies that the entity is accessible (resp. inaccessible) after resolving conflicts. In addition, the accessibility of such entity may be defined by combining many conditions of the access control policy.

4 Policy Reinforcement Mechanism

The query rewriting mechanism has been largely considered for different kind of data [4,5,7,13]. Recently, some authors have adopted this mechanism to reinforce

Algorithm 1: SafeCypher(Q, P, ur)

```
   Input: A Cypher query Q , an ACP P, and a user role ur.
   Output: A safe version of Q, Q^s or ∅ otherwise.
 1 G^P := accessGraph(P);
 2 Q^s := Q;
 3 Add a meta-variable to each free node (resp. relationship) in Q^s;
 4 Extract a set S of Match statements in Q^s;
 5 Define a function L that returns the label of each variable in Q^s;
 6 Define a function T that returns the type of each variable in Q^s;
 7 Let AC be a list of attribute conditions belonging to the Where statement of Q^s;
 8 Let R be a list of statements returned by Q^s;
 9 ACFilters :='true';
10 foreach s ∈ S do
11     sub := isSubQuery(s, Q^s);
12     s' := singleMatchRewriter(s, sub, G^P, ur);
13     if (sub = true) then
14         if (s' = ∅) then Replace s with false in Q^s;
15         else if (s' ≠ s) then Replace s with s' in Q^s;
16     else if (s' = ∅) then return ∅;
17 if (ACFilters ≠ 'true') then
18     if (AC ≠ ∅) then Add ACFilters to the Where statement of Q^s;
19     else Create a Where statement in Q^s with ACFilters as condition;
20 foreach ac ∈ AC with the form "v.attr op val" do
21     rp := attributeReadCheck(G^P, ur, L(v), attr, T(v)) ;
22     if (rp = false) then Replace ac with null in Q^s ;
23 foreach r ∈ R with r = v or r = v.attr do
24     ga := getGrantedAttrs(G^P, ur, L(v), T(v));
25     da := getDeniedAttrs(G^P, ur, L(v), T(v));
26     if (r = v AND (da.contains(*) OR ga = ∅)) then Replace r with v{} in Q^s ;
27     else if (r = v AND ga.contains(*)) then Replace r with v|_da in Q^s ;
28     else if (r = v AND !ga.contains(*)) then Replace r with v{ga} in Q^s ;
29     else if (r = v.attr) then
30         rp := attributeReadCheck(G^P, ur, L(v), attr, T(v));
31         if rp = false then Replace r with null in Q^s;
32 Q^s := optimizeQuery(Q^s)
33 return Q^s;
```

Fig. 3. An algorithm for safe rewriting of Cypher queries.

access control policies over graph databases [9,14]. Given an access control policy P and a user query Q, it is to rewrite Q into a safe one, Q^s, by considering rules of P, so that: for any data graph G, $Q^s(G)$ will traverse and return only accessible entities (nodes, relationships and attributes). We describe hereafter our query rewriting algorithm designed for Cypher queries and Neo4j system.

Our algorithm, referred to as SafeCypher, is given in Fig. 3. It inputs a Cypher query Q, an access control policy P, and the role ur of the user requesting Q. The algorithm produces a safe version of Q, Q^s, or $\emptyset$ if no safe version of Q can be generated. First of all, it is clear that a textual ACP is hard to analysis. That is, we construct a graph representation of our ACP P, called *access graph* and denoted by G^P. Such access graph allows checking easily the accessibility of any entity requested by the query. Due to space limit, we report all details about this structure to the online version. Once the access graph G^P

is constructed (line 1), meta-variables are generated for each free element[5] of the query (line 3). All elements composing Q are extracted (lines 4–8) including: Match statements belonging both to the body of the query and its Where statement; the Return statement; and all attribute conditions defined within the Where statement. Each match statement s is rewritten separately using the procedure singleMatchRewriter (line 12). If s is a subquery[6] of Q and its safe version s' is not empty then s is replaced by s' in the original query, if it is empty then it is simply replaced by *false* in the original query (lines 13–15). Otherwise, if s is not a subquery then the algorithm returns an empty query as a fundamental part of Q cannot be made safe (line 16). After checking all elements belonging to Match statements of Q, their corresponding accessibility conditions are grouped as a conjunctive formulas *ACFilters*. This latter is added to the Where statement of the query to ensure the accessibility of each entity parsed/returned by Q (line 18). If Q has no Where statement, then it is created to enforce the computed accessibility conditions (line 19).

Next, the algorithm examines each attribute condition belonging to the Where statement (lines 20–22) and checks whether the corresponding attribute is accessible. Precisely, for each attribute condition $v.attr\ op\ val$, the algorithm calls the procedure attributeReadCheck in order to check whether the attribute $attr$ is accessible via the element referred to by the variable v. If this is not the case then the attribute condition is replaced with NULL in the query.

On the other side, each return statement rt is rewritten (lines 23–31) in order to check the accessibility of each entity the query Q wants to return. Each return statement can be either a node/relationship (i.e. a variable v) or attributes of this latter (i.e. $v.attr$). The procedure getGrantedAttrs is called to get a list ga of all granted attributes (line 24). Similarly, the procedure getDeniedAttrs is called to get a list da of all denied attributes (line 25). Using these lists, all attributes being returned by Q are checked. If an attribute is inaccessible w.r.t P and it is explicitly returned by Q (i.e. case of $v.attr$) then it is replaced by null at the original query (lines 29–31). Moreover, if Q returns implicitly a list of attributes (i.e. case of v) then this list is sanitized to keep only accessible attributes. We denote by $v_{|da}$ a safe rewriting of v that returns the corresponding entity (node/relationship) without the unauthorized attributes. In addition, $v\{ga\}$ refers to granted attributes of that entity. Thus, $v\{\}$ returns the entity without any of its attributes (lines 26–29). However, Cypher language does not allow to remove a list of attributes from a node making part of the query result while returning the remaining accessible attributes. So, the syntax $v_{|da}$ is implemented using APOC procedures provided by Neo4j. A use-case is given at Example 4.

Notice that adding accessibility conditions to the different elements of Q may lead to a safe but unoptimized version of Q. For instance, the condition null *and* null may be replaced by null. That is why the algorithm calls finally the procedure optimizeQuery to returns an optimized safe version of Q.

[5] I.e. node or relationship with no variable attached to it.

[6] According to our Cypher syntax, a subquery is any Match statement appearing inside an Exists call.

The procedure singleMatchRewriter (Fig. 4) aims to check the accessibility of each entity composing some Match statement in input. It extracts vertex (resp. relationship) tokens composing M, as well as attribute conditions defined over vertices and relationships of M. A token has the form $v : l$ while an attribute condition has the form $\{attr : val\}$. The procedure uses two other procedures: elementTraverseCheck checks accessibility of vertex/relationship; while attributeReadCheck checks accessibility of attribute. If an element (i.e. vertex or relationship) is not accessible then the procedure returns $\emptyset$ (line 7). If such element is conditionally accessible, then its corresponding accessibility condition c will be added to a global filter *ACFilters* (line 8). If the element is unconditionally accessible then no change is done. Furthermore, if an attribute *attr* is not accessible then $\emptyset$ is returned (lines 14–18). The final safe version of M is finally returned (line 19). Given a subquery composed by an Exists call, and a Match statement s and a Where statement w inside it. Then, accessibility conditions related to s are added to w rather than the Where statement of the whole query (lines 10–13). That is why *subACFilters* is used to define accessibility conditions of subqueries only, while *ACFilters* adds accessibility conditions to the Where statement of the entire query.

Example 4. Consider a Cypher query Q made by an administrator and its safe version Q^s computed w.r.t the access policy defined in example 3:

```
Q: Match(hr:HR) where Exists{
       Match(hr)-(x:HAS)->(e:Event)
       Where date='15/08/2020'
   } return hr,e
```

```
Q^s: Match(hr:HR) where Exists{
         Match(hr)-(x:HAS)->(e:Event})
         Where(date='15/08/2020'
         AND x.type='Surgery')
     } return hr{},apoc.map.removeKeys(e,["doc_ids"])
```

The safe version Q^s first adds an accessibility condition, i.e. x.type='surgery', as the administrator can access to only relationships of type *surgery*. The variable e refers to events and thus it is refined by Q^s in order to return only accessible attributes of events: i.e. all attributes excepting docs_ids. The function apoc.map.removeKeys(e,["doc_ids"]), from Neo4j APOC library, is used to remove the unauthorized attributes (doc_ids in this case) from the returned element e. In addition, administrator has a granted traverse for nodes hr but not a granted read, that is no attribute in node hr can be returned. Thus, Q^s replaces hr by hr{} at query result. □

5 Experimental Study

Using real-life data, we next conduct several experiments in order to check efficiency and scalability of our model compared to that of Neo4j.

5.1 Experimental Setting

Datasets. We use the *Stack Exchange*[7] dataset which is a vast collection of anonymized questions, answers and user interactions from various *Stack*

[7] https://archive.org/details/stackexchange.

Algorithm 2: singleMatchRewriter(M, sub, G^P, ur)

```
   Input: A match statement M, a boolean sub, an access graph G^P, and a user role ur.
   Output: A safe version of M, M^s, or ∅ otherwise.
 1 Let VT (resp. RT) be list of vertex (resp. relationship) tokens in M;
 2 Let AC be list of attribute conditions defined over elements of M;
 3 subACFilters := 'true';
 4 M^s := M;
 5 foreach t ∈ (VT ∪ RT) with t = v : l do
 6     ac := elementTraverseCheck(G^P, ur, v, l, T(v));
 7     if (ac = false) then return ∅ ;
 8     else if (ac = c and sub ≠ true) then  ACFilters := ACFilters ∧ ac;
 9     else if (ac = c and sub = true) then  subACFilters := subACFilters ∧ ac;
10 if (subACFilters ≠'true') then
11     if (there exists ac ∈ AC with the form "v.attr op val") then
12         Add subACFilters to the Where statement of M^s;
13     else Create a Where statement in M^s with subACFilters as condition;
14 foreach ac ∈ AC do
15     rp := attributeReadCheck(G^P, ur, L(v), attr, T(v)) ;
16     if (rp = false) then
17         if ac with the form "v.attr op val" then Replace ac with null in M^s;
18         else if ac with the form "{attr : val}" then return ∅;
19 return M^s;
```

Fig. 4. Procedure to safely rewrite Match statements.

Exchange websites. It contains 8 node labels; 7 relationship types representing interaction between nodes; and a vertex set varying from 85K to 1.3M.

Implementation. To ensure the integration of our model within Neo4j, we first implemented a Java application that inputs an access control policy in textual form (i.e. via the syntax given in Sect. 3), and transforms it into an access graph that will be stored within the Neo4j database for rewriting purpose. Secondly, we implemented a Java program that randomly generates access control policies. Finally, we implemented our Cypher query rewriting algorithm in Java and we made its execution possible within the Neo4j browser via the CALL clause. That is, one could simply call our algorithm by inputting any Cypher query. This latter will be first rewritten using our algorithm w.r.t the access graph previously computed, and next its rewritten version will be evaluated over the underlying data graph and the result will be returned to the user.

The experiments were carried out on a machine featuring an 8-core processor Intel i5-1135G7 2.4 GHz, 16 GB of RAM and a NVMe SSD, running Ubuntu 22.04.2. Each experiment were ran 10 times and the mean time is reported.

5.2 Experimental Results

We report hereafter results of our different experiments. The size of a data graph is given by the number of its nodes and edges. The size of a Cypher query is given by the number of nodes and relationships composing all its statements (i.e. MATCH, WHERE and RETURN). Furthermore, the size of an ACP refers to the number of its rules.

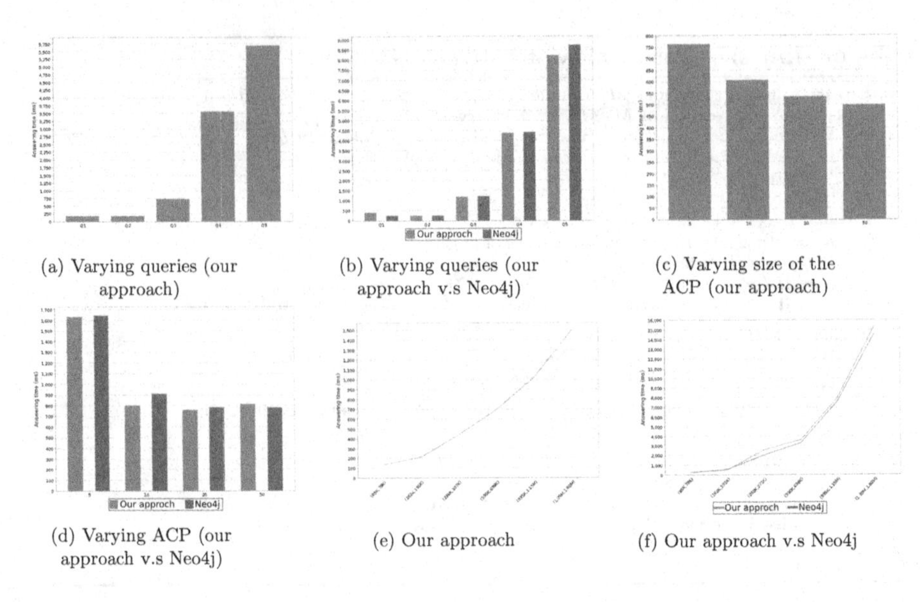

(a) Varying queries (our approach)

(b) Varying queries (our approach v.s Neo4j)

(c) Varying size of the ACP (our approach)

(d) Varying ACP (our approach v.s Neo4j)

(e) Our approach

(f) Our approach v.s Neo4j

Fig. 5. Experimental results

Experiment 1: Efficiency Checking. We experimented the efficiency of our approach by measuring the required times for rewriting Cypher queries as well as times for answering their rewritten versions. We fixed our data graph at 296K nodes and 307K relationships, and we varied sizes of Cypher queries and ACPs.

a) Varying size of the Cypher query: We fixed the size of the access control policy into 50 rules and we varied the size of the Cypher queries. We defined different queries ($Q_1, \cdots, Q_5$) by varying the number of nodes, relationships, attribute conditions, Match statements and path conditions. Details of these queries can be found at the online version. Then, we computed the whole times required to answer the five queries over the underlying data graph. The results reported in Fig. 5-a show that the answering time increases in case of complex queries such as those containing multiple match statements and path conditions.

Next, we compared the efficiency of our model w.r.t that of Neo4j. We first updated our ACP by keeping only unconditional access control rules (as Neo4j does not allow conditional rules), and then we measured the whole time required to answer the five queries used both our approach and Neo4j. The results reported in Fig. 5-b shows that the two models behave in a similar way by varying the complexity of the Cypher queries and demonstrate the same efficiency.

b) Varying size of the access control policy: We considered a Cypher query with 3 Match statements, 7 node and relationship labels and 5 conditions. We varied the size of the ACP by considering different kinds and number of access control rules (i.e. positive, negative and conditional rules). Then, we computed the whole time required to answer the Cypher query using each of the defined ACPs.

Figure 5-c demonstrates that as the number of access control rules increases, the answering time decreases. This is due to the fact that a greater number of access control rules result in more filtering of the rewritten Cypher query's result. Consequently, the result size decreases as more access control conditions are incorporated into the Cypher query. This explains why an access control policy with a smaller (resp. larger) size allows for the retrieval of a larger (resp. smaller) result, thereby influencing the increase (resp. decrease) in answering time.

We updated the ACP by removing all conditional access rules and we computed the answering time of both our approach and that of Neo4j. Results are given in Fig. 5-d. Notice that the differences between both models are minimal.

Experiment 2: Scalability Checking. By fixing the size of the Cypher query (3 nodes, 2 relationships and 1 attribute condition) and that of the ACP (50 rules), we varied the size of the data graph from $(85K, 79K)$ to $(1.35M, 1.80M)$. We answered the Cypher query over each data graph and we reported the corresponding times in Fig. 5-e. Our approach scales well as it takes only 1,500 ms to answer a complex Cypher query over a big data graph composed of $1.35M$ nodes and $1.80M$ of edges, w.r.t a complex access control policy.

Similarly to previous experiments, we removed all conditional rules in order to favor Neo4j, and we compared our answering times with those required by Neo4j (for the same Cypher query and the same ACP). Results reported in Fig. 5-f show that the two models have a high degree of similarity in terms of efficiency.

5.3 Summary of Results

The experiments showed that: 1) our approach scales well with complex Cypher queries, complex access control policies as well as large data graphs. 2) The rewriting process adopted by approach does not make our live much harder since our approach has the same degree of efficiency as Neo4j when considering only unconditional rules. 3) Conditional rules do not decrease the efficiency of our approach as the more conditional rules are added the less is the answering time.

6 Conclusion and Future Works

Our main motivation was to provide a fine-grained access control for the graph databases stored under Neo4j. Therefore, we proposed an access control language that overcomes limits of the Neo4j model and allows specifying attribute-based policies for each entity of the data graph. To enforce our policies, we adopted a rewriting mechanism that transforms arbitrary Cypher query into a safe query which returns only accessible parts of the data graph. Unlike other solutions, our approach seamlessly integrates into the Neo4j database system, offering a practical and efficient solution for enhancing data security in graph databases. We demonstrated the efficiency and scalability of our approach via an extensive experimental study based on real-life data graph. It is worth noting that current works consider only read privileges. The Neo4j access control model takes into account update privileges but several limits arise especially when it comes to understand and resolve conflicts between read and update privileges. We plan to extend our model by addressing these issues.

References

1. Neo4j access control. https://neo4j.com/docs/cypher-manual/current/administration/access-control/. Accessed 10 June 2023
2. Chabin, J., Ciferri, C.D., Halfeld-Ferrari, M., Hara, C.S., Penteado, R.R.: Role-based access control on graph databases. In: SOFSEM, pp. 519–534 (2021)
3. Clark, S., Yakovets, N., Fletcher, G., Zannone, N.: Relog: a unified framework for relationship-based access control over graph databases. In: IFIP, pp. 303–315 (2022)
4. Colombo, P., Ferrari, E.: Efficient enforcement of action-aware purpose-based access control within relational database management systems. IEEE Trans. Knowl. Data Eng. **27**, 2134–2147 (2015)
5. Colombo, P., Ferrari, E.: Towards virtual private NoSQL datastores. In: ICDE, pp. 193–204 (2016)
6. Elliott, A., Knight, S.: Role explosion: acknowledging the problem. In: Software Engineering Research and Practice, pp. 349–355 (2010)
7. Fan, W., Chan, C.Y., Garofalakis, M.: Secure xml querying with security views. In: SIGMOD, pp. 587–598 (2004)
8. Francis, N., et al.: Cypher: an evolving query language for property graphs. In: Proceedings of the 2018 International Conference on Management of Data, pp. 1433–1445 (2018)
9. Hofer, D., Mohamed, A., Küng, J.: Modifying neo4j's object graph mapper queries for access control. In: Pardede, E., Delir Haghighi, P., Khalil, I., Kotsis, G. (eds.) Information Integration and Web Intelligence. iiWAS 2022. LNCS, vol. 13635, pp. 421–426. Springer, Cham (2022). https://doi.org/10.1007/978-3-031-21047-1_37
10. Jin, Y., Kaja, K.: XACML implementation based on graph databases. In: CATA, pp. 65–74 (2019)
11. Mohamed, A., Auer, D., Hofer, D., Küng, J.: Extended authorization policy for graph-structured data. SN Comput. Sci. **2**, 351–369 (2021)
12. Morgado, C., Baioco, G.B., Basso, T., Moraes, R.: A security model for access control in graph-oriented databases. In: QRS, pp. 135–142 (2018)
13. Rizvi, S., Mendelzon, A., Sudarshan, S., Roy, P.: Extending query rewriting techniques for fine-grained access control. In: SIGMOD, pp. 551–562 (2004)
14. Rizvi, S.Z.R., Fong, P.W.: Efficient authorization of graph-database queries in an attribute-supporting rebac model. ACM Trans. Priv. Secur. (TOPS) 1–33 (2020)
15. Valzelli, M., Maurino, A., Palmonari, M.: A fine-grained access control model for knowledge graphs. knowledge graphs. In: ICETE, vol. 2, pp. 595–601 (2020)
16. You, M., et al.: A knowledge graph empowered online learning framework for access control decision-making. World Wide Web, pp. 827–848 (2023)

GRU-Based Forecasting Model for Energy Production and Consumption: Leveraging Random Forest Feature Importance

Alaa M. Odeh[1](✉), Amjad Rattrout[2], and Rashid Jayousi[1]

[1] Department of Computer Science, Al-Quds University, Jerusalem, Palestine
Alaa.odeh4@students.alquds.edu, rjayousi@staff.alquds.edu
[2] Department of Computer Science, Arab American University (AAUP), Ramallah, Palestine
amjad.rattrout@aaup.edu

Abstract. There is huge ignorance of the energy situation in the world. People in different countries are unaware of the amount of consumed energy and how this accelerates the depletion of non-renewable energy resources which might lead to unpredictable consequences. Moreover, the continuous increase in the amount of produced and consumed energy increases the amount of CO2 emission which leads to series pollution in the air. These problems can be solved initially by increasing awareness of the general situation in every country and seeking to solve these problems. In this project, we aimed to develop prediction models for energy production, consumption, CO2 emissions, and share of renewable energy in electricity production. We collected data from several resources and used the random forest feature importance method to select features with high importance scores to produce highly accurate prediction results. We implemented the models using a Gated Recurrent Unit neural network which generated very highly accurate results because of its ability to remember information from the past. In the future, we seek to implement prediction models for each country to increase awareness of energy-related subjects.

Keywords: random forest · gated recurrent unit · Gini importance

1 Introduction

It is unbelievable how the choices humans make can affect their existence, particularly when considering the significant impacts of their choices on climate change, air pollution, resource depletion, energy costs, energy independence, and more [1]. One of the most crucial considerations is to understand how the global energy system works to effectively address these changes and establish boundaries for counterproductive habits. This was a great motivation for data scientists to develop several types of energy forecasting models. These models encompass a spectrum of approaches, which include analyzing historical data to identify patterns in energy production and consumption for renewable and non-renewable resources, developing machine learning algorithms to generate predictions using energy data inputs or integrating both methods to enhance energy predictions

M. Mosbah et al. (Eds.): MEDI 2023, LNCS 14396, pp. 367–380, 2024.
https://doi.org/10.1007/978-3-031-49333-1_26

by leveraging historical data patterns. The choice of the model depends on the data availability, nature of the data, model complexity and interoperability, computational resources, and other forecasting requirements. This guided us in selecting a model that effectively utilizes the available energy data, selecting the most significant features that most influence the energy prediction process, and making the right predictions with minimum errors.

By designing our model, we aim to forecast the amount of consumed and produced energy, the resulting CO2 emissions from non-renewable energy resources, and the share of renewable energy resources in electricity production. Given the numerous features that potentially influence the predicted outcomes, we have utilized the random forest feature importance method to select the most significant features that predominantly affect the forecasted values. Additionally, considering the non-linear relationship within the available data, the substantial volume of data, and the need for continual learning and adaptation, a neural network-based model appears to be the most suitable choice. Specifically, we have developed our model using the Gated Recurrent Unit (GRU), an improved version of the recurrent neural network (RNN). The GRU utilizes two key vectors, namely the reset and update gates, which can be trained to retain relevant information from the past while disregarding irrelevant information [2]. This characteristic proves highly beneficial for our model.

We began by collecting relevant data that we believed would influence the predicted outcomes. This included data on refined oil production, domestic oil consumption, natural gas production and consumption, crude oil production, average CO2 emissions from fossil fuel production, the share of solar, wind, and hydropower in electricity production, temperature changes in the meteorological year, the top 20 countries in energy production and consumption, population data for these countries, and energy production and consumption figures. To preprocess the data, we removed missing values and outliers, consolidated the different datasets into a unified file format, and encoded categorical variables into numerical formats. Subsequently, we utilized the random forest feature importance method to select features for each prediction type. Next, the data was split into training, testing, and validation sets to facilitate the design of the GRU model. Following this step, we defined the key characteristics of the GRU, including the input shape, hidden units, activation functions, learning and optimization algorithm, and training parameters such as the number of epochs and batch size. To assess the accuracy of the prediction process, we employed the mean squared error and the mean average error metrics. Considering all these characteristics enhanced the robustness of our model, resulting in a relatively low loss function and indicating high accuracy in the prediction outcomes.

The detailed model description is explored in the subsequent sections of this paper, which are organized as follows: Sect. 2 addresses the related work relevant to this paper, Sect. 3 discusses the theoretical and mathematical background, Sect. 4 delves into the methodology employed, Sect. 5 investigates the results and outcomes of the model, and finally, Sect. 6 draws conclusions and explores possibilities for further future development of the model.

2 Related Work

Due to the significant impact of the feature selection process of developing predictive models, numerous research studies have focused on identifying key features that greatly influence prediction outcomes. The random forest has been particularly emphasized for its ability to determine feature importance. For example, in [3], authors developed a hybrid forecasting model named RF-CEEMD-DIFPSO-BPNN, which integrated multiple techniques, including random forest (RF), improved grey ideal value approximation (IGIVA), complementary ensemble empirical mode decomposition (CEEMD), the particle swarm optimization algorithm based on dynamic inertia factor (DIFPSO), and backpropagation neural network (BPNN). The importance degree of features was calculated using random forest, and these weights were incorporated into the model to enhance the quality of the training set. While this study intelligently addressed the non-linearity characteristics of photovoltaic (PV) power and achieve higher accuracy in forecasting outcomes, the complexity of the hybrid model poses challenges in understanding, implementing, and interpreting the results.

In [4], researchers explored the use of random forest feature extraction and variable importance in a fault diagnostic scheme for both simple non-linear systems and the benchmark Tennesse Eastman process. They concluded that fault diagnosis using random forest is a promising data-driven approach but further work is required to improve mapping generalization. However, other studies have employed simpler feature selection methods, focusing more on developing high-level prediction models, particularly in the energy field. For instance, in [5], researchers developed a robust multivariate and machine learning model to forecast the biomass high heat value (HHV) based on proximate analysis. Although they used correlation-based feature selection, the results indicated that accurate HHV prediction cannot be achieved by considering individual components of the proximate analysis alone; instead, pairing them as input variables in the developed models is necessary.

In contrast, a study [6] adopted a different approach by establishing an optimal heterogeneous ensemble learning model for building energy prediction using an exhaustive search method. This study evaluated the performance of six machine learning methods in constructing the heterogenous ensemble model. The findings demonstrated the feasibility of the exhaustive search method in identifying the optimal base model. However, it's important to note that the authors focused on a limited data set, which restricts the generalization of the methodology to other types of data sets. Additionally, the ensemble model exhibits high computational complexity and is time-consuming, posing challenges for large-scale models.

Yet, there was enormous research that developed prediction models for carbon dioxide emissions. One notable example is the study [7]. The researchers proposed a novel prediction model for CO2 emissions based on residual neural networks (ResNet) to analyze the structure of energy in different countries in the world. They indicated that traditional neural networks lead to a problem called the vanishing gradient problem which lowers the accuracy of the prediction outcome. Their results revealed that ResNet has as higher a better result than using the traditional CNN. However, ResNet is considered to be a very complex neural network and it was originally designed for image classification so it might be complex to adapt it for time series forecasting. Other methods seem to be more reasonable for time series forecasting models. Such as what the authors used in [8], where they developed a prediction model using an attention mechanism (AM) based gated recurrent unit (GRU) (AM-GRU) to predict gasoline production. They indicated that because of the instability of gasoline equipment operation and the existence of high noise, it is inconvenient to use traditional prediction methods. Their result showed better stability and higher accuracy than models with other used methods.

All of these studies proposed novel methodologies for developing different prediction models. Some of these studies developed complex prediction models which were challenging to follow their steps. Other studies focused on one specific area of the energy prediction model. In contrast, in developing our model we used different methods for wider energy areas. So, our contributions are mainly focused on:

1. Develop forecasting models for non-renewable energy production and consumption, CO2 emissions due to non-renewable energy production and consumption, and share of renewables in electricity production
2. Use the random forest feature importance method to select the most significant features for each developed model which increase the accuracy of the production outcome and minimize the loss function of the designed gated recurrent unit.

3 Theoretical Background

In this section, we will cover the essential concepts necessary to comprehend the entire model. Initially, we will elucidate the characteristics of random forests and their application in feature selection. Subsequently, we will provide a brief overview of neural networks, including their general definition and various types, leading up to the gated recurrent unit and its utilization in constructing the forecasting model.

3.1 Random Forest

Random forest is commonly defined in research papers as a collection of decision trees used for classification and regression analysis. It employs techniques such as bootstrap aggregation and randomization to achieve a high level of predictive accuracy [9, 11, 12]. However, the concept of the random forest was initially introduced by Breiman, L., who drew inspiration from the work of Amit and Geman (1997). The latter characterized numerous geometrical features and explored a random selection of these features for optimal node splitting. Breiman described the random forest as a process that involves generating a large number of trees and then selecting the most popular class [10].

3.2 Random Forest Feature Importance

To prevent overfitting, enhance performance, and gain a more profound understanding of the underlying data-generating processes, we engage in feature selection [16]. There are various types of feature selection methods in machine learning. Such as filter methods, wrapper methods, and embedded methods. The random forest feature importance method is one of the embedded methods which include interactions between features but at the same time maintain reasonable computational costs [15]. Random forest carries out feature selection during classification rules are built [16]. This means that the algorithm selects the most important features to use in the classification process, rather than using all available features. This can improve the accuracy and efficiency of the classification model. Random forest uses two feature importance measures; the Gini Importance and the Mean Decrease Impurity. In the following subsection, we will explain these measures.

Gini Importance

The Gini importance can be depicted as the result of the random forest classifier's implicit feature selection process. It uses a subset of powerful variables to achieve excellent performance on high-dimensional data [17]. Gini importance is calculated using Eq. (1):

$$I_{G(\theta)} = \sum \sum \Delta \mathrm{i}_{\theta}(\tau, T) \tag{1}$$

Mean Decrease Impurity

The Mean Decrease Impurity (MDI) is a metric used to assess the importance of features in a random forest model. It quantifies the reduction in impurity, like Gini impurity, across all decision trees in the random forest when a specific feature is used for splitting. A higher MDI value for a feature signifies its greater significance in classification tasks and provides insights into its contribution to the model's overall performance. MDI can be valuable for feature selection, allowing the removal of less important features to enhance model efficiency and accuracy [18].

3.3 Recurrent Neural Network and Gated Recurrent Unit

Recurrent neural network (RNN) is a type of neural network which is designed to work with sequential data or time series data. It is different from the forward neural network where it has loops to allow information to remain and be passed sequentially from one step to another [19]. And this is what makes it suitable for time series, speech, and natural language modeling. RNN has a key feature which is the ability to sustain a memory of the previous inputs in the sequence. This memory is stored in the form of hidden state vectors and updates at each step of the sequence. These hidden state vectors can be represented by Eq. (2):

$$h_t = g(Wx_t + Uh_{t-1} + b) \tag{2}$$

However, it has been stated that simple RNNs suffered from the vanishing gradient problem because it is difficult for them to capture long-term dependencies [19]. Thus, there were two main modifications to the simple RNN to solve the vanishing gradient problem. They are the Long Short-Term Memory (LSTM) and the Gated Recurrent Unit (GRU). Since our model is implemented using the GRU, we will present a brief explanation of it in the following subsection.

Gated Recurrent Unit (GRU)

GRU is an updated version of the simple RNN. It uses two vectors that determine which information to be passed to the output. These vectors are the update gate and the reset gate. These two vectors can be trained to retain information from the past or remove information that is irrelevant to the prediction [21].

4 Dataset Description

In this section, we will provide a detailed description of the data collection and data preprocessing.

4.1 Data Collection

Since we intend to implement several prediction models, which are non-renewable energy production, non-renewable energy consumption, the share of renewable energy in electricity production, and CO_2 emissions, we collected data from several resources. First, the data needed for non-renewable energy production and consumption prediction models were sourced from the Global Energy Statistical Yearbook. It consists of energy-related statistics on the production, consumption, and trade of oil, gas, coal, power, and renewables, as well as on CO_2 emissions from fuel combustion. This dataset covers 60 countries around the world. However, we aimed to perform forecasting for only the top countries in energy production and consumption, so we used a dataset from IRENA, the International Renewable Energy Agency, which provides the names of the top countries in energy production and consumption. We used this dataset to eliminate some of the countries in the first dataset, which decreased the complexity of the model implementation. Next, we collected the data needed for the share of renewables in electricity production. This includes renewables datasets sourced from Kaggle datasets, the countries' temperature change in the meteorological year from IMF climate change, and the population of each country also from Kaggle datasets.

Data Preprocessing

Initially, we consolidated all the previously mentioned data categories into a single Excel workbook file. The resulting dataset comprises the following features:

- Entity
- Year
- Temperature change in Meteorological year (°C)
- Population
- Electricity from wind (TWh)
- Electricity from hydro (TWh)
- Electricity from solar (TWh)
- Other renewables including bioenergy (TWh)
- CO_2 emissions from fuel combustion ($MtCO_2$)
- Average CO_2 emission factor (tCO_2/toe)
- CO_2 intensity at constant purchasing power parities (kCO_2/$15p)
- Total energy production (Mtoe)
- Total energy consumption (Mtoe)
- Share of renewables in electricity production (%)
- Share of electricity in total final energy consumption (%)
- Oil products domestic consumption (Mt)
- Refined oil products production (Mt)
- Natural gas production (bcm)
- Natural gas domestic consumption (bcm)

- Energy intensity of GDP at constant purchasing power parities (koe/$15p)
- Electricity production (TWh)
- Electricity domestic consumption (TWh)
- Share of wind and solar in electricity production (%)
- Crude oil production (Mt)

The data covers the time interval from 1990 to 2020. However, due to missing values in the year 2020 for certain entities, we excluded the data associated with this year for all entities to ensure that our dataset does not contain any missing or unavailable values. So, the final data file covers the time interval from 1990 to 2019.

However, this data file contains all features needed for the four-prediction model. So, to implement each of the prediction models, we used the random forest feature importance method to extract the most important features for each model. The process will be explained in the methodology.

5 Methodology

We followed a systematic procedure to implement the prediction models. This procedure mainly consists of two phases. The first one involves training the random forest to extract the feature with the highest importance score. The second phase involve training the GRU neural network and evaluate the prediction results. Figure 1 displays these two phases.

- In the initial phase, our goal was to determine the features that exhibited the highest importance scores. To achieve this, we began by partitioning the data from the main file into distinct training and testing sets. Next, we employed the random forest algorithm to train the models and calculate importance scores for each feature. We compiled a list that included the features with the highest importance scores for each model. Subsequently, we created a separate data file exclusively consisting of these selected features. Below is an example snapshot demonstrating the process of generating importance scores for the energy production prediction model.
- In the subsequent phase, we proceeded with the implementation of the GRU neural network for each prediction model. Our objective was to train and assess the neural network using the previously selected features. To accomplish this, we partitioned the data file into separate training and testing subsets. We defined the architecture of the neural network and trained the model using the training data. Following that, we utilized the trained model to generate predictions. Finally, we evaluated the accuracy of the predictions by measuring mean square error (MSE) and mean absolute error (MAE) metrics.

5.1 Model Architecture

The GRU neural network was designed to predict the amount of energy production, consumption, the share of renewable in electricity production, and the amount of CO2 emissions. However, the architecture generally consists of these layers:

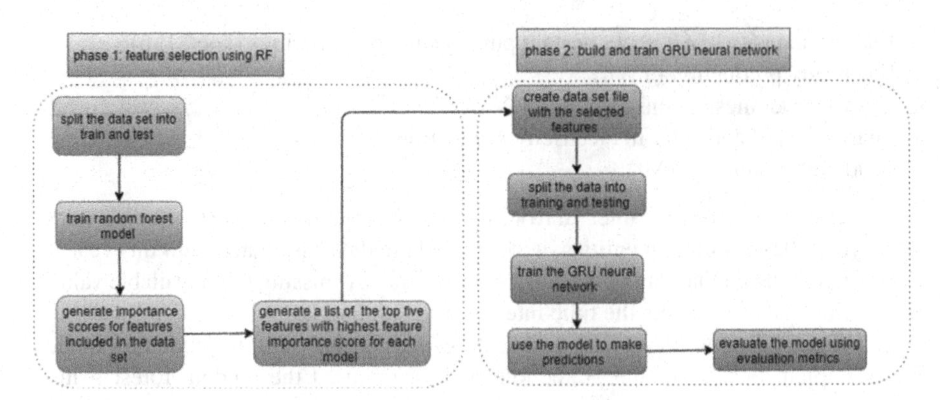

Fig. 1. Model implementation phases

- Input Layer: This layer is where we provide the initial data to the network. It accepts the input data, which for example, in the energy production prediction model, it includes information about oil products consumption, refined oil products production, natural gas production, and population.
- Bidirectional GRU Encoder Layer: This layer processes the input data using a type of recurrent neural network called GRU. The GRU helps capture patterns and dependencies in sequential data by maintaining a memory of past information. The bidirectional aspect means it considers both past and future information at each step, which helps in understanding context and making predictions.
- Repeat Vector Layer: After encoding the input sequence, we repeat the encoded representation for each time step of the output sequence. This allows the decoder layer to have access to the encoded information at each step, enabling it to generate accurate predictions.
- Bidirectional GRU Decoder Layer: This layer takes the repeated encoded representation and decodes it back into a sequence of hidden states. It essentially reconstructs the output sequence based on the encoded information. Similar to the encoder layer, it also uses bidirectional GRU units to consider past and future information.
- Time Distributed Dense Layer: This layer performs a dense (fully connected) operation on each time step of the decoded sequence. It helps transform the hidden states into the final output by considering the relationships between the features and their temporal dependencies.

The model is compiled using the Adam optimizer and the mean squared error loss function. During training, evaluation metrics such as mean squared error and mean absolute error are computed.

6 Results and Discussion

The GRU implementation was carried out using the Python programming language. For each prediction model developed, we generated a comprehensive summary containing key information. This summary encompassed details such as the number of layers within the network, the final loss, mean square error, and mean absolute error values. Additionally, we created visualizations that demonstrated the relationship between the predicted values and the actual values for each trained model. These graphs provided a clear representation of how well the predictions aligned with the actual data.

For instance, Fig. 2 illustrates the energy production prediction model summary:

```
-------------------- Model Summary --------------------
Model: "GRU-Model"
_________________________________________________________________
 Layer (type)                Output Shape              Param #
=================================================================
 Hidden-GRU-Encoder-Layer (B  (None, 64)               7680
 idirectional)

 Repeat-Vector-Layer (Repeat  (None, 5, 64)            0
 Vector)

 Hidden-GRU-Decoder-Layer (B  (None, 5, 64)            18816
 idirectional)

 Output-Layer (TimeDistribut  (None, 5, 1)             65
 ed)

-------------------- Evaluation on Training Data --------------------
Final loss : 0.022314047440886497
Final mean_squared_error : 0.022314047440886497
Final mean_absolute_error : 0.07696016132831573
Final val_loss : 0.0037895343266427517
Final val_mean_squared_error : 0.0037895338609814644
Final val_mean_absolute_error : 0.04964893311262131
```

Fig. 2. Energy production perdition model summary

The provided model summary displays the architecture and structure of the GRU model.

- Hidden-GRU-Encoder-Layer (Bidirectional)
- Repeat-Vector-Layer
- Hidden-GRU-Decoder-Layer (Bidirectional)
- Output-Layer (TimeDistributed)

Also, the provided results indicate that the model achieved relatively low values for the loss, mean squared error, and mean absolute error, both during the validation phase and on the test dataset. This suggests that the model performs well in predicting the target variable.

For further illustration, Fig. 3 displays the predicted amount of produced energy when we use the features with the highest importance score (the crude oil in the case of energy production prediction model):

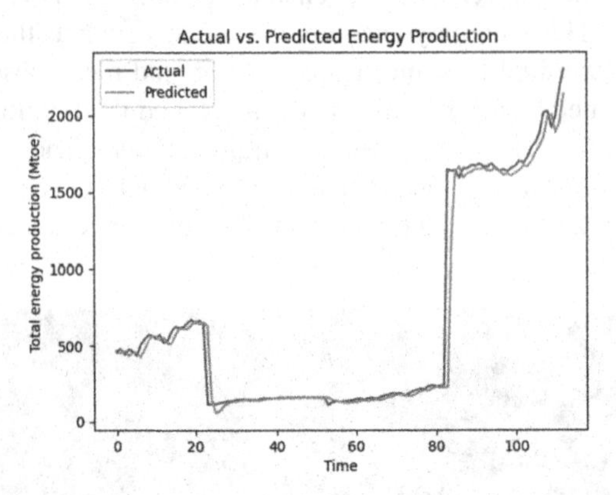

Fig. 3. Actual produced energy vs. predicted produced energy

However, Fig. 4 illustrates the actual amount versus predicted when we use feature with less importance score:

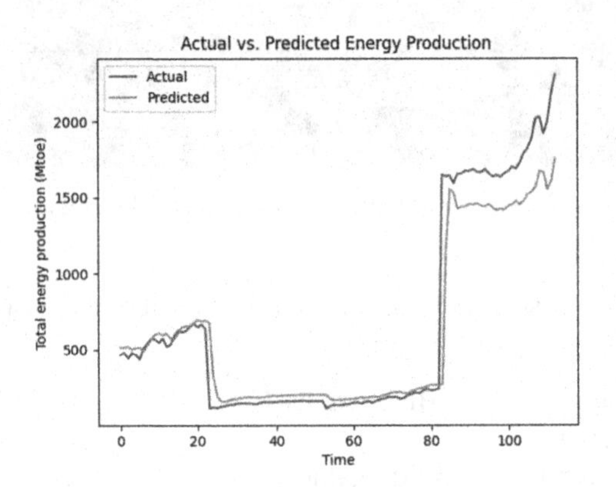

Fig. 4. Actual vs. predicted when using features with less importance score

On the other hand, using all features regardless the importance score generated by the random forest importance score method will produce less accurate prediction, this is illustrated in Fig. 5:

From these results, we can conclude the following;

- The ability of Gated Recurrent Unit neural network to generate accurate time series predicted results.
- Using the random forest feature importance method add more accuracy to the prediction process which indicate the high functionality of random forest feature importance method.

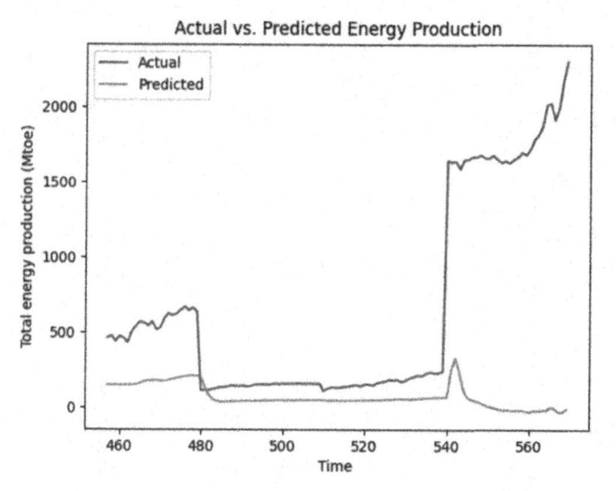

Fig. 5. Actual vs. predicted amount of produced energy when using features regardless of importance score

The actual versus predicted values for the other prediction models (the consumed energy prediction model, the share of renewable in electricity production prediction model, and the CO2 emissions prediction model) are illustrated in the following Figs. 6, 7 and 8.

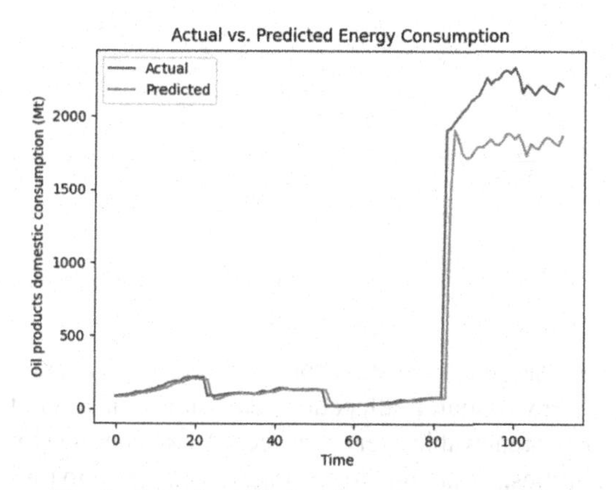

Fig. 6. Actual vs. predicted energy consumption

Through a detailed analysis of the results, it becomes clear that our forecasting models have yielded considerably accurate predictions of energy production, energy consumption, CO2 emissions, and the share of renewables in energy production closely aligning with the actual energy consumption figures. This high degree of precision can be attributed to the careful selection of features using the random forest feature importance methodology, which significantly enhanced the accuracy of our predicted values. Such consistent and reliable predictions serve as a evidence to the efficacy of our chosen methodologies and the expertise applied throughout the project. These findings further highlight the potential of our models to serve as valuable tools in informing

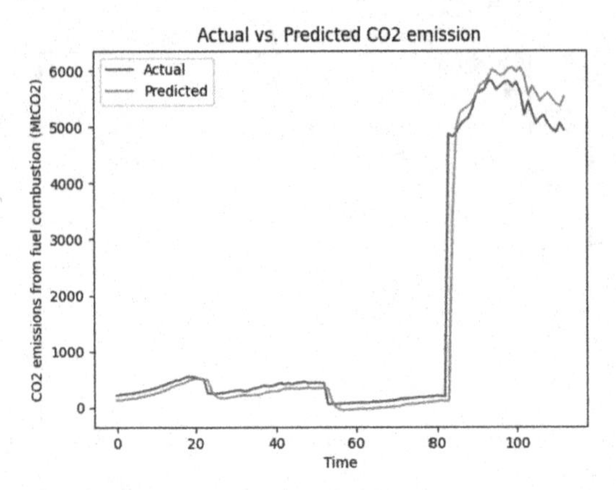

Fig. 7. Actual vs. predicted CO2 emissions

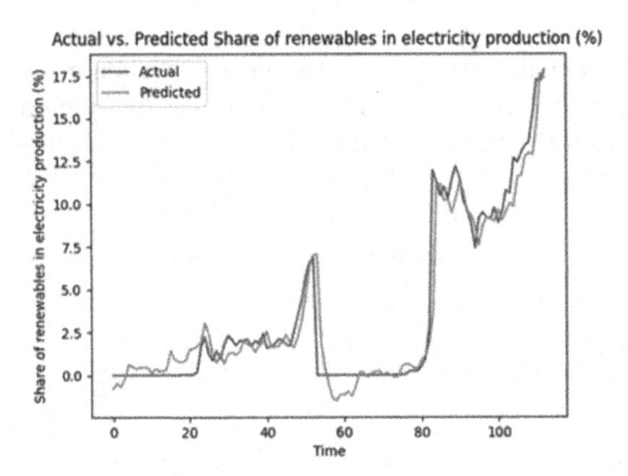

Fig. 8. Actual vs. predicted share of renewable in electricity production

decision-making processes, enabling stakeholders and policymakers to make intelligent choices regarding energy resource allocation, emission reduction strategies, and the promotion of sustainable renewable energy sources. As we continue to refine and expand our forecasting capabilities, we attempt to leverage our expertise to foster a more resilient and sustainable energy landscape.

7 Conclusion

The basic aim of this project was to construct four comprehensive forecasting models all-encompassing energy production, energy consumption, CO2 emissions, and the share of renewables in electricity generation. In order to maximize the accuracy of our predictions, we employed the random forest feature importance method to carefully select the most crucial features based on their importance scores. By utilizing the Gated Recurrent Unit (GRU) architecture, specifically designed for the development of time series forecasting models, we were able to generate significantly accurate prediction values for each target variable.

Moving forward, our future work will be centered around the development of individualized prediction models for each country. This refined approach holds tremendous potential in facilitating the identification of energy resource mismanagement and promoting the adoption of more efficient energy utilization practices at a national level. By adjusting our models to specific country contexts, we can account for the unique characteristics and dynamics of energy production, consumption, emissions, and renewable energy integration within each geographical context. After all, this will empower decision-makers and stakeholders with the insights necessary to implement targeted strategies and policies that drive sustainable energy practices and contribute to a greener and more flexible future.

References

1. Global Energy Outlook 2023: Sowing the Seeds of an Energy Transition. Resources for the Future.https://www.rff.org/publications/reports/global-energy-outlook-2023/
2. Kostadinov, S.: Understanding GRU Networks - Towards Data Science. Medium (2019). https://towardsdatascience.com/understanding-gru-networks-2ef37df6c9be
3. Niu, D., Wang, K., Sun, L., Wu, J., Xu, X.: Short-term photovoltaic power generation forecasting based on random forest feature selection and CEEMD: a case study. Appl. Soft Comput. **93**, 106389 (2020). https://doi.org/10.1016/j.asoc.2020.106389
4. Aldrich, C., Auret, L.: Fault detection and diagnosis with random forest feature extraction and variable importance methods. IFAC Proc. **43**(9), 79–86 (2010). https://doi.org/10.3182/20100802-3-za-2014.00020
5. Dodo, U.A., Ashigwuike, E.C., Abba, S.I.: Machine learning models for biomass energy content prediction: A correlation-based optimal feature selection approach. Bioresource Technol. Rep. **19**, 101167 (2022). https://doi.org/10.1016/j.biteb.2022.101167
6. Wang, Z., Liang, Z., Zeng, R., Yuan, H., Srinivasan, R.S.: Identifying the optimal heterogeneous ensemble learning model for building energy prediction using the exhaustive search method. Energy Build. **281**, 112763 (2022). https://doi.org/10.1016/j.enbuild.2022.112763
7. Han, Y., et al.: Novel economy and carbon emissions prediction model of different countries or regions in the world for energy optimization using improved residual neural network. Sci. Total. Environ. **860**, 160410 (2023). https://doi.org/10.1016/j.scitotenv.2022.160410
8. Liu, J., et al.: Novel production prediction model of gasoline production processes for energy saving and economic increasing based on AM-GRU integrating the UMAP algorithm. Energy **262**, 125536 (2022). https://doi.org/10.1016/j.energy.2022.125536
9. Rigatti, S.J.: Random forest. J. Insur. Med. **47**(1), 31–39 (2017)
10. Breiman, L.: Random forests. Mach. Learn. **45**, 5–32 (2001)
11. Cutler, A., Cutler, D.R., Stevens, J.R.: Random forests. Ensemble Mach. Learn.: Methods Appl., 157–175 (2012)
12. Biau, G.: Analysis of a random forests model. J. Mach. Learn. Res. **13**(1), 1063–1095 (2012)
13. Jaiswal, J.K., Samikannu, R.: Application of random forest algorithm on feature subset selection and classification and regression. In: 2017 World Congress on Computing and Communication Technologies (WCCCT), pp. 65–68. IEEE (2017)
14. Dewi, C., Chen, R.C.: Random forest and support vector machine on features selection for regression analysis. Int. J. Innov. Comput. Inf. Control **15**(6), 2027–2037 (2019)

15. Gupta, A.: Feature selection techniques in machine learning (updated 2023). Analytics Vidhya (2023). https://www.analyticsvidhya.com/blog/2020/10/feature-selection-techniques-in-machine-learning/
16. Qi, Y.: Random forest for bioinformatics. In: Zhang, C., Ma, Y. (eds.) Ensemble machine learning, pp. 307–323. Springer, New York (2012). https://doi.org/10.1007/978-1-4419-9326-7_11
17. Menze, B.H., et al.: A comparison of random forest and its Gini importance with standard chemometric methods for the feature selection and classification of spectral data. BMC Bioinf. **10**(1), 1–6 (2009). https://doi.org/10.1186/1471-2105-10-213
18. Scornet, E.: Trees, forests, and impurity-based variable importance in regression. In: Annales De L'I.H.P, vol. 59, no. 1 (2023). https://doi.org/10.1214/21-aihp1240
19. Li, X., Wang, Y., Basu, S., Kumbier, K., Yu, B.: A debiased MDI feature importance measure for random forests. In: arXiv (Cornell University), vol. 32, pp. 8047–8057 (2019). https://arxiv.org/pdf/1906.10845.pdf
20. Dey, R., Salemt, F.M.: Gate-variants of gated recurrent unit (GRU) neural networks (2017). https://doi.org/10.1109/mwscas.2017.8053243
21. Kostadinov, S.: Understanding GRU Networks - Towards Data Science. Medium (2019). https://towardsdatascience.com/understanding-gru-networks-2ef37df6c

Localizing Non-functional Code Bugs in User Interfaces Using Deep Learning Techniques

Arwa Ahmed(✉), Ahmed Tamer Salah, Ghada Khoriba, and Tamer Arafa

Center for Informatics Science, School of Information Technology and Computer Science (ITCS), Nile University, Giza, Egypt
{a.ahmed2274,aelbardy,ghadakhoriba,tarafa}@nu.edu.eg

Abstract. In the context of modern business digitization, non-functional code issues in User Interface (UI) exert a substantial impact on User Experience(UX). UX-related problems can generate annoyance and potentially result in revenue erosion. However, conventional approaches for detecting and resolving these defects tend to be prolonged, mandate a substantial degree of expertise, and divert precious time and resources away from vital software development tasks, ultimately impairing business operations. A deep learning-based approach automatically detects and localizes non-functional code flaws in user interface (UI) screens. The automated approach reduces the time and effort required to fix non-functional code bugs in UI screens, improving overall quality control and testing and decreasing overall UI testing time. Our model can identify non-functional code flaws by analyzing many UI screens and pinpointing their location on UI screens. We evaluate our proposed approach on a variety of UI screens. The proposed approach achieves better performance in classification by 4% and in localization by 16%.

Keywords: Index Terms-Non-functional bugs · User interface · Machine Learning · Deep Learning · Bug localization · Software engineering

1 Introduction

In recent years, the widespread use of mobile devices has led to an exponential increase in mobile applications, with over 7.26 billion mobile users worldwide [38]. Despite this growth, many mobile applications suffer from visual design issues, which can significantly impact the UX, resulting in reduced usability and user satisfaction [25]. Given the critical importance of visual design in mobile application development, there is a pressing need for more effective approaches to identify and address these issues to improve mobile applications' overall quality and usability.

The Graphic User Interface (GUI), or UI functions as the intermediary between the software applications and its end users, allowing the two layers

M. Mosbah et al. (Eds.): MEDI 2023, LNCS 14396, pp. 381–394, 2024.
https://doi.org/10.1007/978-3-031-49333-1_27

to communicate with one another and giving users a visual way to interact with applications through graphical elements such as widgets, images, and text.

Effective UI development involves skillfully managing user interaction, information structure, and visual elements. A successful UI results in a user-friendly, efficient product that boosts user engagement and loyalty, ensuring the application's success and user satisfaction [15]. Various investigations and studies have revealed a significant correlation between the visual aesthetics of a mobile application and the level of user satisfaction [17].

The importance of mobile applications necessitates thorough testing to ensure the quality of the mobile applications. This study aimed to conduct a pilot study to investigate the occurrence and types of UI display issues in real-world practice, which informed the design of our approach for detecting such issues.

The UI design and development process is inherently prone to errors and can be a time-intensive endeavor, as reported in prior studies [2]. The complexities associated with creating and implementing an effective UI, including the need for iterative design and testing, pose significant challenges for developers and designers seeking to deliver an optimal UX. Identifying and mitigating potential errors during the UI design and development stages can thus play a critical role in ensuring the final product meets user expectations, reduces error rates, and enhances user satisfaction. Numerous research attempts have focused on integrating Human-Computer Interaction (HCI) and AI to prioritize human values and UX in developing AI solutions, as noted in previous studies [25]. This integration can enable the creation of intelligent systems responsive to user needs and preferences while considering their use's ethical and social implications. By integrating HCI principles with AI technologies, developers can create more effective and user-friendly UIs and foster more inclusive and transparent AI systems that align with human values and ethical standards. The continued collaboration between HCI and AI communities can thus play a pivotal role in advancing the development of user-centric AI solutions that promote human well-being.

Human-centered AI (HCAI) is a field that works to construct and create AI systems that consider human necessities and viewpoints. The ultimate goal of HCAI is to develop practical, safe, and ethical technologies for humans. Ultimately, this means that HCAI involves people in its design, development, and testing using actual users, ensuring it meets their needs and is beneficial in some way [2,40].

To ensure the quality of a mobile application, two primary types of user interface UI testing are utilized: automated UI testing and manual UI testing. Our analysis has revealed that a majority of these UI display issues are caused by discrepancies in system settings across different devices, particularly for Android, given the existence of over ten major versions of the Android operating system running on more than 24,000 distinct device models with varying screen resolutions [39]. Although the software can continue to function despite these bugs, they have a detrimental impact on the smooth usage of the application, reducing its accessibility and usability, resulting in poor UX and subsequent loss of users. As such, this study aims to identify and detect these UI display issues [22].

The main contributions of this paper are:

- Issue detection performance: Our developed AI-based solution (deep learning model), specifically VGG-16, demonstrates superior performance in accurately detecting issues within user interfaces, providing valuable insights for UI designers and developers.
- Issue localization performance: Our model showcases impressive capabilities in accurately localizing and pinpointing UI bugs, leveraging the bounding box approach and the heatmap generated by VGG-16, offering a powerful tool for enhancing UI testing and improving overall quality.
- Performance comparison with baselines: We compare the performance of our model with established baselines, showcasing its effectiveness in bug detection and surpassing the majority of the baseline models.

Section 2 discusses related work and research gaps. The proposed methodology is explained in Sect. 3. Section 4 presents the proposed model evaluation metrics. Results and discussion are presented in Sect. 5.

2 Related Work

Two main testing methods are commonly employed in UI testing for mobile applications. Typically employs either manual or automated methods. Automated UI testing has received considerable attention recently, with researchers exploring various approaches. Many studies explored model-based [26,41,42] probability-based [23,24,43] and deep learning-based [19,22,29] approaches. Deep learning techniques have shown promise in modeling user interaction behaviors, which greatly impact modeling complex user interactions characteristic of mobile applications. For instance, Yang et al. [18] utilized a hierarchical recurrent neural network to model list selection tasks effectively. These findings illustrate and highlight the potential of advanced machine learning and deep learning techniques in enhancing the quality and efficiency of mobile application testing.

Traditional methods can be employed to evaluate mobile applications by conducting many evaluation scenarios to cover a broad range of contextual criteria. However, this approach can be costly and time-consuming. Additionally, it is a tedious and challenging task for non-programmers [6,14,33,35], as it requires the analysis of source code and visual design quality of all UIs [40]. Numerous existing studies aim to automatically analyze mobile application design by detecting aesthetic defects and usability code smells, as evidenced by works such as [4,10,28,31,34]. However, such evaluation methods can be costly and time-consuming, as they require an analysis of user interaction behaviors to detect usability issues. The high cost of these methods could be significantly reduced by proposing an automatic prediction of mobile UI design quality as early as possible during the mobile application development phase.

Human testers can find more diverse and complex bugs than automated UI tests, especially those in Fig. 1. However, manual UI testing also has two problems. Firstly, it requires significant human effort. Testers must manually explore numerous pages using various interactive methods while checking for UI display accuracy across different versions and devices with varying resolutions or screen sizes. Secondly, manual test performance is erratic as it highly depends on the skill and experience of the testers, and testers may miss some minor features, especially for these unknown applications [22].

In manual testing, several testers mimic user behavior to explore different application features and find more bugs [5,20,37]. Some researchers streamline manual testing through test cases and prioritization, but these techniques typically rely on specific data [11,21] and testers may miss some minor features, especially for these unknown applications [9,13,27].

With most dynamic UI testing tools, the longer your tests take, the higher your test coverage, the more likely you are to find bugs, and the higher the quality of your application release. However, budget constraints and market pressures force development teams to meet deadlines while balancing testing time with other requirements [16]. It focuses on RAD through frequent iterations and continuous feedback. They use user feedback to uncover the UI display issues, but this passive troubleshooting method may already be hurting application users and leading to a loss of market share [22]. Generally, two main categories of methods are used to evaluate UIs: testing the visual design of mobile applications with access to their source code and evaluating the visual design of mobile applications using UI images [36].

A recent study [38] has revealed that 79% of mobile users are willing to gain a preliminary understanding of common UI rendering issues, [22] conducted a pilot study on 10,330 non-duplicate UI images from 562 crowd-testing tasks for mobile applications. This study aimed to observe any display issues present in these UI images. The results indicate that a non-negligible proportion (43.2%) of these UI images displayed issues that could seriously impact the UX and potentially harm the application's reputation. Additionally, we examined various UI images from randomly selected applications in the commonly-used Rico dataset [3] and found that 8.8% of these applications displayed UI display issues.

The most common categories of UI display issues included component occlusion, text overlap, missing images, null values, and blurred screens [15]. These findings highlight the importance of addressing UI display issues to improve mobile applications' quality and usability.

The pilot study on mobile application UI images revealed that component occlusion (47%) was the most common UI display issue encountered. This issue occurs when textual information or components are blocked or covered by other elements, often in TextView or EditText. It is usually caused by improper element height settings or adaptive issues triggered by larger-sized fonts [22].

Text overlap (21%) was the second most common UI display issue observed. This issue arises when two pieces of text overlap, often caused by adaptive issues between different device models. In contrast to component occlusion, where one component covers part of another component, text overlap involves two pieces of text being mixed together [22].

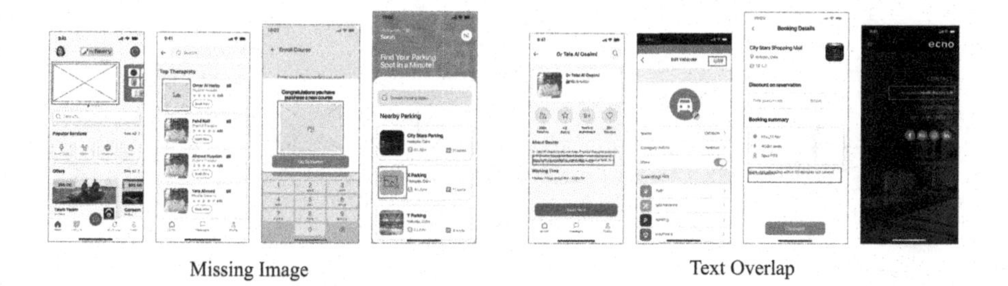

Fig. 1. UI Bug Types

Missing image (25%) was the third most common UI display issue identified. This issue occurs when the image in the icon position does not appear as intended. Possible reasons include incorrect image path or layout position, unsuccessful loading of the configuration file due to permissions, oversized images, network connection issues, code logic errors, or picture errors, among others [22].

Our research focuses on the visual design quality of mobile UIs. We propose a novel model for the automated assessment of UIs in the early stages of mobile application development. Our approach is based on image analysis, allowing for the evaluation and analysis of any mobile application without the need for its source code. As such, we evaluate the UI as an image rather than as source code, enabling a holistic and comprehensive analysis of the entire UI [31]. This approach enables a more efficient and cost-effective evaluation of UI design quality without requiring extensive human intervention or source code analysis.

3 Methodology

3.1 Dataset

We used the Rico dataset [3], which contains over 70,000 UI images from more than 9,000 Android applications spanning 27 Google Play categories, along with corresponding annotations of UI elements such as buttons, text fields, and images. The dataset includes information about the app category, number of downloads, and application rating. The Rico dataset was selected for its comprehensive coverage of mobile application UI design patterns, making it suitable for studying design bugs and developing data-driven design applications. We used this information to generate UIs with two types of non-functional code bugs, which were then used to train a model to correctly localize and predict the presence of buggy UIs, as in Fig. 2. In addition to its comprehensive coverage of mobile application UI design patterns, the Rico dataset includes a collection of unique UIs identified using a novel content-agnostic similarity heuristic. This heuristic enables the identification of UIs that may be visually distinct but functionally similar, providing a more nuanced understanding of design patterns in mobile applications. Moreover, the Rico dataset is large enough to support

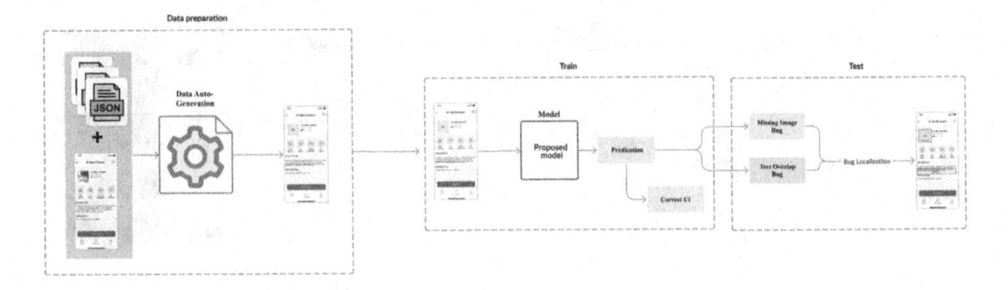

Fig. 2. Proposed Dataset preparation approach

the development of deep-learning applications, making it a valuable resource for exploring advanced machine-learning techniques for analyzing mobile applications' UI. We used an automated method to annotate the UI elements in the Rico dataset. We used a deep learning-based approach to identify UI components such as buttons and text fields and then automatically verified and corrected the annotations to ensure accuracy. The Rico dataset is publicly available and can be accessed at [1]. To train and evaluate our approach, we utilized an updated version of the Rico dataset that includes UI screens with bugs. The Rico dataset, consisting of UI screens and their corresponding JSON files containing annotations, served as the foundation for our work. By leveraging the annotations in the Rico dataset, we could identify and replace specific UI elements with various types of bugs tailored to our approach. This process allowed us to create a comprehensive dataset capturing many potential UI design bugs. By incorporating these bug types into our updated dataset, we aimed to enhance the capability of our model to detect and classify bugs in the UI accurately; the utilization of the Rico dataset as the base for our bug-infected dataset provides several advantages. Firstly, the Rico dataset is widely recognized in the research community and encompasses a diverse collection of real-world UI screens from various mobile applications. Leveraging this rich dataset ensures that our model is exposed to a broad range of UI designs and bug instances, enhancing its generalization capability. Moreover, by aligning our dataset preparation approach with the annotations provided in the Rico dataset, we maintain consistency and comparability with previous studies, allowing for a more robust evaluation of our approach against existing baselines. Overall, the utilization and adaptation of the Rico dataset, combined with our bug injection methodology, provide a solid foundation for training and evaluating our model's bug detection and classification capabilities within the context of mobile UI designs as in Fig. 2.

3.2 Proposed Model Architecture

The proposed model is based on faster R-CNN [7]. The full and detailed architecture is shown in Fig. 3. We used the VGG-16 [32] for better feature extraction.

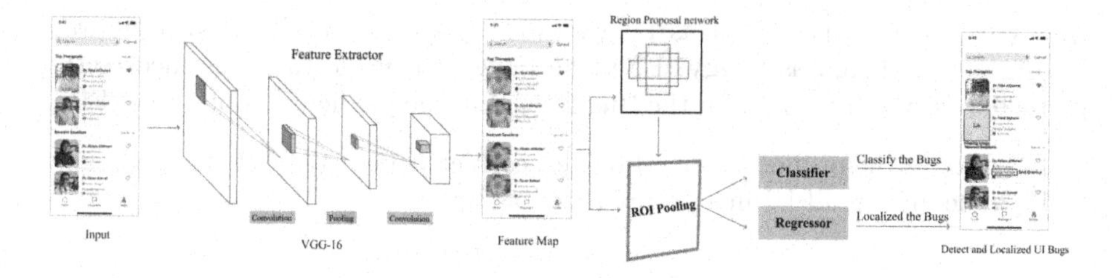

Fig. 3. Proposed Model Architecture

3.3 Region Proposal Network (RPN)

The Region Proposal Network (RPN) [30] is responsible for taking the feature map generated from the CNN as input and then going through a sliding window approach over the feature map to predict the probability of an object being present in this window. The proposed bounding box coordinates are also predicted and ranked by their object's presence score.

3.4 Region of Interest (RoI) Pooling

The Region of Interest (RoI) pooling layer takes in the output of the RPN and the feature maps generated by the CNN backbone. It then extracts a fixed-sized feature map from each proposal by dividing it into a predetermined number of bins and performing max pooling on each bin. This process generates a fixed-sized feature map for each proposal to be passed to the next network stage.

3.5 Object Detection Head

The final stage of the Faster R-CNN model is the object detection head. It takes the fixed-size feature map generated by the RoI pooling layer as input and produces the ultimate object detection output. The object detection head consists of two sub-networks: a classification and a regression sub-network. The classification sub-network predicts the probability that an object belongs to a particular class, while the regression sub-network predicts the coordinates of the object's bounding box. The final object detection output is obtained by combining the classification and regression sub-networks scores and bounding box coordinates.

4 Model Evaluation Metrics

We apply four metrics mostly used for classification to evaluate our proposed model for buggy issue detection and localization accuracy. We evaluate our model accurately as recognition using Precision (P), Recall (R), F1-score, and Accuracy [8]. For buggy issue detection performance, we consider the buggy UI image that was correctly predicted as buggy (TP), the bug-free UI image that was incorrectly predicted as buggy (FN), the buggy UI image that was incorrectly predicted as Bug-free (TN) for the Bug-free UI image predicted as bug-free (FP).

- Precision is the proportion of UI images correctly predicted as having a buggy UI issue among all UI images predicted as buggy.

$$precision = \frac{TP}{TP + FP} \tag{1}$$

- Recall is the proportion of UI images correctly predicted as buggy among all UI images with UI display issues.

$$recall = \frac{TP}{TP + FN} \tag{2}$$

- F1-score (F-measure or F1) is the harmonic mean of precision and recall, combining both metrics above.

$$F1{-}score = \frac{2 * precision * recall}{precision + recall} \tag{3}$$

To assess the effectiveness of our proposed approach in addressing the localization issues, we employed two widely-used evaluation metrics: Average Precision (AP) and Average Recall (AR) [8], which are commonly employed in object detection tasks [12]. Subsequently, we calculated the Intersection over Union (IoU) ratio, which measures the overlap between the predicted and actual buggy regions from its mask. The IoU calculation can effectively address the challenges of coverage of Fig. 1. It was computed as follows: IoU = (intersection of predicted buggy region and real buggy region)/(union of predicted buggy region and real buggy region). To be more detailed, TP was identified as the number of detection boxes with IoU values equal to or greater than 0.5. FP was defined as the number of detection boxes with IoU values less than 0.5, including redundant detection boxes identified within the same ground truth box. Lastly, FN denoted the number of ground truth boxes that were not detected by our system. By applying these evaluation metrics, we conducted a comprehensive assessment of the localization performance accuracy of our system.

5 Results and Discussion

The main target of the model was to classify and localize potential bugs that might be present on the UI. The updated Rico dataset (buggy UIs) was mainly created by adding missing image icons on the images or creating text overlaps by adding a copy of the original text over the text. The updated Rico dataset has an impact on improving the recognition of any bugs in the UI. The updated Rico dataset has an impact on improving the recognition of any bugs in the UI. This approach is illustrated in Fig. 2. The proposed model architecture is shown in Fig. 3, and the detailed model parameters are shown in Table 1. To evaluate

Table 1. Proposed model hyperparameters

Hyperparameter	Value
Backbone	VGG16
Image input size	800 × 1500
Anchor scales	[128, 256, 512]
Anchor ratios	[0.5, 1, 2]
RPN stride	16 pixels
RPN kernel size	3 × 3
RPN feature map channels	512
RPN anchor boxes per point	9
RPN positive IOU threshold	0.7
RPN negative IOU threshold	0.3
RoI Pooling	max pooling
RoI Pooling size	7 × 7
RoI Pooling stride	1
RoI Pooling channels	512
Classification FC layers	4096, 4096
Dropout rate	0.5
Optimizer	Adam
Learning rate	0.0001

Table 2. Buggy issue detection performance for the proposed model

Bug type	Proposed Generated data		
	P	R	F1
Missing Image	0.952	0.864	0.906
Text overlap	0.771	0.702	0.854
Average	0.861	0.783	0.880

the performance of the proposed model, it was compared to several baselines on the Rico dataset with different bug variations (updated Rico dataset), as shown in Table 3.

5.1 Buggy Issue Detection Performance

The proposed model was trained and evaluated on the updated Rico dataset created based on the Rico dataset. The updated Rico dataset contains buggy UIs that include images with missing image icons or text overlaps. The proposed model achieved a promising performance in detecting potential bugs in the UI.

The model's performance in terms of P, R, and F1-score for each bug type is presented in Table 2. Table 2 shows the buggy issue detection performance of the proposed model with the updated Rico dataset. The proposed model achieved an average P, R, and F1-Score of 0.861, 0.783, and 0.880, respectively. The missing image had a P of 0.952, R of 0.864, and F1-score of 0.906, while the text overlap had a P of 0.771, R of 0.702, and F1-score of 0.854. These results demonstrate the effectiveness of the proposed model in detecting and classifying buggy issues on the UI.

5.2 Buggy Issue Localization Performance

The proposed model achieved promising performance accuracy in localizing potential bugs in the UI. The proposed model used the bounding box of the highlighted area of the heatmap in localization, and its performance was compared with the most recent proposed approach. The buggy issue localization accuracy for missing images was 0.88136, and for text overlap, it was 0.76216. Figure 2 and Fig. 3 present the proposed dataset preparation approach and the proposed model architecture, respectively.

5.3 Performance Comparison with Baselines

We compared the proposed model with several baselines using the updated Rico dataset with different bug variations. The baselines include SIFT-SVM, SIFT-KNN, SIFT-NB, SIFT-RF, SURF-SVM, SURF-KNN, SURF-NB, SURF-RF, ORB-SVM, ORB-KNN, ORB-NB, MLP, OwlEye, and Nighthawk. The performance of the baselines and the proposed model evaluated and compared in terms of P, R, and F1-score are presented in Table 3. Table 3 shows that the proposed model achieved a higher P, R, and F1-score than most baselines. The proposed model achieved a P, R, and F1-score of 0.861, 0.783, and 0.880, respectively, outperforming most baselines. The Nighthawk baseline achieved a slightly higher P, R, and F1-score compared to the proposed model, with a P, R, and F1-score of 0.843, 0.837, and 0.840, respectively. This result demonstrates the effectiveness of the proposed model in detecting and localization accuracy of the buggy issues on the UI. Overall, the results indicate that the proposed model with the Updated Rico dataset has achieved a higher performance in detecting and localizing UI buggy issues than other baselines. The updated Rico dataset also proved effective results in improving the recognition of such issues. The proposed model can be a valuable and potential tool for UI designers and developers to be used for UI testing and bug detection to improve the quality of their UIs.

Table 3. Compared with the other baselines with updated Rico dataset with different defect variations

Category	Rico dataset		
	P	R	F1
SIFT-SVM	0.525	0.518	0.512
SIFT-KNN	0.537	0.538	0.537
SIFT-NB	0.564	0.537	0.563
SIFT-RF	0.556	0.563	0.559
SURF-SVM	0.551	0.545	0.548
SURF-KNN	0.560	0.558	0.559
SURF-NB	0.586	0.590	0.558
SURF-RF	0.583	0.585	0.584
ORB-SVM	0.556	0.549	0.552
ORB-KNN	0.564	0.561	0.562
ORB-NB	0.586	0.590	0.588
ORB-RF	0.584	0.586	0.585
MLP	0.611	0.666	0.637
OwlEye	0.790	0.778	0.784
Nighthawk	0.843	0.837	0.840
Our model	**0.861**	**0.783**	**0.880**

6 Conclusion

Based on the experiments, the proposed model successfully detects and localizes potential bugs in mobile applications' UI. The experiment utilized an updated Rico dataset, which includes UI screens with various bug variations, such as missing image and text overlap. The Rico dataset was chosen for its comprehensive mobile application UI design patterns coverage. It includes annotations of UI elements, app categories, download numbers, and ratings. The dataset generated UIs with non-functional code bugs, specifically missing images and text overlap.

The proposed model is based on Faster R-CNN, a deep learning algorithm for object detection and classification. It consists of components such as a CNN backbone, RPN, RoI Pooling, and Object Detection Head.

The proposed model demonstrated strong performance in detecting buggy UI issues, achieving an average Precision of 0.861, Recall of 0.783, and F1-score of 0.880. The model effectively recognized and classified buggy UI elements. The proposed model used bounding boxes to pinpoint areas with missing image and text overlap. The accuracy of localizing missing images was 0.88136, while for text overlaps, it measured 0.76216.

Acknowledgment. This paper is based upon work supported by The Academy of Scientific Technology (ASRT), Egypt. Scientists of Next Generation Scholarship Cycle (8).

References

1. Balakrishnan, R., Baudisch, P., Hinckley, K., Wigdor, D.: Rico: a mobile app dataset for building data-driven design applications. https://www.interactionmining.org/rico.html. Accessed 2 Feb 2023
2. Crease, M., Brewster, S., Gray, P.: Caring, sharing widgets: a toolkit of sensitive widgets. In: McDonald, S., Waern, Y., Cockton, G. (eds.) People and Computers XIV–Usability or Else!, pp. 257–270. Springer, London (2000). https://doi.org/10.1007/978-1-4471-0515-2_17
3. Deka, B., et al.: Rico: a mobile app dataset for building data-driven design applications. In: Proceedings of the 30th Annual ACM Symposium on User Interface Software and Technology, UIST 2017, pp. 845–854. Association for Computing Machinery, New York, NY, USA (2017). https://doi.org/10.1145/3126594.3126651
4. Denden, M., Tlili, A., Essalmi, F., Jemni, M.: Educational gamification based on personality. In: IEEE/ACS 14th International Conference on Computer Systems and Applications (AICCSA), pp. 1399–1405 (2017). https://doi.org/10.1109/AICCSA.2017.87
5. Dyba, T., Dingsoyr, T., Hanssen, G.K.: Applying systematic reviews to diverse study types: an experience report. In: First International Symposium on Empirical Software Engineering and Measurement (ESEM 2007), pp. 225–234. IEEE (2007)
6. Fast, E., Steffee, D., Wang, L., Brandt, J.R., Bernstein, M.S.: Emergent, crowd-scale programming practice in the IDE. In: Proceedings of the SIGCHI Conference on Human Factors in Computing Systems, pp. 2491–2500 (2014)
7. Girshick, R.: Fast R-CNN. In: Proceedings of the IEEE International Conference on Computer Vision, pp. 1440–1448 (2015)
8. Goutte, C., Gaussier, E.: A probabilistic interpretation of precision, recall and *F*-score, with implication for evaluation. In: Losada, D.E., Fernández-Luna, J.M. (eds.) ECIR 2005. LNCS, vol. 3408, pp. 345–359. Springer, Heidelberg (2005). https://doi.org/10.1007/978-3-540-31865-1_25
9. Haas, R., Elsner, D., Juergens, E., Pretschner, A., Apel, S.: How can manual testing processes be optimized? developer survey, optimization guidelines, and case studies. In: Proceedings of the 29th ACM Joint Meeting on European Software Engineering Conference and Symposium on the Foundations of Software Engineering, pp. 1281–1291 (2021)
10. Harrison, R., Flood, D., Duce, D.: Usability of mobile applications: literature review and rationale for a new usability model. J. Interact. Sci. **1**, 1–16 (2013)
11. Hemmati, H., Fang, Z., Mantyla, M.V.: Prioritizing manual test cases in traditional and rapid release environments. In: 2015 IEEE 8th International Conference on Software Testing, Verification and Validation (ICST), pp. 1–10 (2015). https://doi.org/10.1109/ICST.2015.7102602
12. Hemmati, H., Fang, Z., Mantyla, M.V.: Prioritizing manual test cases in traditional and rapid release environments. In: 2015 IEEE 8th International Conference on Software Testing, Verification and Validation (ICST), pp. 1–10. IEEE (2015)
13. Hemmati, H., Sharifi, F.: Investigating NLP-based approaches for predicting manual test case failure. In: 2018 IEEE 11th International Conference on Software Testing, Verification and Validation (ICST), pp. 309–319. IEEE (2018)

14. Ines, G., Makram, S., Mabrouka, C., Mourad, A.: Evaluation of mobile interfaces as an optimization problem. Procedia Comput. Sci. **112**, 235–248 (2017)
15. Jansen, B.J.: The graphical user interface. ACM SIGCHI Bull. **30**(2), 22–26 (1998)
16. Joorabchi, M.E., Mesbah, A., Kruchten, P.: Real challenges in mobile app development. In: 2013 ACM/IEEE International Symposium on Empirical Software Engineering and Measurement, pp. 15–24. IEEE (2013)
17. Lee, M., Kent, T., Carswell, C.M., Seidelman, W., Sublette, M.: Zebra-striping: visual flow in grid-based graphic design. Proc. Hum. Factors Ergon. Soc. Ann. Meet. **58**(1), 1318–1322 (2014). https://doi.org/10.1177/1541931214581275
18. Li, Y., Bengio, S., Bailly, G.: Predicting human performance in vertical menu selection using deep learning. In: Proceedings of the 2018 CHI Conference on Human Factors in Computing Systems, pp. 1–7 (2018)
19. Li, Y., Yang, Z., Guo, Y., Chen, X.: Humanoid: a deep learning-based approach to automated black-box android app testing. In: 2019 34th IEEE/ACM International Conference on Automated Software Engineering (ASE), pp. 1070–1073. IEEE (2019)
20. Liu, Z., Chen, C., Wang, J., Huang, Y., Hu, J., Wang, Q.: Owl eyes: spotting UI display issues via visual understanding. In: Proceedings of the 35th IEEE/ACM International Conference on Automated Software Engineering, pp. 398–409 (2020)
21. Liu, Z., Chen, C., Wang, J., Huang, Y., Hu, J., Wang, Q.: Guided bug crush: assist manual GUI testing of android apps via hint moves. In: Proceedings of the 2022 CHI Conference on Human Factors in Computing Systems, CHI 2022. Association for Computing Machinery, New York, NY, USA (2022). https://doi.org/10.1145/3491102.3501903
22. Liu, Z., Chen, C., Wang, J., Huang, Y., Hu, J., Wang, Q.: Nighthawk: fully automated localizing UI display issues via visual understanding. IEEE Trans. Softw. Eng. **49**(1), 403–418 (2022)
23. Machiry, A., Tahiliani, R., Naik, M.: Dynodroid: an input generation system for android apps. In: Proceedings of the 2013 9th Joint Meeting on Foundations of Software Engineering, pp. 224–234 (2013)
24. Mao, K., Harman, M., Jia, Y.: Sapienz: multi-objective automated testing for android applications. In: Proceedings of the 25th International Symposium on Software Testing and Analysis, pp. 94–105 (2016)
25. Margetis, G., Ntoa, S., Antona, M., Stephanidis, C.: Human-centered design of artificial intelligence. Handb. Hum. Factors Ergon., 1085–1106 (2021)
26. Mirzaei, N., Garcia, J., Bagheri, H., Sadeghi, A., Malek, S.: Reducing combinatorics in GUI testing of android applications. In: Proceedings of the 38th International Conference on Software Engineering, pp. 559–570 (2016)
27. Nakagawa, T., Munakata, K., Yamamoto, K.: Applying modified code entity-based regression test selection for manual end-to-end testing of commercial web applications. In: 2019 IEEE International Symposium on Software Reliability Engineering Workshops (ISSREW), pp. 1–6. IEEE (2019)
28. Palomba, F., Di Nucci, D., Panichella, A., Zaidman, A., De Lucia, A.: Lightweight detection of android-specific code smells: the adoctor project. In: IEEE 24th International Conference on Software Analysis, Evolution and Reengineering (SANER). IEEE (2017)
29. Pan, M., Huang, A., Wang, G., Zhang, T., Li, X.: Reinforcement learning based curiosity-driven testing of android applications. In: Proceedings of the 29th ACM SIGSOFT International Symposium on Software Testing and Analysis, pp. 153–164 (2020)

30. Ren, S., He, K., Girshick, R., Sun, J.: Faster R-CNN: towards real-time object detection with region proposal networks. In: Advances in Neural Information Processing Systems, vol. 28 (2015)
31. Shirazi, S., Alireza, N., Henze, A., Schmidt, R., Goldberg, B., Schmidt, H.: Insights into layout patterns of mobile user interfaces by an automatic analysis of android apps. In: Proceedings of the 5th ACM SIGCHI symposium on Engineering Interactive Computing Systems, pp. 275–284. Association for Computing Machinery (2013)
32. Simonyan, K., Zisserman, A.: Very deep convolutional networks for large-scale image recognition. arXiv preprint arXiv:1409.1556 (2014)
33. Soui, M., Chouchane, M., Gasmi, I., Mkaouer, M.W.: PLAIN: PLugin for predicting the usAbility of mobile user INterface. In: Proceedings of the 12th International Joint Conference on Computer Vision, Imaging and Computer Graphics Theory and Applications. SCITEPRESS - Science and Technology Publications (2017)
34. Soui, M., Chouchane, M., Mkaouer, M.W., Kessentini, M., Ghedira, K.: Assessing the quality of mobile graphical user interfaces using multi-objective optimization. Soft. Comput. **24**(10), 7685–7714 (2020)
35. Soui, M., Ghedira, K., Abed, M.: Evaluating user interface adaptation using the context of use. Int. J. Adapt. Resilient Auton. Syst. (IJARAS) **6**(1), 1–24 (2015)
36. Soui, M., Haddad, Z.: Deep learning-based model using densnet201 for mobile user interface evaluation. Int. J. Hum.-Comput. Interact. 1–14 (2023)
37. Su, Y., Liu, Z., Chen, C., Wang, J., Wang, Q.: OwlEyes-online: a fully automated platform for detecting and localizing UI display issues. In: Proceedings of the 29th ACM Joint Meeting on European Software Engineering Conference and Symposium on the Foundations of Software Engineering, pp. 1500–1504 (2021)
38. Taylor, P.: Forecast number of mobile users worldwide 2020–2025 (2023). https://www.statista.com/statistics/218984/number-of-global-mobile-users-since-2010/
39. Wei, L., Liu, Y., Cheung, S.C.: Taming android fragmentation: characterizing and detecting compatibility issues for android apps. In: Proceedings of the 31st IEEE/ACM International Conference on Automated Software Engineering, pp. 226–237 (2016)
40. Xu, W.: Toward human-centered AI: a perspective from human-computer interaction. Interactions **26**(4), 42–46 (2019)
41. Yang, S., et al.: Static window transition graphs for android. Autom. Softw. Eng. **25**, 833–873 (2018)
42. Yang, W., Prasad, M.R., Xie, T.: A grey-box approach for automated GUI-model generation of mobile applications. In: Cortellessa, V., Varró, D. (eds.) FASE 2013. LNCS, vol. 7793, pp. 250–265. Springer, Heidelberg (2013). https://doi.org/10.1007/978-3-642-37057-1_19
43. Zeng, X., et al.: Automated test input generation for android: are we really there yet in an industrial case? In: Proceedings of the 2016 24th ACM SIGSOFT International Symposium on Foundations of Software Engineering, pp. 987–992 (2016)

Author Index

M. Mosbah et al. (Eds.): MEDI 2023, LNCS 14396, pp. 395–396, 2024.
https://doi.org/10.1007/978-3-031-49333-1

Printed in the United States
by Baker & Taylor Publisher Services

Printed in the United States
by Baker & Taylor Publisher Services